Oval Track and Other Permutation Puzzles

And Just Enough Group Theory to Solve Them

Published and Distributed by
The Mathematical Association of America

ISBN 0-88385-725-1

Library of Congress Catalog Card No. 2003108186

Printed in the United States of America

10 9 8 7 6 5 4 3 2 1

Oval Track and Other Permutation Puzzles

And Just Enough Group Theory to Solve Them

by

John O. Kiltinen

Northern Michigan University

Published and Distributed by

The Mathematical Association of America

In memory of my father,

Chester S. Kiltinen

CLASSROOM RESOURCE MATERIALS

Classroom Resource Materials is intended to provide supplementary classroom material for students—laboratory exercises, projects, historical information, textbooks with unusual approaches for presenting mathematical ideas, career information, etc.

101 Careers in Mathematics, 2nd edition edited by Andrew Sterrett

Archimedes: What Did He Do Besides Cry Eureka?, Sherman Stein

Calculus Mysteries and Thrillers, R. Grant Woods

Combinatorics: A Problem Oriented Approach, Daniel A. Marcus

Conjecture and Proof, Miklós Laczkovich

A Course in Mathematical Modeling, Douglas Mooney and Randall Swift

Cryptological Mathematics, Robert Edward Lewand

Elementary Mathematical Models, Dan Kalman

Environmental Mathematics in the Classroom, edited by B. A. Fusaro and P. C. Kenschaft

Geometry From Africa: Mathematical and Educational Explorations, Paulus Gerdes

Identification Numbers and Check Digit Schemes, Joseph Kirtland

Interdisciplinary Lively Application Projects, edited by Chris Arney

Inverse Problems: Activities for Undergraduates, Charles W. Groetsch

Laboratory Experiences in Group Theory, Ellen Maycock Parker

Learn from the Masters, Frank Swetz, John Fauvel, Otto Bekken, Bengt Johansson, and Victor Katz

Mathematical Evolutions, edited by Abe Shenitzer and John Stillwell

Mathematical Modeling in the Environment, Charles Hadlock

Mathematics for Business Decisions Part 1: Probability and Simulation (electronic textbook), Richard B. Thompson and Christopher G. Lamoureux

Mathematics for Business Decisions Part 2: Calculus and Optimization (electronic textbook), Richard B. Thompson and Christopher G. Lamoureux

Ordinary Differential Equations: A Brief Eclectic Tour, David A. Sánchez

Oval Track and Other Permutation Puzzles, John O. Kiltinen

A Primer of Abstract Mathematics, Robert B. Ash

Proofs Without Words, Roger B. Nelsen

Proofs Without Words II, Roger B. Nelsen

A Radical Approach to Real Analysis, David M. Bressoud

She Does Math!, edited by Marla Parker

Solve This: Math Activities for Students and Clubs, James S. Tanton

MAA Service Center
P.O. Box 91112
Washington, DC 20090-1112
1-800-331-1MAA FAX: 1-301-206-9789

Preface

Development history

This book and the accompanying puzzle software grew to the present form over several years of classroom use. In the early 1980s, I, like many others, had fun with *Rubik's Cube* and the spate of permutation puzzles which followed it, particularly enjoying the concrete representation they gave for the theory of permutation groups. I brought my collection of puzzles to my abstract algebra class, and discussed with the students how group theory provides exactly the right mathematical structure for understanding and solving these puzzles.

One of the puzzles of this genre to come out in the later 1980s was called *Top Spin*, marketed by Binary Arts. It has twenty numbered disks that rotate freely on an oval-shaped track. In addition, there is a circular turntable at the top of the track that has room for four disks. By rotating this turntable through 180° with four disks on it, one can reverse the order of the disks. By sliding the disks around the track and using the turntable, you can scramble the disks and then try to solve the puzzle.

This puzzle has the virtue of being purely a permutation puzzle, in contrast to *Rubik's Cube*, which has an overlay of issues involving the orientation of the moving pieces. Described in permutation group terms, *Top Spin* is a concrete realization of the subgroup of the symmetric group S_{20} which is generated by the twenty-cycle $(1, 2, 3, \ldots, 20)$ and the product $(1, 4)(2, 3)$ of two disjoint transpositions.

Group-theoretically, the puzzle is simple to describe, but from a practical standpoint, it is nontrivial to solve. This makes it an excellent object of study for students of abstract algebra, giving them a concrete representation of a nontrivial and fruitful application of the theory of permutation groups.

Applying the theory to solve the physical puzzle presents challenges to one's memory, however. There are certain sequences of steps that one repeats frequently that must be memorized, and there is the need to remember *ad hoc* sequences of as many as ten steps so that you can do them in reverse later on. This represents a serious obstacle to persisting with study of the puzzle for all but the most serious aficionados. This is where the computer comes to the rescue.

The idea occurred to me to present Top Spin on the computer and to have the computer take over the memory functions. The computer also makes it possible to offer variations on the puzzle that would be impossible to realize in plastic and which expand the exploration opportunities.

The HyperCard development environment for the Macintosh operating system gave ready access to the right set of graphic interface and programming tools to do the job. I set about developing the computerized version of the puzzle, and then was able to incorporate applications of permutation groups to solving puzzles as a regular unit in my abstract algebra course. The computer gave ready access to the puzzles, as well as useful enhancements for studying them.

After developing the Top Spin idea into my *Oval Track Puzzles* package, I created three other HyperCard stacks to round out this computerized exploration of the theory of permutation groups. One of these called *Transpose* simply provides an environment for exploring the fundamental role of transpositions in the theory of permutation groups. The other two implement other puzzles, which expand on the learning that takes place through working with the Oval Track Puzzles. These three puzzles are now combined into a single package called *Multi-Puzzle*.

Some years after developing the software as HyperCard stacks, I converted it to MetaCard, which makes it possible to use the puzzles not only under the Macintosh operating system but also on computers using the Windows operating system.

This book came after the software was developed. It offers instruction on how to use the puzzle software, copies of which accompany the book on a CD ROM. The book also introduces just enough of the theory of permutation groups to allow the reader to apply this theory in solving the puzzles, and presents numerous problems related to the puzzles and solutions to many of them. In addition, there is a section in which I show how the group theory package of the commercial symbolic mathematics software system, *Maple,* and a freeware group theory system called *GAP* can be used to provide useful information related to the puzzles.

The intended audiences

The book is written with abstract algebra students and their instructors in mind. For this audience, the book can serve as a supplement to a standard text, providing deeper exploration of one of the standard topics covered in the typical abstract algebra course.

The book and the puzzle software can be a source of individual research projects for students in abstract algebra and related courses. The exercises at the end of each chapter offer many suggestions for explorations that can be undertaken with the puzzles. These suggestions can guide students to make nontrivial discoveries regarding permutation groups.

However, I have written the book so that it can also be read independently by anyone with approximately a post calculus level of mathematical maturity who has the tenacity and interest to want to master the puzzles. It is my hope that the book will be of interest to people who enjoy recreational mathematics.

If a community of recreational players of the puzzles develops, I will be delighted. At the local level, this community potentially could be a high school, college, or university mathematics club that adopts playing with the puzzles as one of its activities.

Approach to the mathematics

I believe that, for most people, it is very important to have concrete, visual ways of thinking about abstract concepts in order to understand them fully. The theory of permutation groups is certainly an abstract subject, but the puzzles upon which this book is based gives a very useful way to relate them to models that one can see, manipulate, and understand.

This belief in concrete models shapes my approach to the group theory presented in the book. I relate all of the group theory ideas such as inverses, conjugates, commutators, and parity to the puzzles. Rather than giving formal, notationally based proofs of theorems, I state facts, and give explanations of them based upon reasoning about the puzzles, or other visual models.

For example, one can think of a permutation on a finite set by identifying the elements of the set with a collection of points and representing the permutation by a set of arrows between points, with exactly one arrow going out and one arrow coming in at each point. If we think this way, we could think of two permutations as being represented by arrows of different colors, say red and blue. Making our model more tangible, we could think of these arrows as being made of flexible plastic tubing. We could then think of composing a red and a blue permutation as corresponding to duct-taping the pointy end of a red arrow to the blunt end of a blue arrow at every point. Thinking in this way, the associativity of composition becomes rather transparent, and the rarity of permutations commuting under composition is obvious.

My approach to the mathematics reflects this belief in visualization. The informal style is intended to make the book a more comfortable companion to a formal textbook in an abstract algebra course, as well as a less intimidating source of abstract information for the recreational reader.

About the exercises

There are exercises at the end of each chapter that guide the reader through additional exploration of the chapter contents and through the features of the puzzles under discussion. They vary in their objectives from ones that focus on learning more about the software features, to others that give routine practice with some of the mathematical ideas, to yet others that suggest open ended explorations. This means that some will require just a few minutes of effort while others will require much more time.

The reader is urged to at least read through the exercises to pick up additional insights, and to work on many of them in order to become better at solving the puzzles and more knowledgeable about the mathematics behind them.

How to read and use this book

This book is a cross between a software manual, a textbook, and a recreational mathematics book. The first four chapters are written to familiarize you with the basic things that the puzzle software can do. They also ease you into some of the notation that will be needed later, and give some glimpses into the kinds of questions that we will be exploring later using some mathematics. After that, we start getting into some mathematics, and later, we get more intensively back to the puzzles and show how the mathematics allows us to understand them.

The book is not designed to be read straight through. Ideally, the reader will intersperse sessions reading it with time spent working with the puzzles. Jumping around between chapters is encouraged. You may wish to pursue a solution for one of the puzzles, for example, and return to the theory later when you appreciate the need for it.

The CD ROM, web site, and upgrades

The CD ROM that comes with this book contains the version of the software that was current at the time of the printing of the book. Upgrades of the programs that correct problems and add additional features are planned.

Register your copy of the software to get your name and e-mail address onto a distribution list for information about updates. Registered owners will receive upgrade announcements as they become available.

Information about upgrades in the software will be posted on the web site `www.nmu.edu/mathpuzzles`. This web site also includes additional information of interest about these puzzles, including ideas and tricks provided by users of the puzzles. There are also links to other resources on the web for people who enjoy puzzles of this type.

A final thought

It is interesting to observe that there are so many provably possible arrangements of the pieces on puzzles such as *Rubik's Cube* and our Oval Track Puzzles that one can argue convincingly that most of them have never been seen by any human being. In spite of this, it is possible for our human intelligence, applying the tools of mathematics, fully to understand them and restore them to order from any scrambling whatever.

For this author, making discoveries about these puzzles has been both exhilarating and humbling. The exhilaration comes from solving challenging problems. The humility comes from recognizing that the capability of human intelligence (mine) to solve problems like this is not a solo act, but requires "standing on the shoulders of giants." The mathematical knowledge needed to solve these puzzles results from millennia of human exploration of mathematical challenges.

I invite the reader to share with me not only the challenge of applying some mathematics to solving some engaging puzzles, but also, through mathematical discovery, to share the satisfaction of being part of one of humanity's most enduring intellectual enterprises.

Acknowledgments

I wish to express my appreciation to Northern Michigan University, for the support it provided to this project. Specifically I am very appreciative for the sabbatical leave during the 2000–01 academic year, which provided the time to write the book and to do some major upgrades on the software, and for a Faculty Grant, which provided extra funding.

I am indebted to Jacqueline Landman Gay of Hyperactive Software in Minneapolis, Minnesota for her technical expertise that contributed to getting the software to a form in which I was satisfied to offer it to the public. I also want to thank Tuviah Snyder of Diskotek in Brooklyn, New York, who assisted with the initial conversion of the Oval Track Puzzles from HyperCard to MetaCard. Scott Raney of the MetaCard Corporation provided lightening fast technical support for my development of the software on the rare occasions that problems with MetaCard came up. Bill Ritchie, the founder and president of Binary Arts, has been supportive of my software implementation and generalization of one of his puzzles every since I contacted him over ten years ago.

Numerous students and several colleagues who have used the software have made suggestions for its improvement, and I want to thank them. I cannot name them all, but a few stand out for their contributions to improving the software and to discovering useful puzzle solution ideas. Brent Sauvé made a nice discovery regarding an improved 3-cycle for the Hungarian Rings. Michael Kowalczyk made some fundamental discoveries about efficiency in expressing permutations in terms of transpositions, and found a vastly more efficient transposition for one of the puzzles. He also did some important torture testing of the software as it was nearing its final form, and made some very insightful suggestions for improvement.

My colleague, Barry Peterson, wrote some exhaustive search programs that found an improvement on Kowalczyk's transposition method and got us to optimality for one interesting problem. Another colleague, Randy

Appleton, provided technical help with designing the online registration system. My Freshman Fellow during 2000–01, Raymond Hedberg, did some important library research and software testing and made some helpful suggestions for improvement. Bryan Langer was one of my most eagle-eyed proofreaders.

Several people from the MAA have made noteworthy contributions to this project. Specifically I want to thank the editorial board for the Classroom Resource Materials series, and the two editors, Andrew Sterrett and Zaven Karian, who have overseen the project. The two staff members with whom I have worked most closely, Elaine Pedreira Sullivan, Assistant Director for Publications, and Beverly J. Ruedi, Electronic Production Manager, have been very helpful and patient, and have made this a better book.

Given that music is my second passion following mathematics, it was natural for me to want to incorporate it into the software side of the project. I thank the makers of that music. Organist Eddy Layton gave permission to use the clips of his organ music that I found on the Web. Paul Niemisto of the Boys of America Finnish brass septet, and Antti Hosioja of the Finnish folk music group, Tallari, have given permission to use clips of their music as the opening serenades for the two puzzle packages. Finally, I want to thank composer, Peter Hamlin, who got brought into the project too late for his work to get other mention than this in the book. He wrote some "commutator music" for me that presents musical interpretations of the commutator concept that is so fundamental to solving permutation puzzles. He also wrote a "Commutator Music Machine" computer program that allows anyone to experiment with making music using commutator ideas. Information about his work is available on the CD ROM as well as on my puzzles web site.

I want to thank my wife, Pauline, for her support and understanding throughout the software development and book writing process. Anyone who has worked on software knows that it can be an all-consuming task for long periods of time. Pauline's patience, and willingness to pick up the slack on other tasks being neglected during these times has made my job easier.

My mother, Eva Kiltinen, also has had a role in this book coming into being. Ever since I began my academic career thirty-six years ago, she has mentioned occasionally her dream that I would write a book. I always dismissed the idea; not wanting to write one unless I felt that I had something to say that had not been said as well or better by others. The combination of ingredients in this project seemed sufficiently original, so finally her dream became mine. I thank her for that.

John Kiltinen
Marquette, Michigan
June 10, 2003

Contents

Software Installation and a First Tour

0

Installing the Permutation Puzzles Package onto your computer is a simple process. Everything you need to know to install and run the Puzzle Driver application is found in the next few pages. This chapter will also whet your appetite by explaining the menu.

What You Need

This book comes with a CD ROM in a sleeve inside the back cover. This CD, readable by computers using either the Windows or the Macintosh operating system, contains everything you need.

System Requirements

Operating System: Windows: Windows 95 or later.
Macintosh OS Classic: System 7.1 or later.
Macintosh OS X Native: Any version of OS X

A 200 mhz or faster processor is recommended. The software runs on computers with slower processors, but it will be sluggish.

Hard Drive Space: Windows: 3.5 megabytes
Macintosh: 5.4 megabytes

Installation

To install the Permutation Puzzles Package software, you must first insert the CD ROM into the CD drive of your computer. After opening the Permutation Puzzles Package disk, you will find a folder that contains the puzzle software and a few other items. Among these is a "Read Me" text file that contains the details about the installation process. Double-click on the icon for this file to open it. You will have on your screen the straightforward instructions for installing the software.

On the Windows platform, an installer program handles the details of the installation. On the Macintosh platform, it is just a matter of dragging things to your hard drive. For the Macintosh platform, there is a version of the package that runs natively under the newest OS X operating system, and another that runs under older versions of the Macintosh operating system. Choose the one that is right for your operating system.

Puzzle Package Contents

Your Permutation Puzzles folder contains the following items:

Puzzle Driver is the formal name of the application program that runs the puzzles ("PuzzleDriver.exe" on the Windows platform and "Puzzle Driver for Mac" on the Macintosh). It opens a small window from which you can navigate to the puzzle and resources windows and from which you can set some controls that affect the behavior of the puzzles.

OTPuzzles.pzl is the Oval Track Puzzles component. You open it from the Puzzle Driver window by clicking on the button labeled "To Oval Track." This opens a second window, which presents 19 different versions of the Oval Track Puzzle.

MultiPuz.pzl gives you three other puzzles. The file name is a shortened form of the full name, Multi-Puzzle. You open Multi-Puzzle the same way as you open Oval Track Puzzles.

Resources.pzl offers a window that is the place to go for accessing and storing additional resources for all of the puzzles.

PuzzlePrefs.dat contains your preferences.

Hint: You need to keep all of these components together in a single folder, and you must not change the names of the separate files. The software looks for these files to be in the same folder, and looks for them by their original names.

Starting the Software

Starting the software is simply a matter of double-clicking on the Puzzle Driver icon. (The Windows installer will put a shortcut for Puzzle Driver onto your desktop, which you can also double click.) Once the small Puzzles Home window opens, as shown in Figure 0.1, click on the buttons to open the windows for the puzzles, the resources window, or the documentation window.

Registering the Software

When you start Puzzle Diver for the first time, after you click the "OK" button on the credits screen, the program takes you to a screen that encourages you to register your software. When you send in your registration information, you become entitled to technical support and you will receive announcements of new enhancements or other updates. Thus, you are encouraged to register.

Clicking on the "Register Now" button opens a new window that allows you to enter your contact information. After you have done so, click on the "Press to Continue" button, which takes you to a screen in which you confirm your information.

Once your contact information is accurate, the easiest way to register is over the Internet, assuming that you have an Internet connection active. If you do, just press on the "By Internet Now" button. When you do, your information is sent over the Internet to a collection point, and you will get a confirmation within seconds. If the Internet option does not work for you, the registration screen gives you the option of either printing your information for mailing or faxing, or of copying your information for sending by e-mail.

The registration reminder screen also gives you the options of indicating that you intend to do so, but later, or that you never intend to register the software. If you register or indicate that you never plan to do so, the program will stop reminding you. If you say you will register later, it will continue to remind you every time you start the program.

A Bonus Program

Along with the Permutation Puzzles folder, you will find on the CD ROM another application program. It is called "ParityDemo.exe" in its Windows version and "Parity Demo" as a Macintosh application.

This program demonstrates the proof of the Parity Theorem that is presented in Chapter 8. It gives you hands-on experience with using the algorithm that underlies the theorem, offering you a more concrete understanding of why the theorem is true.

This program is independent of the puzzle package. It is up to you to decide whether or not you load it onto your hard drive. It takes an extra 4.7 megabytes of hard drive space for the Windows version and an extra 5.9 megabytes for the Macintosh. To run the program, drag it to a convenient location on your hard drive and double-click its icon. See Chapter 8 for more about it.

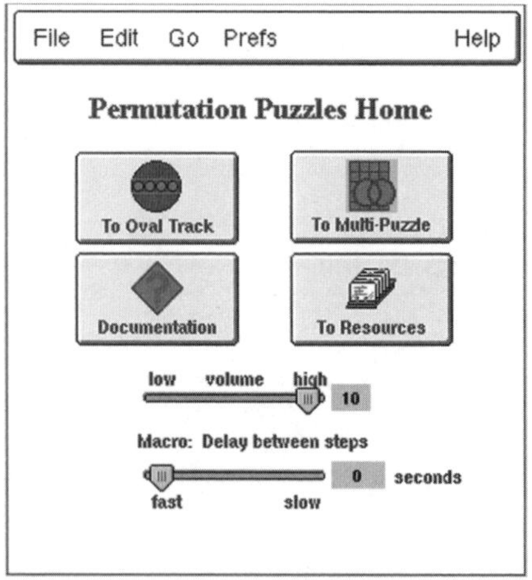

Figure 0.1. The Puzzles Home window opened by Puzzle Driver.

Uninstalling the Package

If you need to remove the Permutation Puzzles package from your computer for some reason, the process is as simple as dragging the Permutation Puzzles folder to the trash. This is all you need to do on either the Windows or Macintosh platform. However if you installed the software using the Windows platform installer, then an uninstaller was created that can be used.

A Tour of the Puzzles Home Window

Once you have installed your software, it is time to start Puzzle Driver and begin getting familiar with the Puzzles Home window (Figure 0.1). It contains four navigation buttons and two slide controls. In the Windows version, it also contains a menu bar at the top as shown in the figure. In the Macintosh version, you do not have this menu bar within the window.

Instead, Puzzle Driver for Mac puts these menu items onto the main Macintosh menu bar at the top of your computer's screen.

Buttons and Controls

Each of the four **Navigation** buttons opens another window. The top two take you to the windows for the puzzles.

The **To Oval Track** button takes you to the collection of Oval Track puzzles. There are eighteen fixed variations on the same basic theme, plus one "build your own" version in which you can create an unlimited number of variations for yourself. The basic theme involves a set of numbered disks on an oval-shaped track. You can rotate the disks around the track. In addition, you can change the positions of a few of the disks by pressing a button. The object is to have the computer scramble the disks and then you get them back in order using the tools available to you.

The **To Multi-Puzzle** button opens a window that presents three different puzzles: Transpose, Slide, and Hungarian Rings. These follow the same theme of moving numbered disks, or in the case of the Hungarian Rings Puzzle, colored disks also, under certain constraints, but the type of constraints are different from those of the Oval Track Puzzles.

The **Documentation** button opens text-based documentation for all of items of the Permutation Puzzles Package.

The **To Resources** button gives you access to tools for storing and retrieving resources for use with the puzzles. Using Resources, you can store files of two types: puzzle configurations and macro programs.

This means that you can store an unlimited number of procedures for making useful changes to the puzzles and move them back and forth to the macro buttons on the puzzles. The macro buttons allow you to perform a sequence of moves by pressing a single button.

You can also store and retrieve configurations for the puzzles. For example, we have pre-loaded certain scramblings of the puzzles into the Resources collections and these are referenced in the exercises. You can load one of these configurations into your copy of the puzzle and then do the exercise. Since we know what the scrambling is, we can include specific hints and a solution that we know in advance will work.

The Resources window also gives you the capability of creating your own configurations for the puzzles. This will allow you to invent an unlimited number of variations on the basic themes.

Sound and the **Volume** Control**:** The Permutation Puzzles make use of sound. In part this is just for the fun of it, but the sound can also serve a useful function. For example, when you are rotating the disks on an Oval Track Puzzle or on the Hungarian Rings, the computer will sound a pitch a number of times equal to the number of positions that you are moving the disks. This provides confirmation that you have done the maneuver that you intended and have not hit the wrong button or hit the right one too lightly for the computer to have gotten the message. The sounds of the moves can become part of your experience with the puzzles and can aid in remembering things.

The package gives you control of the volume from within. You can of course control volume externally either using operating system controls or a physical volume control if you have external speakers. The volume can be set on a scale from 0 to 10 with 0 being off. You can set it by moving the controller along the sliding scale. Either drag the controller to where you want it or click on the slide on the side of the controller to which you want it to move. This second method will change the setting in increments of one unit. You can also control the sound from the Prefs menu.

The **Macro Delay** Control**:** The bottom slide control is used for setting the "Macro Delay" for the puzzles. Using it, you can control the speed at which your macro buttons run the sequence of steps that have been programmed into them.

All of the puzzles but one have macro buttons that you can program. These allow you to perform a useful sequence of moves with a single press of a button. If you have the macro delay set to zero, when you press a macro button, the computer will perform the sequence of steps as fast as it can. Unless you have a very old and slow computer, it will go too fast for you to follow the individual steps.

There will be times when you want to slow the process down so that you can follow the pattern of steps in a macro program. The macro delay control allows you to do this. It builds a short delay between steps into the execution of a macro program. You control how long this delay lasts on a scale from zero to three seconds. Slide the controller along the scale to change the setting. When you release the mouse button after making a change, the computer will demonstrate the length of the delay by showing a progress bar and bracketing your delay time between two beeps. As an alternative to dragging the controller along the scale, you can click on the scale on the side of the controller toward which you want to move. This will cause the controller to move in quarter second increments.

A Tour of the Menu Bar

The menu bar that is shown at the top of the window in Figure 0.1 appears also in the windows for Oval Track, Multi-Puzzle, and Resources, but not in the documentation window. For the most part, the functions available to you from the menu bar are standard ones and thus will be clear to you based upon experience with other software packages. However, there are a few aspects of how the menu items perform that are specific to this puzzle package, so it will be worthwhile to give a brief rundown of the menus here.

Platform Differences: Before describing the specific menu items, let us mention that the menus are handled somewhat differently on the Windows and Macintosh platforms. On the Windows platform, each window has its own menu bar if one has been put there. On the Macintosh platform, a standard menu bar is always at the top of the screen, and menu items there are modified according to what application's window is the current front window.

This package has its menus where you expect them according to the standard practice for the platform. The figures in this chapter have the Windows look (but were generated on a Mac), but the way the menus behave will be the same on each platform.

The **File Menu** is shown in Figure 0.2. It offers standard items that you expect on a file menu.

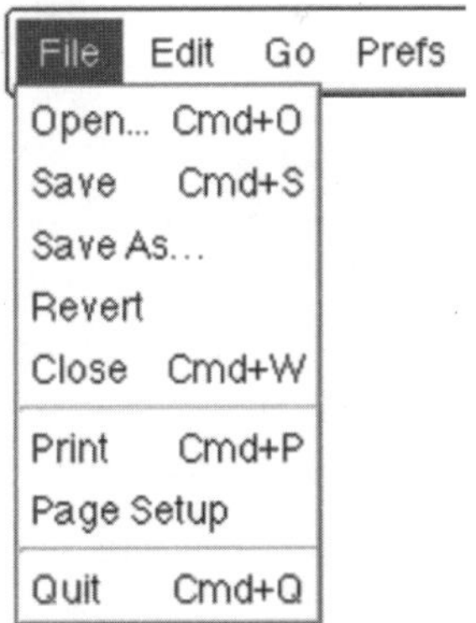

Figure 0.2. The File menu.

The **Open** command is an alternate way to open the windows for the puzzles. You can also use it to open copies of the puzzles that you may create. This gives you, for example, a way to have two copies of a puzzle open side by side for comparing approaches or doing side-by-side demonstrations.

The **Save** command saves changes in the current versions of the puzzles. If Puzzle Driver is to the front, this command will be inoperative because you cannot save changes in application programs.

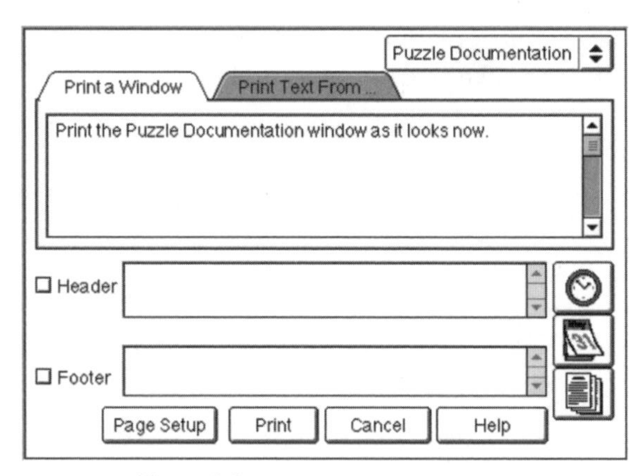

Figure 0.3. The print dialog window.

The **Save As ...** command allows you to save copies of the puzzles. Warning: Give the copies you make different names from the names of the original copies unless you want to replace the originals.

The **Revert** command restores the last saved version of the puzzle that is to the front, removing any changes that you have made to the currently open puzzle.

The **Close** command causes the current front window to close without quitting from the Puzzle Driver program.

The **Print** menu item opens a print control dialog box that is part of the Permutation Puzzles Package. From within this dialog box, you can do printing of two types: (1) screen shots of the currently open windows, and (2) the contents of various text fields. The controls of this print dialog are easy to figure out, but there is also a Help button, which walks you through the steps for setting up a print job.

The print tools built into the Permutation Puzzles package are rather basic, and do not give you fine control over formatting. The tools are designed to enable you, for example, to print a usable but not elegant, copy of the online documentation, or to record a screen shot. If you want to print materials generated by the package with more control than the package offers, you can usually copy what you need from the puzzle window and paste it into a document from an application program that is designed to produce well-formatted printed matter.

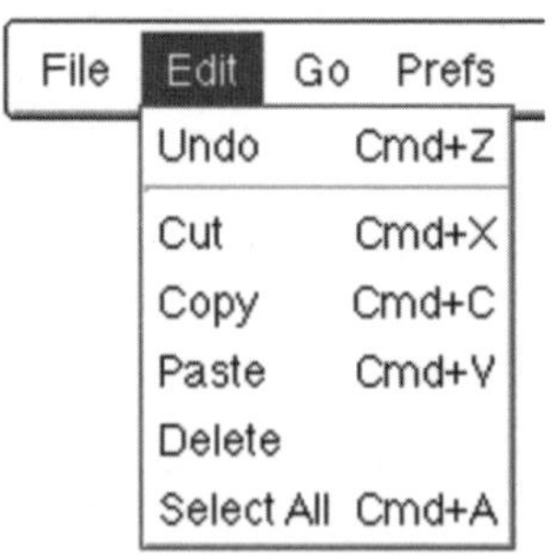

Figure 0.4. The Edit menu.

The **Page Setup** menu item opens the standard Page Setup dialog box for your operating system.

The **Quit** command shuts down the Puzzle Driver program and also closes any puzzle windows that may be open.

You use the **Edit** menu for a limited number of text handling functions in the Permutation Puzzles Package. The menu is shown in Figure 0.4. Most of the time, all but the Copy command are deactivated on this menu. The entire menu is active only when you are using the Note Pad of one of the puz-

Go Prefs

Puzzles Home Cmd+H

Multi-Puzzle Transpose
Multi-Puzzle Slide
Multi-Puzzle Hungarian Rings

OT 1: [1,4] [2,3]
OT 2: [4,3,2,1]
OT 3: [3,2,1]
OT 4: [5,4,3,2,1]
OT 5: [1,2] [3,4]
OT 6: [1,11] [4,14]
OT 7: [5,3,1]
OT 8: [1,3] [2,4]
OT 9: [4,2,1]
OT 10: [1,3,5,7]
OT 11: [7,4,2,1]
OT 12: [11,7,4,2,1]
OT 13: [1,5,7,3]
OT 14: [1,6,5,3,4,2]
OT 15: [1,3,4,2,5]
OT 16: [1,4,3,5,2,6]
OT 17: [1,6] [2,5] [3,4]
OT 18: [1,11,7,13,4,9]
OT 19: Build Your Own

Resources Cmd+R

Figure 0.5. The Go menu.

zle windows. When you are using the Note Pad, these menu items function as they do in any word processor to allow you to manipulate the text of the Note Pad.

The **Copy** command is always available. You can use it to copy text from many of the text fields that appear in the puzzle windows. As in other programs, you highlight the text and use Copy to place your selection on the clipboard. From there, you can paste it into documents of other programs such as a word processor.

The **Go** Menu (Figure 0.5) offers one of several alternative methods of navigating through the Permutation Puzzles windows. When you click on Go on the menu bar, you open a long list of the various puzzles, including all three that are in the Multi-Puzzle window and all nineteen versions of the Oval Track Puzzle. Move the cursor to any of these and release the mouse button and that puzzle is brought to the front. You can also go to the Puzzles Home or Resources window from this menu.

The package offers other ways of navigating using buttons on a button bar in the puzzle window. We introduce the button bar in Chapter 1.

The **Prefs** Menu is shown in Figure 0.6. From this menu you control the musical introductory tour, the tool tip hints, and how your puzzle saves changes and settings. You can also set the color scheme and the volume from this menu. The menu includes access to its own help screen.

To get all of the details, go to the Prefs menu and select the Prefs Menu Help command. This takes you to a screen in the Puzzles Home window that shows the menu. Moving the cursor over a menu item opens a field with an explanation of it.

Figure 0.6. The Prefs menu.

The **Help** Menu: On the **Windows platform**, the Help menu has two items on it. One is the About item which takes you back to the credits screen in the Puzzles Home window. The "Help..." item takes you to the text-based documentation from the Puzzles Home window, and, from any of the puzzle windows, to the graphic-based help screen for that window. On the **Macintosh** platform, these items are in the standard places on the Macintosh menu bar at the top of your screen. When any window of the Permutation Puzzles Package is the front window, the About item of the Apple icon menu takes you to the credits screen in the Puzzles Home window. The "Help..." item of the Help menu takes you to the text-based documentation if Puzzles Home is in front and to the graphics-based help screen if a puzzle is in front.

Exercises

0.1 Getting familiar with the documentation. Click on the documentation button to open the text-based documentation window. (a) Resize the window by dragging the lower right corner. (b) With the Puzzle Home documentation showing, click on the Search button, and enter the word "macro." Check the "Match whole words only" box, and find all occurrences of this word by repeatedly pressing the "Find Next" button. How many do you find? (c) Click on the "Rings" button to bring up the Hungarian Rings Puzzle documentation. Scroll down the Table of Contents on the left until the entry "9. Scrambling" shows, and click on this line. Read the text in red in Section 9, and answer this question: How can you avoid the confirmation dialog box that comes up when you press the scramble button? (d) Close the documentation window by clicking in the close box in the upper right corner (Windows) or upper left corner (Mac).

0.2 Getting familiar with the macro delay. (a) In the Puzzles Home window, try different settings on the macro delay slider control. Drag the pointer along the slide track. Then drag it all the way to the left so that it shows a delay of 0 seconds. Try clicking to the right of the pointer on the slide track to advance the setting in quarter second intervals. (b) To see the effect of the macro delay, click on the "To Oval Track" button to open the Oval Track Puzzles window or bring it to the front if it is already open. (It will go to version 1 upon launching, but if you've had it open to another version, go back to version 1.) Now click on the macro button labeled "B" and watch as the computer performs an eight-step procedure. (This assumes that the program that is in button B when the puzzle is new is there. Otherwise whatever program you have changed to will run.) Go back to Puzzles Home, change the macro delay, return to Oval Track 1, and try button B again. See how the delay affects the speed of the operation of the program.

0.3 Try out the Prefs Menu Help. Move the cursor to the Prefs menu, open it, and select the bottom menu item, "Prefs Menu Help." The Puzzles Home window will then display a picture of the Prefs menu. Slide the browser hand over this graphic of the menu to bring up explanations of each menu choice as the browser passes over it. (You do not need to hold the mouse button down while doing this.) Press on the "Return" button to bring the usual Puzzles Home screen back.

0.4 Setting the volume. In the Puzzles Home window, experiment with using the upper slide control to set the volume. Do this by dragging the slide control and also by clicking on the slide track to one side or the other of the control.

0.5 Getting familiar with the ToolTip hints. Go to the Prefs menu and select the "Tool Tips on" command. Open the menu again to see that this menu item now has a check mark showing to the left of it. When the check mark is showing, "tool tip" hints will show as you move the browser over various objects in the puzzle package windows. How many different objects in the Puzzles Home window have tool tip messages?

0.6 Practice with printing. Go to the "File" menu and select the "Print" command. This opens a print dialog window. (a) Click on the "Help" button. Then click on the arrows that replace it to go through a description of the six steps involved in setting up a print job. (b) If you have a printer, try printing out an image of one of the puzzles, or some of the documentation.

0.7 Another help mode. When you have any of the three windows for Oval Track Puzzles, Multi-Puzzle, or Puzzle Resources open, you will see in the lower right corner of the window a button with an "i" icon at the right end of the button bar. Pressing this button takes you to a graphic help screen within the window. This screen looks like the puzzle or resources screen. By clicking with your browser on any of the objects on the screen, you will open a text field that gives you an explanation of that item. You return back to where you came from by pressing the "Go Back" button which is where the "i" button had been. (a) Go to one of these graphic help screens and try clicking on a few objects. (b) Since the graphic help screen looks like the puzzle, it is possible to mistakenly think that you are if fact looking at a puzzle screen. See if you can identify three visual cues that this is the help screen and not a puzzle. (c) Describe the transition that takes you from a puzzle screen to the graphical help screen.

0.8 What's your favorite color? The Prefs menu gives you the option of picking from among four colors for your puzzle windows backgrounds. When you move the browser over the menu item named "Color scheme," a submenu appears which offers the options: Tan, Blue, Green, Yellow. Try out the four colors to see which one you prefer. What works for you will depend not only on your taste in colors but also on the sort of color reproduction your monitor offers.

0.9 It's time to Go! Experiment a bit with the "Go" menu item. Try using it to go to various puzzle and resources windows. What happens if you select a puzzle that belongs to a window that is not currently open?

An Overview of Oval Tracks

Now that you have installed the software that came with this book, let's get acquainted with it. We will start with Oval Track Puzzles. Go to the folder into which you installed the puzzles and find the Puzzle Driver program (PuzzleDriver.exe if you are using Windows, or Puzzle Driver for Mac if you are using Mac). Double-click the icon to start Puzzle Driver. Then click the "OK" button on the credits screen, and through the registration prompts if you've not yet registered to get to the Puzzles Home window. Then click on the "To Oval Track" button.

When the program is running, you will see a screen like the one shown in Figure 1.1. There are numbered disks in numbered positions on an oval-shaped track. You can move the numbered disks around the oval track, keeping them in the same order. You can also make one other change in the relative positions of a few of the disks. On the first puzzle that you see, you can reverse the order of the four disks on the circular "turntable." The program offers more versions, each having a different way of reordering a few of the disks. For example, Figure 1.2 shows puzzle number 7, which replaces the turntable with a *three-cycle* that moves the disks in spots numbered 1, 3, and 5. You can define your own rule for changing a few disks in the "Build Your Own" version (number 19).

The object is to mix the disks up, and then try to get them back into the original order. Alternatively, you can try to put them into whatever arrangement you would like.

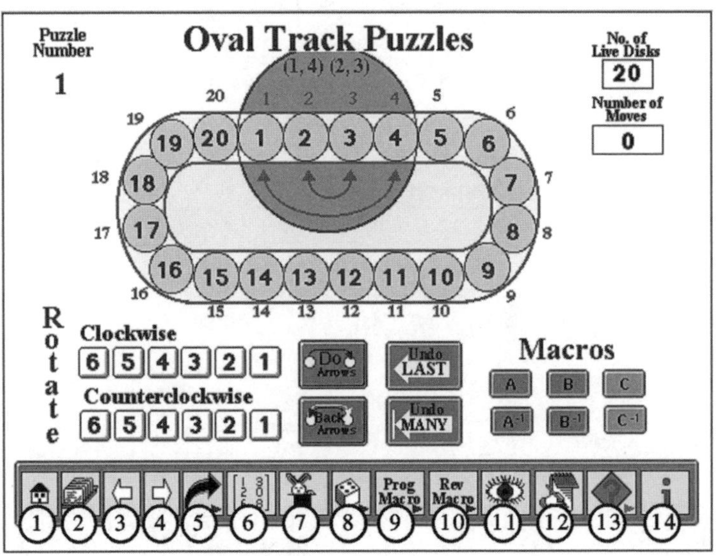

Figure 1.1. Oval Track Puzzle Number 1.
Numbers on buttons keyed to explanations below.

1.1 How the Puzzle Works

Let's begin by familiarizing ourselves with how the puzzle works. We will start with how you move the disks around by dragging and pressing buttons, then describe the use of the other free standing buttons below the oval-shaped track, and finally describe what each of the buttons on the lower button bar will do.

Moving Disks Around the Track

The numbered disks can be rotated on the oval track in two ways. By clicking on one of the numbers in the lower left corner, you can rotate the disks by that number of places either clockwise or counterclockwise. Alternatively, you can use the mouse to drag any disk to a new location. When you drag one disk to a new location, they all rotate in the way they must to move the disk you dragged.

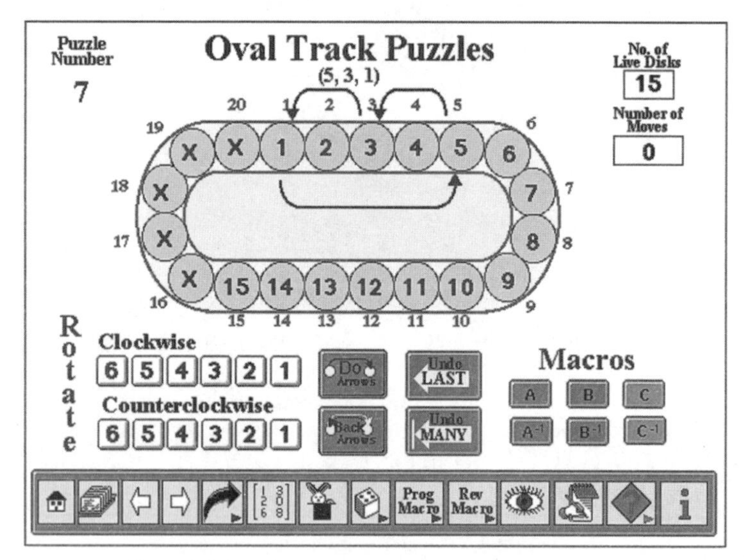

Figure1.2 Oval Track Puzzle No. 7 with 15 disks active.

The Do Arrows Operation

The "Do Arrows" button causes the disks in locations that have red arrows pointing out from them to move to new locations following the arrows. For example, on the first puzzle, the disks in locations 1 and 4 trade places, and those in locations 2 and 3 do the same. The "Back Arrows" button reverses what the "Do Arrows" button does. In those cases where the "Do Arrows" button takes many applications to bring things back to where they started, "Back Arrows" can be a time-saver. (If "Do Arrows" just causes some disks to trade places, then the "Back Arrows" operation does the same thing.)

Backing Up

Sometimes after making a few moves, you discover that it was not what you really wanted. If this happens, you can reverse your series of moves by pressing the "Undo Last" button once for each move that you want to reverse. Each time you do this the most recent move will be undone, and the counter will decrease by one. If you want to undo many moves, press the "Undo Many" button. A dialog box will ask you how many moves you want to undo. Enter this number and click "OK" and the computer will reverse that many moves. You may also hold down the Alt key (Windows) or Option key (Mac) while pressing the "Undo Many" button to have the computer undo 10 steps without asking.

The Macro Buttons

In working with the puzzle, you will discover certain series of moves that you will want to use over and over again because they are so useful. It is possible to program such series into one of the "Macro Buttons." These are the buttons labeled A or B or C. Once you have programmed one of these buttons, you will only need to press that button, and the computer will automatically do your sequence of moves. The button A^{-1} or B^{-1} or C^{-1} will perform your sequence of moves in the reverse order. This is useful if you have programmed button B, say, to do a "3-cycle" that takes the disk in spot 1 to spot 2, the one in spot 2 to spot 3, and the one in spot 3 back to spot 1. If in solving the end game you needed to move not from 1 to 2 to 3 and back to 1 but rather 3 to 2 to 1 and back to 3, you get this by simply pressing button B^{-1}. You could also do the job with button B pressed twice, but it would take twice as many steps.

You begin to program a macro button by pressing one of the buttons on the button bar. We will describe this process later.

The Button Bar

The button bar across the bottom of the window provides access to procedures that you will use less often than the buttons above that you use to do manipulations of the puzzle. We will describe what each of them does, working our way from left to right. The numbers used here correspond to the ones shown for the buttons on Figure 1.1.

1. The **Home** Button (house icon) brings the Puzzles Home window to the front. You use it when you want to go to the home window to change a setting, for example.

2. The **Resources** Button (cardfile icon) opens a new window, (or brings it to the front if it is already open), which you use to store and retrieve macro programs and configurations for the puzzles. Read Appendix A to learn about Puzzle Resources.
3. The **Previous** Button (left arrow icon) takes you to the version of the Oval Track Puzzle that comes one place earlier in the numbering sequence.
4. The **Next** Button (right arrow icon) takes you to the version of the Oval Track Puzzle that comes one place later in the numbering sequence.
5. The **Go Anywhere** Button (vaulting arrow icon) opens a pop-up menu that lists all of the puzzles. By selecting one of them from the menu, you can go directly to it, in contrast to going step by step as you do with the next and previous buttons. The menu allows you to go to the puzzles in the Multi-Puzzle window, as well as to any in the Oval Track Puzzles window. You can also use it to go to Resources or Puzzles Home.
6. The **Reset Live Number** Button (number array icon) is the one to use if you want to restart the puzzle and at the same time reset the number of active disks. The number of active disks can range from the maximum of twenty down to the smallest number, which leaves room for the "Do Arrows" to operate. You will discover changes, sometimes minor and sometimes dramatic, in how some of these puzzles work as you change the numbers of active disks. When you press this button, the computer will confirm your intent to restart, and then will open a dialog box where you indicate how many disks you want to be active.
7. The **Reset Current** Button (rabbit-out-of-a-hat icon) is the one to press to start over with the current configuration. When you press it, you will return each numbered disk to its original position and reset all of the counters.
8. The **Scramble** Button (die icon) scrambles the numbered disks. When you press this button, you will get a pop-up menu offering four options. These are a complete scramble, which randomly scrambles all of the currently active disks, an end game scramble, which scrambles only the disks within the range of the "Do Arrows" move, and two options for a "Buttons Scramble." Figure 1.3 shows the results of a typical complete random scramble.

   A *buttons scramble* is one that is done by having the computer perform a series of presses on randomly chosen rotation buttons and the "Do Arrows" button, but without showing the steps. How scrambled the disks become will depend on how many times the computer presses the buttons. If it is just a few, you can solve the puzzle quickly. If the number is large, solving the puzzle might be as challenging as a full random scramble. In all cases, a buttons scramble is always solvable, while a random scramble may or may not be.

You will use end game scrambles to test out and refine your methods for getting the last few disks into their right places. This is the most difficult part of solving this puzzle, so having access to end game scrambles will allow you to focus attention where it will be needed most.

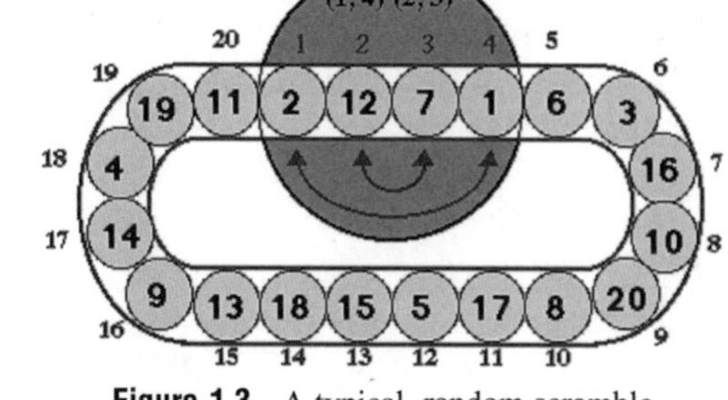

Figure 1.3. A typical random scramble.

9. The **Program a Macro** Button starts the process of programming a sequence of steps into one of your macro buttons. When you press it, you get a pop-up menu that offers you a choice of which button to program, namely *A*, *B*, or *C*. When you have made your selection, a field with instructions appears, and a button that gives you one last chance to abort the process if you have changed your mind. If you want to proceed, you press the Program a Macro button again, which has changed color to green and now says "Begin." After this, you perform the sequence of steps you want programmed into the button and the computer will record your steps. When you are done, press the "Record a Macro" button again, which now is red and says "Done." That is all there is to it. You can repeat the sequence by pressing the button you have programmed.
10. The **Review a Program** Button opens a pop-up menu that lists the three-macro buttons. Choose one, and you will be shown a field that displays the current program recorded in that macro button, together with an explanation of the coding system used. You will also see two buttons that can be used to copy the program to the clipboard for moving it to another document such as a word processing file or to move the program into your Resources file for long-term storage. See Section A.4 of Appendix A for details regarding storing macros.
11. The **View Moves** Button (eye icon) displays a field that shows the sequence of moves that you have made since the last time you reset the puzzle. As was the case with the Review a Macro button, when you press the View Moves button, you also make the two copy buttons visible, which allow you to move the record to another document or turn it into a macro button sequence and store it for future use. Section A.4 of Appendix A describes this procedure for storing macros.

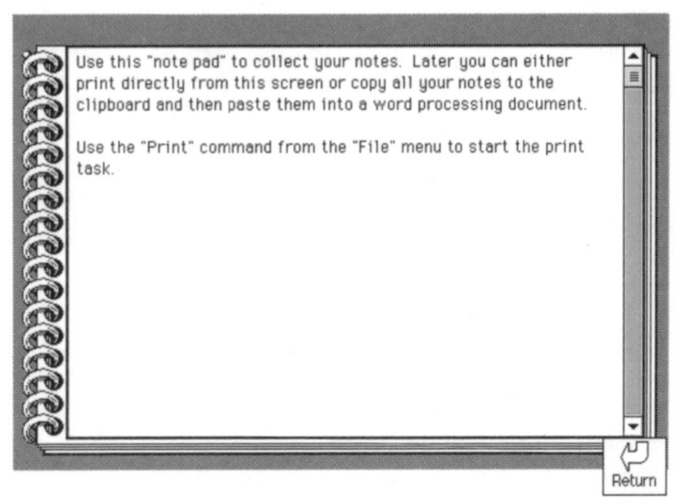

Figure 1.4. The Note Pad screen.

12. The **Note Pad** Button (note pad icon) takes you to a "Note Pad" in the same window. Figure 1.4 shows what it looks like. You may use it to collect a written record of your discoveries about the puzzles as you are playing with them. It offers you limited word processing functionality such as Copy, Paste, and Select All. Access to these functions are available from the menu at the top of the window (Windows) or at the top of your desktop (Macintosh), following the standard conventions.
13. The **Documentation and Ideas** Button (question mark icon) offers you a choice of going to the on-line, text-based documentation for the Oval Track Puzzles, or opening a window which gives you some ideas about the current version of the puzzle you are looking at.
14. The **Graphical Information** Button ("i" icon) takes you to a graphically-based help screen. This screen looks like the puzzle window, but when you click on any item on this screen, you open up an explanation of it.

1.2 The Opening Game

You will want to spend a bit of time trying things on the puzzle to get familiar with it. After a few minutes of this, you will be ready to start learning how to solve the puzzle. So go ahead and press the "Scramble" button, and choose the option "Complete Random." Start doing moves to bring the disks back into their original order. Practice this a few times, by putting five to ten disks into the proper sequence.

In order to work from the same scrambling, let's consider the one that is shown in Figure 1.3. (Note that you can get this scrambling of the puzzle from the Resources window, load it into your puzzle and duplicate the steps described here as you read. See Exercise 1.1 for details.) The numbered disks are quite mixed up here. This sort of scramble was the computerized equivalent of dumping all of the disks off of the track, and then randomly picking a first one to put into spot 1, randomly picking a second to put into spot 2, etc., until all disks were returned to the track. There are 20 ways of picking the first disk to return, 19 ways of picking the second, 18 ways of picking the third, etc. The total number of ways in which this process could have been done is, according to one of the most basic principles of combinatorial mathematics, the product of these numbers of choices at each step. That means that there are 20 times 19 times 18 times … times 2 times 1, or 20 factorial ways of scrambling the disks. The standard notation for factorials puts an exclamation point after the largest number. We note that 20! = 2,432,902,008,176,640,000. This is about 2.4 quintillion, a very large number indeed. But not to worry. Whichever of the several quintillion of the possibilities the computer confronts us with, we can begin to solve it.

Let's start by getting the low numbers in order and working our way up. (There is no reason why we have to do it this way; we could equally well start at the high end and come down.) We observe that disks 1 and 2 have ended up rather close together, but there are two other disks between them and disk 2 is on the wrong side.

Our first task will be to get these disks, numbers 1 and 2, together. Let's call them our *focus disks* for now. We will bring our focus disks together using the only maneuvers available to us: moving all the disks one or more positions around the track, and using the "Do Arrows" button to switch the disks in spots 1 through 4 according to the arrows.

If we press the "Do Arrows" button first, we will have solved our orientation problem. That is, disk 1 is now on the left and disk 2 on the right. The result is shown in Figure 1.5. Note that disks 1 and 2, which were in spots 4 and 1 respectively before have now traded places, while disks 7 and 12 have traded places between spots 2 and 3. We still have to do something to get the two disks that separate disks 1 and 2 out from between them.

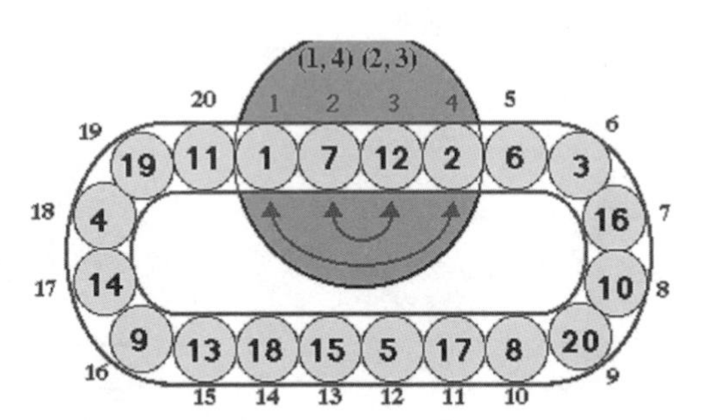

Figure 1.5. The scrambling of Figure 1.3 after performing a "Do Arrows".

The fact that there are two disks between our focus disks is better for us than if there were one disk in between. It means that it will take us fewer steps to get them together. The key insight for this first "turntable" version of the puzzle is that *three disks between two focus disks is just a turntable away from getting them together.*

Our intermediate goal will be to get three disks between our focus disks 1 and 2. To get started, let us shift everything two positions counterclockwise. The result is shown in Figure 1.6a.

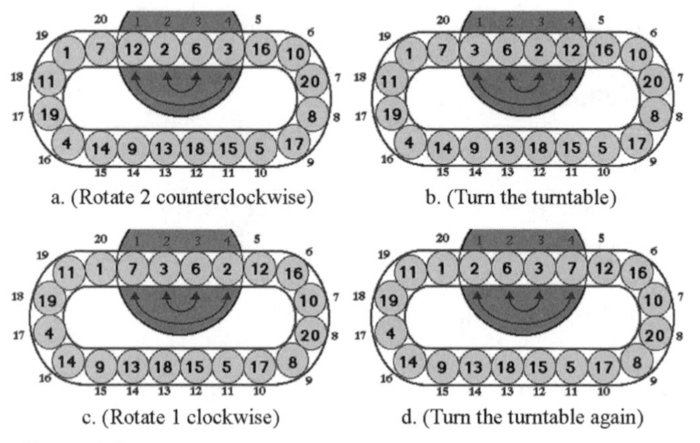

a. (Rotate 2 counterclockwise) b. (Turn the turntable)

c. (Rotate 1 clockwise) d. (Turn the turntable again)

Figure 1.6. Four more steps to get the focus disks 1 and 2 together.

Do you see where this is headed? With focus disk 1 off on the left of the turntable, and with focus disk 2 in spot 2 on the turntable, if we now perform a "Do Arrows," what happens to disk 2? It is moved to spot 3, which means that there will be three disks between the focus disks, as shown in Figure 1.6b. Next, rotate the disks one position clockwise, as shown in 1.6c, and finally, turn the turntable again. The result, as shown in Figure 1.6d, is what we wanted. We have gotten our focus disks 1 and 2 together. It has taken us five steps.

Our focus now shifts to getting disk 3 next to disk 2. Since there is one disk between them, it will take two applications of the turntable's ability to widen the gap by one to get the new focus disks 2 and 3 with three disks between them. When we do, we can bring them together. Getting disk 3 next to disk 2 will take six steps. We show these steps in Figure 1.7.

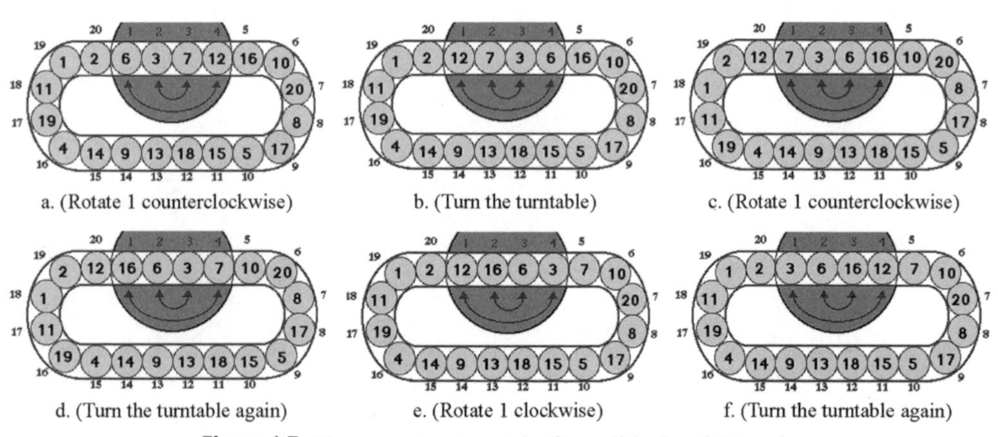

a. (Rotate 1 counterclockwise) b. (Turn the turntable) c. (Rotate 1 counterclockwise)

d. (Turn the turntable again) e. (Rotate 1 clockwise) f. (Turn the turntable again)

Figure 1.7. Six more steps to get the focus disks 2 and 3 together.

So far so good. We have gotten three disks together, and we are starting to see how to do it in general. We confidently look for disk 4 with the goal of bringing it next to disk 3.

Perhaps you are dismayed when you see how far disk 4 is away from disk 3, that is, if you measure from 3 to 4 in a clockwise direction. However, we quickly realize that this distance does not prevent us from bringing 4 next to 3; it only means that we will need more steps.

Let's start out by doing a rotation to bring disk 4 into spot 4. How many positions is that? Twelve counterclockwise or eight clockwise, but you do not really need to know that. This is a time when the dragging method of rotating the disks proves its value. You just drag disk 4 to spot 4, and you do not need to concern yourself with how many positions that is, either clockwise or counterclockwise.

After getting disk 4 into spot 4, turn the turntable. This will move disk 4 to spot 1. You have moved it three positions closer to disk 3. Now rotate the disks three positions clockwise to get disk 4 back into spot 4 and turn the turntable again. You have gotten disk 4 three places closer to disk 3. Keep this up until disk 4 gets down within the last few positions away from our goal. From here, the process is like what we described earlier.

This isn't so bad after all, is it? We are not having to use much more than common sense to guide us. We could keep this up until …. Until when, exactly?

Well, until we get down toward the last few disks. Try a few scrambles and follow these ideas about getting things back into place. Occasionally, you will be lucky and the last few will fall into place of their own accord, but most of the time you will find that getting the last few into place will be very difficult. Precisely, the "last few" for the first version of "Oval Track" is four disks, which is the size of the turntable. When we get to this stage, we say that we are in the *end game*.

For the scramble we started with, one way of working the disks brings the puzzle to the state shown in Figure 1.8 in 66 steps. (The 66 steps that get you to this stage are available in the Resources window for this puzzle. See Exercise 1.2 for the details on how to load these steps as a macro so you can see this one way of doing the opening game.) This is not necessarily the least number of steps to the end game, and it is definitely not the case that every procedure one might adopt will lead to the same end game. What we see here is one way that it worked out for the author.

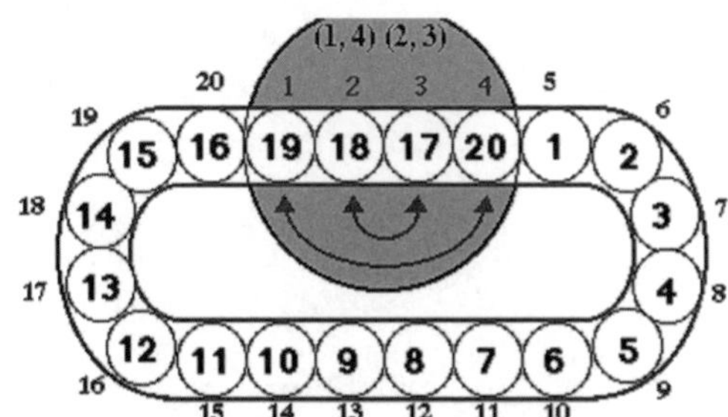

Figure 1.8. An end game for the scramble of Figure 1.3 after 66 steps.

If you were to continue with the methods we have been using focusing on disks 16 and 17, you could bring them together, but the problem would be that disks 1 and 2, which have been under control for so long now, would get brought back into the jumble. Each time we get one more high number right, we bring one more low number into the jumble of four disks.

When you get down to these last four disks, you are into the "end game" for the puzzle. It is at this point that you need mathematics to guide the solution strategy. We will defer consideration of the end game until after we have developed the mathematical ideas that we need.

1.3 Notation and a Bit of Algebra

In order to study these puzzles, it will be very helpful to have a notation for recording information. Since there are two basic operations we can use to move the disks, we can build a very helpful notation with two basic symbols. Let us define that

(1.1) R denotes a one-position clockwise rotation of the disks, and
T denotes a turn of the turntable, or more generally, pressing the "Do Arrows," button to move disks following the arrows.

We can now denote any series of maneuvers by stringing together a sequence of these two symbols. For example, the sequence $R\,R\,R\,T\,R\,R\,T$ represents rotating the disks three positions clockwise, then performing a "Do Arrows," then rotating two more positions clockwise, and then performing another "Do Arrows."

How would we represent a one position counterclockwise rotation in this very simple notation? Well, if all 20 of the disks are active, we note that rotating one position counterclockwise is the same as rotating 19 positions clockwise. We could denote a one position counterclockwise rotation by writing down 19 R's.

Such a notation would be cumbersome, and it is easy to see how to improve it. We will add notations for the *inverses* of the two basic moves.

(1.2) R^{-1} denotes a one-position counterclockwise rotation of the disks, and
T^{-1} denotes pressing the "Back Arrows," button to move disks following the arrows in the reverse direction.

We observe that the inverse of T for the first version of the Oval Track Puzzle gives us nothing new. The arrows for this version with the turntable are bidirectional, so to follow them in the opposite direction is the same as following them in the forward direction. That is, $T^{-1} = T$. However, for some of the later versions, the inverse of T

will be different from T, and having a separate notation for it will be helpful, both for making more compact representations and for theoretical clarity.

With the introduction of superscripts to denote inverses, our notation is looking like multiplication. We will take this one step further and introduce exponentiation. For multiplication, exponents are used to represent repeated multiplications. Here, they can represent repeated successive applications of a maneuver. For example, we can represent $R\,R\,R$ more compactly by R^3. The meaning of negative integer exponents ought to be clear also. We will understand R^{-3} to denote the application of R^{-1} three times. That is, $R^{-3} = (R^{-1})^3$. More generally,

(1.3) For any positive integer n, $R^{-n} = (R^{-1})^n$.

The analogous definition applies to T raised to negative integer powers.

With the introduction of integer exponents, both positive and negative, our system is nearly complete and about as efficient as we can get. For completeness, we introduce the integer 0 as an exponent. We define

(1.4) X^0 denotes zero applications of the maneuver X, where X is either R or T.

Applying R or T zero times clearly makes no change in the disks. We will call a procedure that makes no changes *the identity procedure*, and will denote it by I sometimes. The use of the definite article *the* is no grammatical oversight here. Even though we can think of not doing any number of things to change the disks, the act of not doing any of them produces the same outcome as moving nothing. It is the uniqueness of this outcome that justifies the use of the definite article. As often happens with mathematical notations, we can pack most of this insight into a single string of equalities: $R^0 = T^0 = I$.

To illustrate the use of our notation, let us look back to the steps we performed in the last section to start the solution of the scrambling presented in Figure 1.3. We can describe all of these steps with the expression

(1.5) $$T\,(R^{-2}\,T\,R\,T)\,(R^{-1}\,T\,R^{-1}T\,R\,T)$$

We have used the pairs of parentheses to group the steps according to the way they were illustrated. The result of applying the first (leftmost) T is what is shown in Figure 1.5. Following the steps within the first pair of parentheses, we get to the arrangement displayed in Figure 1.6d, and following those within the second pair, we get to Figure 1.7f.

Parentheses will never be necessary in this notation to eliminate ambiguity. That is, if we write down a string of these symbols, we can group them in any way we want without changing the outcome of an application of the maneuvers they represent to the puzzle. After all, what do the parentheses really do? They indicate some way of grouping the steps, which may or may not be of some aid in interpreting the string. We used them in (1.5) to group the steps to conform to the graphical illustrations. At some point, you might find it helpful to divide a string into two groups, separating them at a point where you paused in working a puzzle to check your e-mail or to go and get a cup of coffee. Where you took this pause had no effect on what the sequence did to the disks. The effect comes from which steps are in the sequence and what their order is.

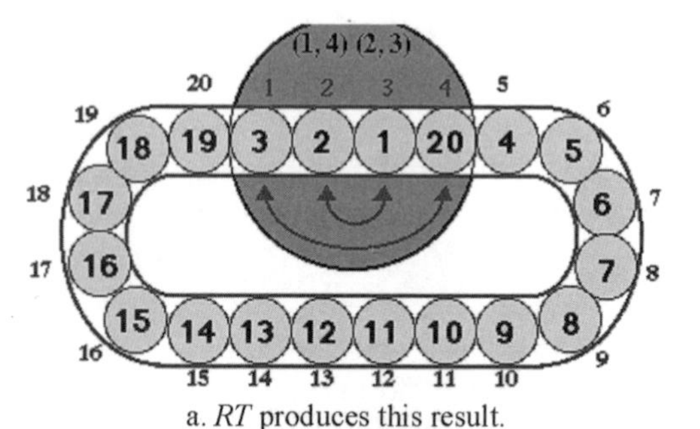

a. RT produces this result.

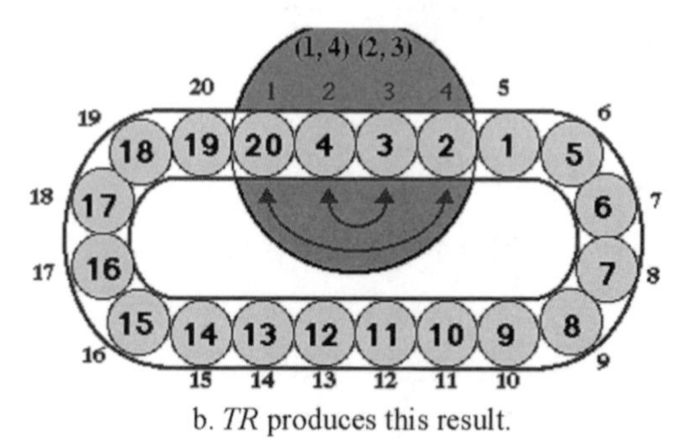

b. TR produces this result.

Figure 1.9. The maneuvers R and T do not commute.

This matter of the order of the steps is worthy of mention at this point. In contrast to the fact that any grouping imposed makes no difference, the order in which these maneuvers on the puzzle are applied does make a difference. In most cases, if we perform two steps A and B first in the order AB and then in the order BA, the outcomes will be different. That is, in general, $AB \neq BA$. Figure 1.9 illustrates this in the case of R and T, showing that $RT \neq TR$. The figure shows the result of starting with the puzzle having everything in its right place and then applying RT (on the top) or TR (on the bottom). We see that the results are different. While the outcome is the same for most of the disks, specifically those numbered 5 through 19, the places where disk 20 and disks 1 through 4 end up is different.

We summarize this observation by saying that on the Oval Track Puzzle the rotation R and the turntable T *do not commute* with each other. We will say more about commutativity later.

1.4 Heuristics

This book is meant to be read with a computer on and the puzzles up and running, and with frequent periods of setting the book aside to test out ideas on a puzzle. If you were doing this while in §1.2, you perhaps had thoughts such as, "it ought to be possible to do this more efficiently." Part of the fun of working with these puzzles is going beyond learning how to solve them to finding better ways to do it. Formal mathematics will not be as useful for this as *heuristic* methods will be. The word "heuristic," as a noun, has evolved to refer to a procedure that is discovered through experience and which is believed to do the job. As an adjective, "heuristic" means derived from experience as opposed to having been proved.

We will see in later chapters that the issues that arise in the end game will be settled by rigorous mathematical methods. These methods will also allow us to settle many global questions about what is possible and what is not for many of the puzzles. However, when it comes to issues of efficiency that live in the opening game, rigor will give way to heuristics.

As you work with the puzzles, you will want to give thought to how you can make use of features of each particular one to solve it most efficiently. For example, in §1.2, we discussed what to do next with the puzzle in the state shown in Figure 1.7f. The next step was to move disk 4 next to disk 3. The procedure we described to do this was kept simple so as not to overwhelm you.

However, you could ask yourself if one could not get the job done quicker by moving disk 4 in a clockwise direction to where it needs to go. Disk 4 needs to move 6 positions in this direction instead of the 14 positions that it must move going counterclockwise. Since 6 is significantly smaller than 14, going clockwise ought to take fewer steps.

Another option that we could have considered is to take disk 4 around counterclockwise, but on the way to bringing it into position, we could join it up with disk 5 and move the two together. After all, they both need to go the same way and are not too far away from each other. You will want to explore methods that work on two or more disks at a time. Each puzzle will present differing opportunities for efficiencies of this sort.

Exercises

1.1 Practice with the Figure 1.1 scrambling. With Oval Track Puzzle 1 open on your computer, click on the Puzzle Resources button on the button bar (the button that is second from the left with the file cards icon). This takes you to the Puzzle Resources window. In the upper left field labeled "Locally Resident Presets," click on the first entry titled "1. Exercise 1.1 …." Then in the dialog box that appears, press the "Load" button. This will load the scrambling shown in Figure 1.3 into your copy of the puzzle. Now carry out for yourself the procedure described in this section for starting to solve this scrambling. You may use the "Undo Many" button to return to the scramble as often as you wish to try different approaches.

1.2 More on the Figure 1.3 scrambling and loading a macro. Go to the Puzzle Resources window for Oval Track 1 and in the upper right field labeled "Locally Resident Macros" click on entry number 5 titled "A 66-step opening game for the scrambling in Figure 1.3." In the dialog box that comes up, click on the button A. This will load this 66-step procedure as a program into macro button A of your puzzle. Load the scrambling of Figure 1.3 again as described in Exercise 1.1. Now press macro button A. The computer will go through the 66 steps that result in the end game situation shown in Figure 1.8. Set the macro delay in the Puzzles Home window to slow down the process to the point that you can start seeing the strategy being used.

1.3 Practice with another preset scrambling. Go to the Puzzle Resources window for Oval Track 1 and load the Exercise 1.3 preset configuration. This is a second completely random scrambling of the puzzle. As in Exercise 1.1, solve this scrambling as far as an end game that leaves only disks 17 through 20 out of order.

1.4 Practice with a buttons scramble. Load the Exercise 1.4 preset configuration from the Puzzle Resources for Oval Track 1 into your puzzle. This is a buttons scramble for this puzzle with $k = 5$. This means that to produce the scramble, the computer did five rotations and "Do Arrows." The puzzle ought to be solvable with the count going to at most 10. See if you can solve it, and if you can keep the count to 10 or less.

1.5 A second preset buttons scramble. Load the Exercise 1.5 preset configuration from Puzzle Resources for Oval Track 1 into your puzzle. This is another five-fold buttons scramble, but this time with 19 disks active. See if you can solve this one keeping the count to 10 or less.

1.6 A preset scramble for version 2. Go to version 2 of the Oval Track Puzzle and from there, press the Puzzle Resources button. This will take you to the Puzzle Resources window for Oval Track 2. You will see as the first entry in the "Locally Resident Presets" field one labeled as Exercise 1.6. Click on this line and load the configuration into your puzzle. This is a totally random scrambling of the puzzle. Solve it, starting with disk 1 and working up, as far as an end game, that is, until all but disks 17 through 20 are in order. Pay attention to how your strategy differs between version 2 and version 1.

1.7 A macro that reduces the OT 2 scramble to an end game. The second macro program included in the Puzzle Resources for Oval Track 2 is a 59-step procedure for reducing the scrambling of Exercise 1.6 to an end game. Load the same scrambling you used in Exercise 1.6 into your puzzle, and then load the Exercise 1.6 partial solution from the Locally Resident Macros collection into macro button *A*. Run this macro slowly and watch how the disks are put into order. Do you see any heuristics at work as the solution unfolds? Can you think of any ways to improve the procedure?

1.8 Practice with yet another version of Oval Track. Go to version 4 of the Oval Track Puzzle. Then go to its Puzzle Resources window and load the first preset configuration that is labeled as being for Exercise 1.8. This is the same scrambling as you studied in Exercise 1.6 on Oval Track 2. Reduce the scramble to an end game with this new "Do Arrows," and compare the way it goes with your results for Exercise 1.6 with the same scrambling and Oval Track 2 "Do Arrows."

1.9 A macro to reduce the last scramble. The second macro program included in the Puzzle Resources for Oval Track 4 is a 54-step procedure for reducing the scrambling of Exercise 1.8 to an end game. Load the same scrambling you used in Exercise 1.8 into your puzzle, and then load the Exercise 1.8 partial solution from the Locally Resident Macros collection into a macro button. Run this macro slowly and watch how the strategy it follows, making use of the Oval Track 4 "Do Arrows," compares with the strategy used by the Exercise 1.7 macro which used the Oval Track 2 "Do Arrows."

The Transpose Puzzle: An Introductory Tour

All of the puzzles, other than the Oval Track Puzzles introduced in Chapter 1, are bundled together in one file called “Multi-Puzzle.” Return to the Puzzles Home window opened by Puzzle Driver, and click on the Multi-Puzzle button to open these other puzzles. After the computer goes through its musical introductory tour (if you have this feature turned on), you will be brought back to the Transpose Puzzle (Figure 2.1).

2.1 What It Does and How

The Transpose puzzle presents you with a box with 24 numbered compartments. Each compartment contains a disk with a number on it, except for some at the bottom that have X’s indicating that they are out of service. Transpose gives you some experience with the fundamental role of transpositions. A *transposition*, defined informally for now, is a swapping of two disks between two boxes. That any scrambling of disks to boxes can be obtained by a sequence of transpositions is a basic idea in the branch of mathematics that underlies the study of these puzzles. After some playing with Transpose, you will become thoroughly convinced that this is so.

If you have already explored the Oval Track Puzzles and read Chapter 1, you will know what to expect from most of the buttons on this puzzle. Thus, our description of how this puzzle works can be briefer.

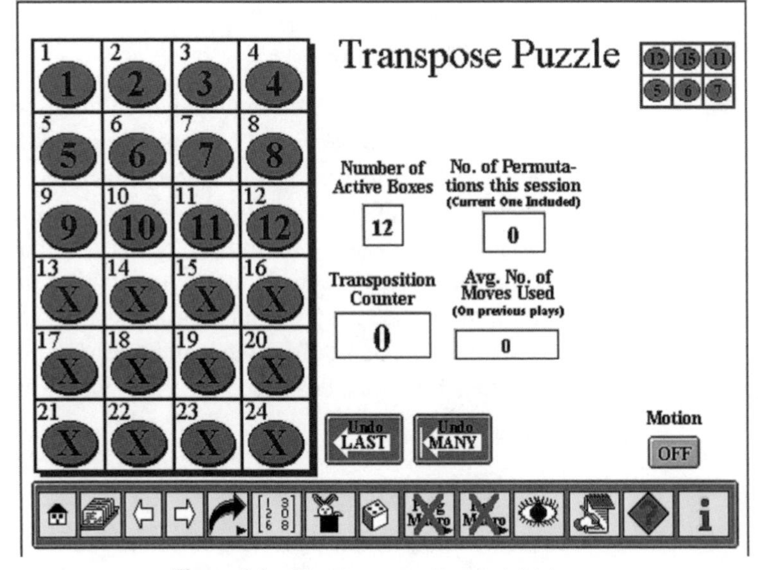

Figure 2.1. The Transpose Puzzle initial screen.

Moving Disks By Dragging

There is only one way to move disks on Transpose. That is by dragging a disk from one box to another. Move your mouse to put the browser hand onto a disk and then press down the mouse button. While holding the mouse button down, "drag" the hand to another box. When you have it over the desired box, release the mouse button. When you do so, the computer will perform a swap of the disks in the two boxes. That is, the disk in the "source box" where you started the drag will be moved to the "destination box" where the drag ended, and the disk from the destination box will be moved to the source box. The computer will sound a quick sequence of three tones to confirm that the swap has been made and will add one to the count number in the "Transposition Counter."

The disks that have X's on them are not available for swapping. If you try to drag a disk with an X on it, or if you try to drag a disk with a number to a box that contains a disk with an X, the computer won't do anything.

The Undo Buttons

Transpose has two Undo buttons which function exactly as they do on the Oval Track Puzzles. Pressing the "Undo Last" button will cause the computer to undo the last recorded transposition that you have done. It will also erase the record of this transposition and reduce the "Transposition Counter" by one. It confirms its action with a three tone audible report at a lower pitch than what it uses to confirm transpositions.

The "Undo Many" button can be used to backtrack many steps. When you press this button, you get a dialog box asking you how many steps you want to undo. Either keep its initial suggested value of 10 or enter the number of steps you want to undo and press the "OK" button or the return key.

When you press "OK," the computer will undo the number of steps that you have indicated. It will also perform the other functions described for the "Undo Last" button involving erasing the transpositions from the record and updating the counter, and give an audible confirmation.

The Motion Toggle

You will notice that there is a button in the lower right part of the window that is labeled "Motion." It is shown in Figure 2.1 in its "OFF" mode. If you click on this button, you switch it between off and on modes, which you recognize by the label changing between "OFF" and "ON" and by its changing colors between red and green.

When the motion toggle is in off mode, the swaps of disks between compartments are very rapid. The numbers just jump from one compartment to the other. When you have the motion toggle on, the swap is shown by having the numbers glide from one box to the other. There may be times when you want to see the swap in slow motion. When this is the case, just turn the motion toggle on. When you want to go as fast as possible, turn the motion toggle off.

The Button Bar

The buttons on the button bar for the Transpose Puzzle will present you with no surprises assuming that you have become familiar with the Oval Track Puzzles already. There are only some minor differences in their behavior because of the unique features of this puzzle. For this reason, rather than give a complete description of what the buttons do, we will only discuss the differences.

First, you will note that the buttons relating to macros are crossed out. That is because there are no macro buttons for Transpose. The puzzle is so simple that you really do not need macros for anything.

Let us mention again the distinction between the two reset buttons near the middle of the button bar. The one on the left with the number array icon is the one to use if you want to change the number of active disks. The one to its right with the rabbit-out-of-the-hat icon will reset everything with the current configuration of the puzzle. This means if you have loaded a configuration from the Resources window, perhaps one that uses letters rather than numbers, then you will reinitialize the puzzle with this alternative configuration.

The Objective

The objective of the Transpose Puzzle is to have the computer present many random scrambles of the disks for you to put back into their right places. The only way to do this is to perform a sequence of transpositions. A little experience will convince you that it is not only possible but also quite easy.

Once you are convinced that you can always unscramble any scrambling that the computer presents, you should also see that you ought to be able to go the other way and get from the identity permutation to any other permutation (scrambling) using transpositions.

It is easy to see why this claim is true. Think of it this way. Someone gives you an arbitrary permutation to express with transpositions. Imagine that the Transpose puzzle's scramble button had presented you with this permutation as a scrambling. You would know how to solve it using transpositions, returning the puzzle to the identity permutation state. The computer has recorded the sequence of transpositions that you used. If you run this sequence in the reverse direction, which you can do by repeatedly pressing the "Undo Last" button until you have undone the whole string, you get back to the original scrambling. Voila! You have produced the scrambling using transpositions.

2.2 Notation for Transpositions

A transposition on this puzzle can be identified by a pair of numbers, namely the numbers of the two boxes involved in the transposition. We will adopt the notation of representing a transposition with source box a and destination box b by the ordered pair (a, b). Thus, for example, the notation (2, 5) represents the transposition, which switches the disk from box 2 to box 5 and the one from box 5 to box 2.

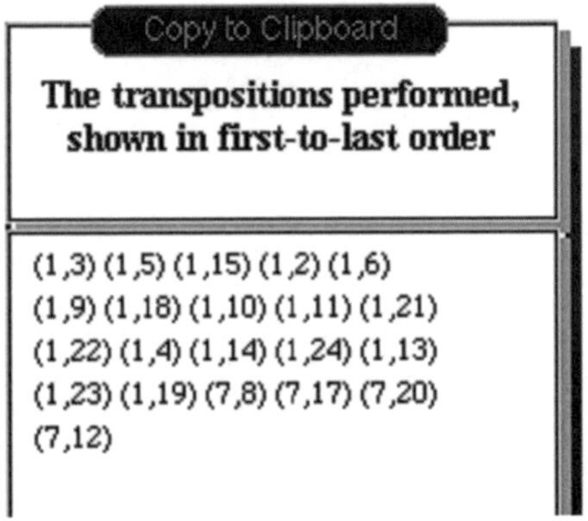

It is important to remember that we are using box numbers in this notation, not disk numbers. One could use disk numbers, but box numbers are going to be preferable for reasons that will become clear later.

Copy to Clipboard

The transpositions performed, shown in first-to-last order

(1,3) (1,5) (1,15) (1,2) (1,6)
(1,9) (1,18) (1,10) (1,11) (1,21)
(1,22) (1,4) (1,14) (1,24) (1,13)
(1,23) (1,19) (7,8) (7,17) (7,20)
(7,12)

Figure 2.2. The "View Moves" window of Transpose.

Return to the "View Moves" button now, the one with the eye icon. Pressing this button will allow you to open a window that displays a listing of all the transpositions that you have done on the puzzle since the last time it was initialized with the scramble or one of the reset buttons. See Figure 2.2 for a typical listing of a series of transpositions in the ordered pair notation that we have adopted.

This listing is for the solution of a certain scramble with all 24 compartments active. It took 21 transpositions to solve the puzzle in this case. The first swap performed was between boxes 1 and 3, and the last was between boxes 7 and 12.

Another important thing to note about the listing is that, reading from left to right, you are reading in first-to-last order. We are writing things this way because it is more convenient and natural, but it goes against the more common practice. If you read about the ideas we are discussing in other mathematics books, you are more likely to find the notation written so that as you read from left to right for a sequence of transpositions, you are reading from the last performed to the first performed. In this other system, to read from first to last, you must read from right to left. It is not very difficult to move back and forth between these two systems, but it is important to be aware of the difference so that you know which one is in use in the various sources you might be reading.

2.3 Heuristics for Transpositions

The Transpose Puzzle will present some heuristic issues, but of a simpler sort than with the Oval Track Puzzles. Heuristic thinking arises, for example, in considering various algorithms for getting all the disks back into their right boxes after you have had the computer scramble them. There are several that ought to come to mind after you have had a bit of experience.

In the case of such a simple puzzle, it will be possible to identify a "best" approach that requires the fewest number of transpositions. As you work with this puzzle, you might want to think about what a best algorithm might be, and indeed if there could be more than one best algorithm.

Exercises

2.1 Two scrambles for practice. Load the first preset configuration, labeled Exercise 2.1(a) from the Transpose Puzzle Resources window. It presents a scramble of twelve disks. Can you get all of the disks to their home boxes with 8 transpositions? Can you do this in more than one way? Now load the Exercise 2.1(b) configuration. Can you restore each disk to its home box in 9 transpositions? [Hint: To try out several strategies for solving the same scramble, after having solved it, use the "Undo Many" button to restore the original scramble. Another way to restore the scramble is to press the "Restore" (rabbit out of a hat) button. Hold down the alt key (Windows) or option key (Mac) to avoid the confirmation dialog and have the computer do the restore immediately.]

2.2 Practice with cycles. There are three preset configurations in the Transpose Puzzle Resources window that you will use for this exercise. They are labeled Exercise 2.2, (a), (b), and (c). They present three of the fundamental types of rearrangements of the disks called cycles. (These will be discussed formally in Chapter 5.) Load these configurations into your Transpose puzzle and perform transpositions to restore the disks to their home boxes. Pay attention to how the three arrangements are similar and how they differ. Look for a pattern in how you restore the disks, and describe how the pattern is modified each time the length of the cycle is increased by one. [Hint: You may find it helpful to look for patterns in the listing of transpositions you used. Press the "View Moves" (eye icon) button to open a field that shows this list. Press in the list field or on the "OK" button to hide the list field.]

2.3 How many transpositions needed? In Exercise 2.1, you looked at two pre-defined scrambles and for patterns in how you solved them. To continue this exploration, have the computer scramble the disks and then you put them back into their starting places. (a) Use the "Scramble" (die icon) to do this. Hold down the option key (Mac) or alt key (Windows) if you want to avoid the confirmation dialog. Do this several times. Make note of how many transpositions it takes to do the job. Is it always the same for each scramble for a given number of active disks? (b) See how the number of transpositions needed changes as you change the number of active disks. Use the "Reset Active Number" (number array icon) button to do this. Identify several algorithms for doing the job.

2.4 Comparing solution algorithms. Continue the exploration discussed in Exercise 2.3. After restoring the disks to their home boxes using one of the algorithms you have found, make note of how many transpositions you used. Then return to the initial scrambled state and restore the disks using another algorithm. How do the algorithms compare in terms of how efficient they are? See the hint in Exercise 2.1, which describes two ways to restore a scramble after you have solved it one way so that you can try a different way.

2.5 A formula for the number of transpositions? Think about whether it is possible to determine in advance how many transpositions you will need to return all the disks home for a given scrambling of the disks. Could you devise a formula that gives this number? That is, can you give a formula for $T(\alpha)$ = the least number of transpositions to restore a scrambling α to the identity arrangement?

2.6 How many transpositions on average? For a given number n of active boxes on the Transpose Puzzle, let μ_n denote the average number of transpositions needed to restore all disks to their home boxes following scrambling of the disks. Can you determine a formula for μ_n?

2.7 Using Transpose for anagrams. (a) It is possible to use the "build your own" tools of the Puzzle Resources window to create configurations for the Transpose Puzzle that involve letters of the alphabet instead of numbers. If you configure Transpose with a word or phrase, you can then rearrange the letters on the puzzle to look for anagrams, that is, permutations of the letters that spell other words or phrases. To illustrate the possibility, load the Exercise 2.7(a) preset of Transpose Puzzle Resources, which loads the author's name into the puzzle. Rearrange the letters to spell out the phrase J O I N / K I L N / T H E N. Can you find any other anagrams?

(b) Load the Exercise 2.7(b) preset, which presents three four-letter girls' names that do not repeat any letters: MARY, BETH, and LONI. See what anagrams you can make with these letters.

(c) Build your own phrase to load into the Transpose Puzzle. To do this, go to the Transpose Puzzle Resources window and click on the "Build Your Own" (hand and pen icon) button on the lower button bar. Enter your configuration into the data entry grid that appears. Press the "Load" button to move the configuration into the puzzle. Now see what anagrams you can get from your phrase.

2.8 A Checkerboard Problem. (a) Load the Exercise 2.7 preset configuration from the Transpose Puzzle Resources window into your puzzle. It configures the puzzle with the letters A and B in a four by four alternating checkerboard pattern. Scramble the puzzle and then restore the original pattern. Do this several times. (b) Can you predict accurately how many transpositions you will need to restore the checkerboard pattern? (c) Can you give a formula for the average number of transpositions that you will need to restore the checkerboard pattern following scrambles?

The Slide Puzzle: An Introductory Tour

Let's now take a look at the second of the three puzzles presented by Multi-Puzzle, namely the Slide Puzzle. To get to it, start Multi-Puzzle, and then use one of the navigation methods we have already discussed. Assuming you have the Transpose Puzzle showing, you can use the "Next" button on the button bar, the one with the right-pointing arrow, to move to the Slide Puzzle (Figure 3.1).

3

Note that the screen for the Slide Puzzle looks very similar to the one for Transpose. To help you distinguish between them, the background of the grid for Slide is a light blue in contrast to the light yellow of Transpose. In addition, the two screens have different animated logos in their upper right corners, which demonstrate how the disks move when you first go to the screen.

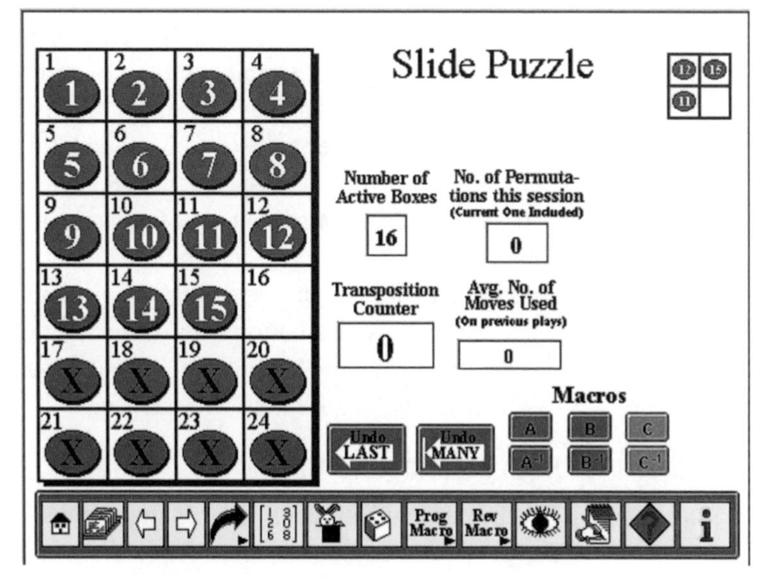

Figure 3.1. The Slide Puzzle in its initial configuration.

3.1 What It Does and How

This puzzle is based upon the very familiar "15" puzzle that you probably owned in your youth, or maybe still do. It is the plastic puzzle that has a square frame that holds 15 numbered disks. The fifteen pieces and one blank are locked into a four by four square frame.

Any piece that is adjacent to the empty square can be moved into the empty spot. Doing a series of moves like this, you can maneuver the empty spot all over the puzzle, and in the process can rearrange the plastic pieces.

The Slide Puzzle presents this familiar puzzle, together with many variations on it that your plastic version could never give you. Its interface window looks very much like that of the Transpose Puzzle, and for the most part will function as you expect.

Moving a Disk into the Blank Spot

The standard "15" puzzle has one blank spot, but you can initialize this computerized version with either one or two blanks. If you use a standard configuration rather than one that you have created from the Resources window, the two blanks will start out in the highest numbered active boxes. If you are creating a configuration from the Resources window, use an asterisk to indicate where you want a blank spot. You are free to create as many as you want it this case, but the computer will show all but two as disks with asterisks on them rather than as blank spots.

A disk can move into the blank spot if it is in a spot that is adjacent to the blank, either horizontally or vertically. To make the move, simply put the browser hand onto the disk you want to move and click the mouse button. When you do, the computer will move the disk into the blank spot and the blank into the box in which you just clicked. The computer will sound two quick tones to confirm that the swap has been made and will add one to the count number in the "Transposition Counter."

If you click on a disk that is not adjacent to a blank, nothing will happen. When you are using two blanks, a numbered disk may be adjacent to both of them. If you click on that disk, the computer will open a dialog box presenting you with the numbers of the blanks and asking you to choose the one to which you want to move the disk you clicked. Press the button with the number of the blank that you want to use.

The disks that have X's on them are out of action. You cannot move a blank into any of these boxes. If you click on an X'ed-out box, nothing will happen.

Reconfiguring the Puzzle

This puzzle can be reconfigured in many more ways than the others so here are the options in some detail.To start the reconfiguring process, press the button with the number array icon. The rabbit-out-of-the-hat button is for starting over with the current configuration.

First, you get a dialog box that asks how many active disks you want. The number window will be initialized with the number of active disks on the previous version of the puzzle. This number will be selected and highlighted so that you can either start typing to replace it or just press "OK" if this is the value that you want to use again. Pick the number that you want and then press the "OK" button. There is also a "Cancel" button if you change your mind about going forward with the reset.

When you press "OK," the next thing you see is a window that asks how many blank spots you want. The choices are "1" or "2" or "Cancel." If you choose two blank spots, the puzzle will be easier to solve than if you choose one. If you choose "Cancel," you exit from the whole reset process.

Once you have clicked in the dialog box to record the number of blanks that you want, you are presented with a dialog box that asks if you want to cross out any boxes. This gives the opportunity to create a version of the puzzle that has some of the boxes inactive, thus serving as obstacles. You may want to experiment with obstacles to see what effect they have on solvability.

Tell the computer where you want to place obstacles by entering the numbers of those boxes that you want to be put out of action. Enter the numbers separated by spaces. If you do not want to create any obstacles, leave this field blank.

Once you have made your choices regarding obstacles, press the "OK" key. When you do this, the computer brings up a dialog box asking for your choice regarding "wrapping." The puzzle allows you to, in effect, bring the two sides or to bring the top and bottom together. You can do this with either or both or neither. As a consequence, you can set the puzzle up as though it were on the surface of a vertical or horizontal cylinder or a doughnut-shaped torus. If you choose to have side to side wrapping, for example, the positions on the left edge will be regarded as adjacent to those on the right edge. This means that if the blank is in box 16, since boxes 16 and 13 are adjacent, you can click on the disk in box 13 and move it to box 16.

When you have a wrapping mode turned on, the computer will display a pair of curving arrows that signal this, and will also show this in writing under the main title line.

When vertical wrapping is turned on and there are up to three rows at the bottom of the grid that are out of action, then the computer will interpret the lowest row that has active spots in it as adjacent to the first row. For example, if you have configured the puzzle with the first four rows active, then row four is adjacent to row 1. When horizontal wrap is turned on and the fourth column is completely out of service, then column three will be regarded as adjacent to column one.

Once you have made your wrapping choice, the computer will redraw the puzzle display with your wishes taken into account. It will also reset the counters "No. of Permutations this Session" and "Transposition Counter" to zero, and will reset the "Avg. No. of Moves Used" to zero as well. The computer will set the "Number of Active Boxes" display field to the number of active boxes that you set at the start of the reset process, minus the number of boxes that you chose to cross out. It will play an audible fanfare to confirm its action.

(a) A 3-by-3 version of the puzzle. (b) With 3 obstacles and side-to-side wrap.

Figure 3.2. Variations of the Slide Puzzle that are possible.

To illustrate the process of reconfiguring the puzzle, let's run through an example. Suppose that you would like to play a three by three version of the puzzle, that is, an "8" version of the familiar "15" puzzle. You can do this by selecting "11" when the computer asks "How many active boxes?", then select "1" for the number of blank spaces, and finally entering "4 8" when asked "How many cross outs?" The result will be Figure 3.2(a).

You can create a configuration with a barrier in it (Figure 3.2(b)). This one is obtained by using 24 as the number of active boxes, choosing two blank spaces, and putting obstacles in boxes 10, 15, and 18, and turning side-to-side wrap on. You may wish to experiment with variations like this one to learn the effects of obstacles and wrapping on the solvability of these puzzles.

The "Undo" Buttons

The "Undo" buttons of the Slide Puzzle behave in the same way that they do for Transpose.

The Macro Buttons

These are programmed and used the same way as the macro buttons of Oval Track. Any sequence of steps can be programmed into one of three buttons, labeled *A*, *B*, and *C*. When the sequence is recorded for a button, it can be reproduced by simply pressing that button, or played in a reverse order by pressing the inverse version of the button.

There is one subtlety about running programs that applies to the mechanics of the Slide Puzzle but not to Oval Track. That has to do with the unique role of blank squares. A blank square must be in the same place it was when a sequence was recorded in order for the macro button to work. If the blank is not where it needs to be when you press a programmed button, the computer will present a dialog box explaining that it cannot run the program. Similarly, in order for the inverse of a program to run, a blank must be in the spot where it was left when the button was programmed. If you push the inverse of a programmed button and a blank is not where it needs to be, you will get the same dialog box explaining the problem.

Another issue is that the macros you have programmed into the buttons remain there when you reconfigure the puzzle. You can program a macro with one configuration and then change to another where the macro cannot work. For example, you may have added an obstacle in a location that the macro uses, or you may have had a wrapping option on when the program was created but that wrapping option is off in the new configuration. When this happens and you try to run the macro, the computer will detect the problem, stop, show a dialog box explaining the problem, and then reverse any steps it did in trying to run the macro.

Why are there macro buttons for the Slide Puzzle but not for Transpose? It is because there is a clear need for them with Slide but not with Transpose. As we work with Slide, we will see that there will be a need to run a rather lengthy sequence of steps in a reverse direction at times. Sometimes these sequences can be more than 30 steps.

This is clearly more than you would want to commit to memory, and, although it would be possible, writing down every step for sequences of such length is tedious. Having the computer do the remembering for us will be very helpful.

3.2 Notation of the Slide Puzzle

We will use the same ordered pair notation for the Slide Puzzle that we are using for Transpose. Each move is a transposition involving the contents of two boxes. We will represent a transposition involving boxes a and b with the blank in box a by the pair (a, b). For example, if the blank is in box 11, then we can do transpositions involving it and the numbered disks in the adjacent boxes 10, 12, 7, and 15. If we move the blank from box 11 to box 15 and the disk in box 15 to box 11, we will represent this by the pair (11, 15). The blank is now in box 15 and is adjacent to box 19. Thus, the transposition (15, 19) is one that can follow (11, 15).

For a sequence of ordered pairs to represent moves that are possible on the Slide Puzzle, certain restrictions must apply. One can perform the transposition (a, b) when and only when (i) a blank is in box a, and (ii) box b is adjacent to box a and is currently active. If a pair is going to represent a move that follows (a, b), then it must be of the form (b, c).

A listing of the moves you make is recorded by the computer and can be viewed by pressing the "View Moves" button (eye icon). This listing will look like what you see with the Transpose Puzzle. Figure 3.3 gives a typical display of a sequence of moves on the Slide Puzzle. The first move performed, according to this listing, is the transposition (16, 12). This tells us that the blank spot started out in box 16, and after the first step, it was moved to box 12 while the disk in box 12 slipped over to box 16.

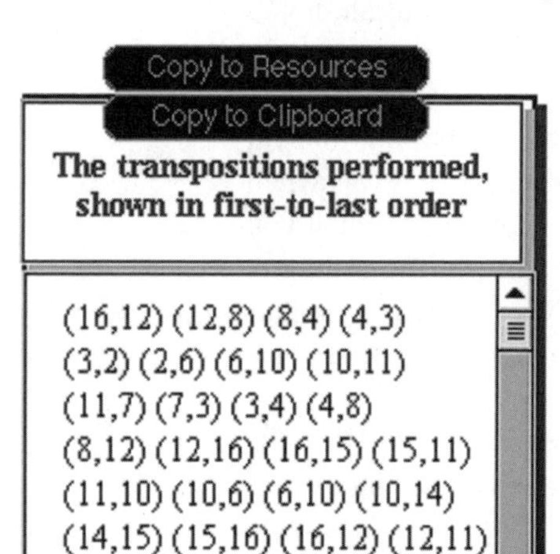

Figure 3.3. The listing of a certain 32-step procedure. (Note: Right entries of each pair match the left ones of the next pair reflecting movement of the blank)

Note in the listing that the right entry in each pair is the same as the left entry in the next pair. If there is only one blank, this will always be the case. It reflects the fact that the blank box must be involved in every transposition, and that we always list the number of the box where the blank is on the left of the ordered pair.

3.3 Heuristics for the Slide Puzzle

The heuristic-to-theory ratio for the Slide Puzzle is the highest for any of the puzzles of this package. You will understand what this means much better after you have had experience with all of the puzzles. Theory will tell us what is possible and what is not, but will not be as involved in solving end game problems as with the others. The opening game is the main issue, and looking for efficient strategies will provide some of the major fun with the Slide Puzzle.

As you are studying strategies for the puzzle, you will want to make use of the "Avg. No. of Moves Used" register, which will display the average number of moves that you used on each run of the puzzle, excluding the current one, since the last time you reset the puzzle. The number displayed in this register will not have much meaning if you are just dallying with the puzzle, but it will be of interest if you are running a series of tests of a certain strategy and solving the puzzle as efficiently as possible after every scramble.

Let's say that you want to get an idea how many steps, on average, it takes to solve the "8" puzzle shown in Figure 3.2a. Reset your puzzle to this configuration. When you do this, the average register is reset to zero. Now scramble the disks and try to solve the puzzle. It is possible that you cannot solve it completely. If you cannot, at least you can get everything into its right place except for the highest two numbered disks, 9 and 10 in this case. We will get back to this point after we have looked at the necessary group theory. If you run a series of trials and each time either solve the puzzle or get it to the point that only disks 9 and 10 are switched, and if you are following your strategy as efficiently as you can, then this register will give you a measure of its efficiency. For example, the author ran a series of five scrambles with the "8" puzzle. For three of them, the puzzle was totally solvable and for the other two, the puzzle ended up with the last two disks out of place. On average, it took 43.2 transpositions to solve the puzzle or get it to the "as close as possible" state.

Note that if you are keeping track of the average number and you want to include a final solution in the average, you will need to update the average by doing another scramble. For example, with the five trials just discussed, the author got the fifth trial included in the average by doing a sixth scramble, even though he did not intend to solve this one.

Another reason for a high heuristics-to-theory ratio for the Slide Puzzle is the possibility of building barriers into the puzzle. If you get interested in exploring the puzzle with barriers in place, you will discover that they may or may not make a difference in the degree of solvability of the puzzle depending on geometrical factors. Even when a set of obstacles does not prevent the puzzle from being solved, it may have a major impact on what strategy is needed, causing the average number of steps needed to increase many-fold. You will find that questions like these are going to be more in the realm of heuristics than theory.

Exercises

3

3.1 Practice with the configure button. The configure button (number array icon) is on the lower button bar. To gain a sense of the options this button gives you for configuring the puzzle, use it to put the puzzle into the configurations listed here.

	Number of active boxes	Number of blanks	Cross outs (obstacles)	Wrapping options turned on
(a)	24	2	6 11 16	Top to bottom
(b)	11	1	4 8	Left to right
(c)	19	1	7 10 17 18	Top to bottom
(d)	20	1	7 10 15	None

3.2 Working on efficiency. The more you work with the Slide Puzzle (as well as the others), the more efficient you get at solving them. Configure your Slide Puzzle in the basic "15" configuration as shown in Figure 3.1. Scramble the disks with the scramble (die icon) button. Then solve the puzzle completely, or until the two highest numbered disks are the only ones out of place. Make note on the Transposition Counter of how many steps it took you. Repeat this process several times and see if your efficiency improves.

3.3 Practice with programming a macro button. Have your Slide Puzzle configured in the basic "15" mode. Then program a procedure into button A which moves disk 13 into spot 14, disk 14 into spot 15, and disk 15 into spot 13. We call such a rearrangement a 3-cycle. Follow these steps to program the button. (i) Press on the "Prog Macro" button on the button bar. Select "Program Button A" from the pop-up menu. (ii) The "Prog Macro" button now is green and says "Begin." Press on it to start the recording process. (iii) Perform the steps to accomplish the desired rearrangement. When you are done, press on the red "Done w Prog" button (that is what the "Prog Macro" button now says). The process is now complete. Press on button A to see the steps you performed while programming the button now get repeated automatically. [Note: To accomplish the particular 3-cycle described above, get disks 13, 14, and 15 together with the blank in a two by two square, say in the lower left corner. Then run them around this square with disk 13 going to where 14 was, disk 14 going to where disk 15 was, and disk 15 going to where disk 13 was. Then reverse the steps you used to arrange the disks this way to get them back to their final destinations.]

3.4 Getting familiar with how macros behave under changes in configuration. The macros that you program into the macro buttons A, B, and C remain there until you change them. Even if you change the configuration on the Slide Puzzle, the macros do not change. However, a change in a configuration can prevent of macro program from being performed. For example if a macro program moves the blank through spot 10 and a new configuration of the puzzle has spot 10 out of action, the macro is blocked from working. (a) Program a macro into button A that passes the blank through spot 10. Then reconfigure the puzzle with spot 10 crossed out. Now press button A and see what happens. (b) A macro program that was created with a wrapping option turned on cannot run under a configuration that has this wrapping option turned off. To verify this, configure the puzzle with left to right wrapping turned on. Now program button A with a macro that has disks jumping from one edge to the other. Then reconfigure the puzzle with left to right wrapping turned off, press button A, and see what happens. (c) Macro programs that are created under a set of constraints can run when these constraints are removed. For example, a macro created with no wrapping options on will still run when the wrapping features are turned on. Verify this by programming a macro button with no wrapping, then reconfiguring the puzzle with wrapping turned on, and then running the macro.

3.5 Practice with designing your own configurations. Configure the Slide Puzzle so that it is like the basic "15" puzzle, except that it uses the letters A through O in place of the numbers 1 through 15. To do so, follow these steps. (i) Press the Resources button (card file icon) on the button bar of the Slide Puzzle to bring the Puzzle Resources to the front. (ii) Press on the "Build your own" button (hand and pen icon) of the Puzzle Resources window to make the data entry display visible. (iii) Press the "Clear" button to the left of the display if necessary to clear out the entry locations. (iv) Click in the upper left box of the data entry display and type in the first letter, A. Use the tab key to move from one box to the next and type in the rest of the letters. After you have typed in the letter O, tab to the 16th box and type in "*". (v) Click on the Wrapping Choice button on the upper right of the data entry display and choose the wrapping option that you want from the pop-up menu. (vi) Press the "Load" button to start the process of loading this configuration into your puzzle. (vii) When the dialog box comes up that asks you for a name for your configuration, enter the name you want to give it, or keep the default name, and press "OK." *Voila*! Your configuration will be loaded.

The Hungarian Rings Puzzle: An Introductory Tour

The final puzzle included in the Multi-Puzzle collection is the Hungarian Rings Puzzle. You get to it from either of the other two Multi-Puzzle versions by pressing either the “Next” or “Previous” arrow buttons. When you do, the screen transforms into what is shown in Figure 4.1.

Like version one of the Oval Track puzzle, the Hungarian Rings puzzle is modeled on an actual plastic puzzle. This puzzle came on the market in the post-Rubik’s Cube 1980s. It has two circular tracks with spherical plastic balls in them. There is room for twenty balls on each track, and given that the tracks intersect in two points, two balls are on both tracks. This means that there are 38 balls all together.

The balls are in four different colors. The numbers of each color are ten, ten, nine, and nine. Initially, the balls of like color are grouped together as is represented by the letters in Figure 4.1, which on your computer screen will also be in different colors. The object is to mix them up by a series of rotations of the two rings, and then get them back into this initial configuration. Later on, you can look for other interesting arrangements of the colors.

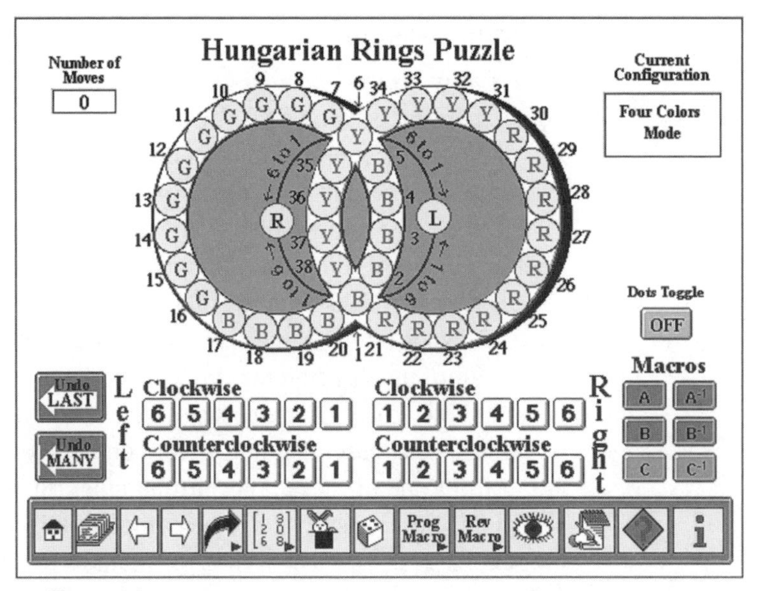

Figure 4.1. The Hungarian Rings Puzzle initialized in four-color mode.

4.1 What It Does and How

The Hungarian Rings Puzzle is implemented on the computer much like the Oval Track Puzzles, but it differs in a few respects. Like Oval Track, it has disks that move around on a track. However where Oval Track has one oval shaped track, the Hungarian Rings has two ring shaped tracks. Both puzzles have two fundamental ways to move disks. On Oval Track, one of these, the rotation, moves all of the disks and the other, "Do Arrows," just moves a few disks. In contrast, on the Hungarian Rings Puzzle, the two basic moves are rotations of disks on either the left or right ring. Each of these moves many disks—one more than half of the total number.

The various buttons on the two puzzles correspond by name, even though they are differently placed. If you have spent any time with Oval Track, then the way Hungarian Rings operates will present no surprises.

The similarities will make Hungarian Rings feel familiar when you first start to use it. However, you will discover that the differences will make solving the puzzle significantly harder.

Moving Disks Around

The two ways to move disks are to do a rotation on either the left ring or on the right ring. You can do these rotations in either of two ways: by pressing the number buttons on the bottom of the screen or by dragging.

The number button method is preferable for moving disks according to a specified pattern that you are using to accomplish a certain objective. You see from Figure 4.1 that they are grouped according to which of the rings you want to change and whether you want to move the disks in a clockwise or counterclockwise direction. You have buttons that will move all of the disks up to six positions either way on either ring.

To drag the disks on one of the rings, put the browser hand onto one of the disks that you want to move. Drag it to the spot on the same ring where you want it to go. When you release the mouse button, the computer will move all of the disks on the ring in such a way that the disk you dragged goes where you indicated. Note that you cannot drag a disk from one ring to the other. If you try such a drag, the computer will simply ignore your action.

One special case of a drag—from one of the intersection locations to the other—needs a different procedure. A drag from one of these points to the other, that is, from spot 1 to spot 6 or vice-versa, could be on either of the rings. If you do such a drag, the computer will present you with a text window explaining the problem and a dialog box asking you which ring you want to rotate. By pressing "Left" or "Right" in this dialog box, you can complete the rotation. You can also press the "Neither" button to have the computer take no action.

A way of doing a drag from one intersection point to the other in a single step which avoids the dialog box is to use either of the "drop-off points" labeled "L" and "R." Say you want to drag from spot 6 to spot 1 using the left ring. Start a drag at spot 6 and end it at the "L" drop-off point. The computer will immediately move all the disks on the left ring so that the one that was on spot 6 moves to spot 1.

You may wonder why the drop-off point labeled "L" is on the right and the one labeled "R" is on the left. This is because the short arc of the left ring between the intersection points is to the right of the short arc of the right ring between these points. If on a physical puzzle you were to do a drag on the left ring between the intersection points, the drag would in fact start out going to the right. The drop-off points were placed to simulate this tactile aspect of a physical puzzle.

Hungarian Rings uses an audible report of moves you make on the puzzle. It plays a number of quick notes that equals the number of positions that you move. For the left ring, the tone is a percussive piano-like sound and for the right ring it is a softer flute-like sound. A clockwise rotation will produce a higher pitch and a counterclockwise rotation a lower pitch on either ring.

Resetting in Either of Two Modes

This puzzle can be set in two basic modes: "Numbers" and "Four Color." You choose the mode you want using the "Reset" button (number array icon) on the button bar. When you press this button, you will open a pop-up menu, which offers the two different modes. When one is selected, the puzzle is reinitialized in the mode that you chose.

If you select the "Four Color" option, you will initialize the puzzle in four-color mode as shown in Figure 4.1 above. Even though the illustration is not in color, the display on your computer screen will show the letters on the disks in different colors (if your computer has a color monitor). The B's will be in blue, the G's in green, the R's in red, and the Y's in yellow. Having the different letters for the different colors gives a way to play this puzzle, even if one cannot distinguish colors. When one does have color, the letters provide an additional visual cue for the differences between the four sets of disks.

When you first initialize in color mode, the computer will also display some colored dots around the border of the rings. The colors on these dots match the colors of the disks on the rings, and serve as a guide to where the colors were originally, something you might find helpful after you have scrambled the disks and want to recall where

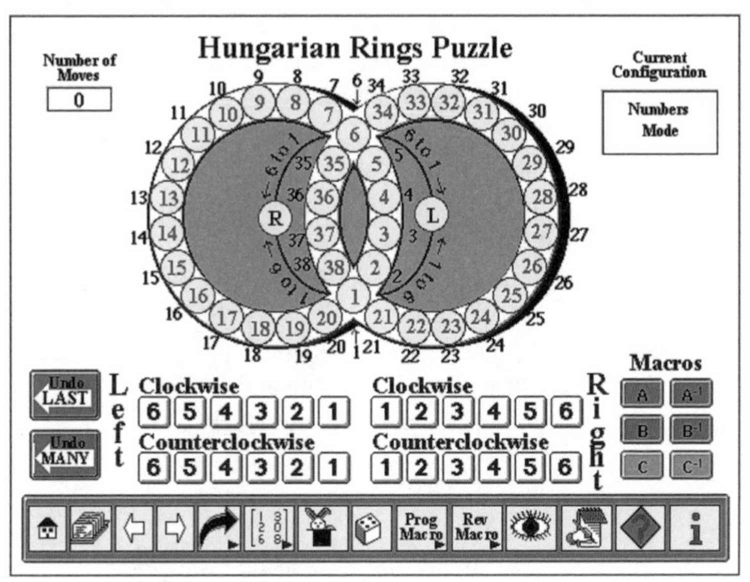

Figure 4.2. The Hungarian Rings initialized in number mode.

they were initially. When you are in color mode, a "Dots Toggle" button will be visible on the right side of the window. Use it to make the dots appear and disappear as you wish.

When you choose the "Numbers" option from the "Reset" button, the computer responds by giving you disks that have the numbers 1 through 38 on them. The puzzle is initialized with each disk in the spot with the same number that is on the disk, as shown in Figure 4.2.

When the puzzle is initialized, the "Number of Moves" counter is reset to zero.

The two different modes for the puzzle require very different approaches to solving it. This will particularly be the case for the end game.

The Other Buttons

All of the other buttons of Hungarian Rings function like they do on the other puzzles. Note that we have macro buttons again. For this puzzle, they are indispensable.

4.2 Notation of Hungarian Rings

The notation for the Hungarian Rings puzzles is very similar to that of the Oval Track Puzzles. We use two fundamental symbols, L and R to represent the building block moves of a one-position clockwise rotation of the disks on the left ring, denoted by L, or on the right ring, denoted by R. Multi-position rotations on either side are indicated by means of superscripts on these two symbols. Positive superscripts will denote clockwise rotations and negative ones will represent counterclockwise rotations. As with all of the puzzles, we will write expressions from left to right indicating first to last order.

To illustrate, consider the notation $L^5R^5L^{-5}R^{-5}$. This denotes a five-position clockwise rotation on the left ring, followed by a five-position clockwise rotation of the right ring, then five positions counterclockwise on the left and finally five counterclockwise on the right. If you initialize the puzzle in the number mode and try these four steps, you will find that it produces a limited change: The disks in spots 1 and 25 will have traded places and the ones in spots 6 and 11 have done the same. (Perhaps there is a way that we can make use of this later. Hint, hint.)

Since getting the software to render exponents presented technical problems, Multi-Puzzle modifies the notation just described for the purposes of displaying moves that you have made or have programmed into a button. The exponents are brought down and enclosed in a pair of parentheses. Thus, the four-step series given as an example in the previous paragraph is represented in the form $L(+5)R(+5)L(-5)R(-5)$ in the field that opens if you press the "View Moves" button. In this field, you also get each step numbered. If you view the moves programmed into a button by pressing the "Review Macro Program" button, you will find the same notation used, except without the steps being numbered.

4.3 Heuristics for Hungarian Rings

In the previous chapter, we said that the Slide Puzzle has a high heuristic-to-theory ratio. In stark contrast, you will find that the heuristic-to-theory ratio for Hungarian Rings is very low. It is not quite so low in color mode, but in number mode, you will find that the mathematical theory takes on much greater importance. Also, you will find the heuristics interwoven with the theory to a much greater extent.

Let us explore the opening game heuristics just a little, thinking about the puzzle in numbers mode. Suppose that you scramble the disks and try to get them back into their right places. A sensible strategy is to work on getting the disks that are on the extreme left of the left ring or on the extreme right of the right ring into place first. The reason is that these ones are far away from the action, which takes place at the intersection points. Once you get them into place, perhaps you can just leave them there while you work on the remaining disks that are nearer the "front lines."

Do a scramble and try this out. You will see that things go fine for about ten disks per side. You can get ten disks on the left ring into place, say about number 9 through 18, and then get ten on the right ring into place, perhaps number 23 through 32. After this, it starts getting harder, because you need to make rotations of about seven or eight positions on the rings to work on the remaining disks. When you do this, say focusing attention on the lower intersection point, you are also making chances that you probably do not want at the upper intersection point.

Once you have gotten the first twenty disks into place, you will find that common sense has taken you about as far as it will. Maybe the luck of the scramble will allow you to get a few more disks into place just with common sense, but for all practical purposes, you are at the end game where some mathematical ideas are very much needed.

The end game for Hungarian Rings comes when the puzzle is just a bit over half solved. Contrast this with the first version of the Oval Track Puzzle, which you can get 80 percent solved with common sense and need theory for only the final 20 percent.

Exercises

4.1 Practice with moving the disks. This exercise guides you through some practice with moving disks on the Hungarian Rings. (a) Practice moving the disks on both the left and right rings by dragging a disk from its current location to where you want it to go. Do some of the drags following the curvature of the ring and others going straight from the source to the destination, noting that it is not necessary to follow the curvature. (This method is the preferred one when you are doing opening game tasks of getting the first disks in order.) (b) Now practice moving disks by pressing the rotation buttons. Try these sequences of moves using the buttons: (i) $(L^5R^5)^4$; (ii) $(RL)^{10}$; (iii) $(LR^{-1}L^{-1}R)(L^2R^{-1}L^{-1}R)^3L$. Then repeat these sequences by dragging. No doubt you agree that repetitive sequences of moves such as these, which get used during the end game, are easier to perform by pressing buttons.

4.2 Moving disks between the intersection points. Sometimes you want to drag a disk from one of the intersection locations to the other. This requires different procedures to specify which ring should be rotated. (a) Do a drag from spot 1, the lower intersection location, to spot 6, the upper one. Read the explanation field that opens at the top of the screen, and make a choice of which ring to rotate in the dialog box that opens. Practice this method of dragging between the intersection points several times, doing it from spot 6 to spot 1 as well. (b) Now practice doing drags from one of the intersection points to either of the drop off points labeled "L" and "R." These drop off points allow you to do a drag between intersection points and specifying which ring at the same time. Do enough of these drags to the drop off points to be sure you are certain how they work.

4.3 Doing shuffles. A *shuffle* is a frequently used sequence of moves. You alternate one-position rotations of the two rings in the same direction. For example, $L\,R\,L\,R\,L\,R\,L\,R$ is a four-fold, left first clockwise shuffle. Practice doing shuffles, both left and right first and both clockwise and counterclockwise. Do them with the puzzle initialized in both the numbers mode and color mode. Pay attention to how the disks move, and describe the patterns.

4.4 Order of procedures. The *order* of a sequence of moves on the Hungarian Rings Puzzle, or on any of the others, is the number of repetitions of the sequence it takes to get back to the identity arrangement of the disks if you start from the identity arrangement. (The identity arrangement is the one that has every disk in its home location.) (a) Determine the order of the basic shuffle moves $(L\,R)$, $(R\,L)$, $(L^{-1}\,R^{-1})$, and $(R^{-1}\,L^{-1})$. That is, for example, what is the smallest positive integer k such that $(L\,R)^k$ is the identity arrangement? (b) Determine the order of $(L^2\,R^2)$. (c) Determine the order of $(L^5\,R^5)$.

4

4.5 Reading notations. These examples are intended for use on the puzzle initialized in the four-color mode. (a) Carry out the procedure defined by the expression given here to get an arrangement that switches the place of the set of green disks with the set of red disks and similarly switches the blues and yellows.

$$(L\,R)^3\,(L\,R^{-1}\,L\,R\,L^{-1}\,R)\,(L\,R)^2\,(L^{-1}\,R\,L\,R^{-1})\,(L\,R)^8\,L \tag{4.1}$$

[Hint: Recall, for example, that $(L\,R)^3$ is a way to compress the notation $L\,R\,L\,R\,L\,R$.]

(b) Reinitialize the Hungarian Rings in the four-color mode. Here is a 39-step procedure that takes you from this initial arrangement to one which alternates the reds and yellows on the left ring and alternates the greens and the blues on the right ring.

$$(L^2\,R^2)^9\,L^2\,(R\,L)^5\,(R^{-1}\,L^{-1})^5 \tag{4.2}$$

4.6 Saving a sequence of steps as a macro. If you have just done one of the procedures in Exercise 4.5, the computer will have kept a record of the steps you performed. You can turn this sequence into a macro program by performing the following steps. (i) Press the "View Moves" (eye icon) button on the button bar. (ii) When the field that displays the sequence of moves appears, there will be two buttons above it. Press the button that says "Copy to Resources." (iii) A dialog box asks you to name to this procedure. Enter an appropriate name and then press the "OK" button in the dialog box to complete the process of converting and storing the sequence as a macro program. (iv) Press the "OK" button on the button bar where the eye icon button was to close the display field. (v) Go to the Puzzle Resources window and scroll down to the bottom of the "Locally Resident Macros" field to see the entry for your new macro program. (vi) Click on this entry to start the process of loading it into one of the macro buttons if you wish.

4.7 Actual versus apparent order. When the puzzle is in four-color mode, the procedure in display (4.1) that switches the red and green sets of disks and the blue and yellow sets appears to be of order two. If you apply the procedure twice, the four sets of colors all return to their initial locations. However, the *individual* disks do not return to their own starting positions. To verify this, program this procedure into one of the macro buttons, say button B. Reinitialize the puzzle into numbers mode, and run the procedure twice by pressing the macro button you programmed. You will see that many of the disks have not returned to their home locations. What is the actual order of this procedure? That is, how many times do you have to press button B before all the disks return home?

Permutation Groups: Just Enough Definitions and Notation

5

If you have spent enough time with the puzzles while reading the first four chapters, you have a reasonably good understanding of the opening game of each of the puzzles. That is, you can scramble a puzzle and proceed to get most of the pieces back into place in a strategic and purposeful way.

You have also learned that there is an end game part of solving the puzzles that is qualitatively different. For the Oval Track Puzzles, the end game starts when you have a contiguous group of disks remaining to be put into place that is as wide as the span of the "Do Arrows" move of the puzzle. For Hungarian Rings in number mode, the end game starts when you have gotten about 10 or 11 disks in order onto the outer edges of each of the rings. In color mode, after you get all 10 of the green disks together on the left ring and all 10 of the red ones together on the right ring, you are already edging into the end game. In either mode, you have an end game that involves almost half of the disks. For the Slide puzzle, there is no clearly identifiable end game.

After getting a handle on the opening game, perhaps you have spent some time trying to solve end game problems. It is sometimes possible to do this based on a bit of luck and trial and error. In spite of the occasionally lucky break, spending some time with the end game is certain to convince you that there is a need for some new type of knowledge in order to finish the puzzles with skill.

Thus, if the first four chapters and your initial experiences with the puzzles have done their intended job, you feel a "need to know" some theory of permutation groups.

As an aside, and at the risk of belaboring the point, we observe that creating a need to know is a basic theme of human interaction. Educators use a fancy term for leading students to the point of wanting to learn a new thing – "anticipatory set." Philosophers lead their students through a thought experiment that creates a dilemma, and a need to "equilibrate," or restore an equilibrium. Comedians look for just the right straight line that sets up the need for their punch line. Novelists take their readers through a careful process of exposition and development leading to the crisis, which creates the need for the resolution that comes in the climax. The composer of a great work of music creates dramatic tension through dissonance, which creates a need for the final consonant release.

Perhaps you agree that these puzzles create a need to learn a bit of the mathematics that can relieve the frustration of trying to solve them. So let's set about doing just that.

5.1 The Permutations on a Set

Each of the arrangements of the disks in locations on any of the puzzles can be thought of as a permutation. By a *permutation*, we mean a bijective function from some set S to itself.

Surely you know that a *function* from S to S is a rule that gives back an element of S for every element of S. We will use Greek letters such as α, β, and γ to denote functions. If x is some element of S, the common way to denote the element that a function α returns when x is the input is $\alpha(x)$. However, given that we are adopting left-to-right notation, we will denote the image of x under the action of α by $x\alpha$.

When we say that a function α from S to S is *bijective*, what we mean is that α maps distinct elements of S to distinct elements of S, and that every element of S has something mapped to it by α. The first part of this means that if x and y are two different elements of S, then $x\alpha$ and $y\alpha$ will also be different. The second part means that if we take any element y in S, there must be an element x in S such that $x\alpha$ is y.

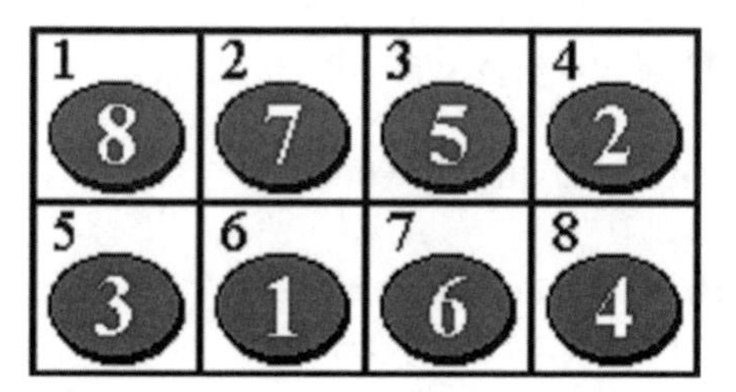

Figure 5.1. A scrambling (permutation) on the set $\{1, 2, \ldots, 8\}$.

We can interpret arrangements of the disks on one of the puzzles, say with spots numbered 1 through n active, as permutations on the set of integers $\{1, 2, \ldots, n\}$. To illustrate, let's use the Transpose Puzzle with 8 active disks. Take, for example, the scrambling of the disks shown in Figure 5.1. We will identify this scrambling with the permutation α, which maps the numbers 1 through 8 as follows.

$$\begin{array}{ll} 1 \to 6 & 5 \to 3 \\ 2 \to 4 & 6 \to 7 \\ 3 \to 5 & 7 \to 2 \\ 4 \to 8 & 8 \to 1 \end{array}$$

The fact that disk number 1 has gone to box number 6 will be represented in the associated permutation α by defining α so that it maps 1 to 6. That is $1\alpha = 6$. More precisely, the equation $1\alpha = 6$ will be interpreted as saying that this permutation moves the disk that was in box 1 (which initially was disk 1) to box 6.

The fact that a scrambling of the disks in the boxes numbered 1 through 8 gives rise to a function from the set $\{1, 2, \ldots, 8\}$ to itself corresponds to the fact that each of the disks numbered 1 through 8 is in one of the boxes having these numbers. Since no box has more than one disk, distinct disks have gone into distinct boxes. This corresponds to distinct elements of $\{1, 2, \ldots, 8\}$ mapping under α to distinct elements of this same set. We observe that every box contains a disk. This corresponds to α having the property that for every y in $\{1, 2, \ldots, 8\}$, there is an element x in $\{1, 2, \ldots, 8\}$ such that $y = x\alpha$.

These observations, taken together, explain why a scrambling does indeed give rise to a permutation. If you consider any scrambling of any one of the puzzles, you will see that the same things will apply. Have you ever come up with a case in which some box had two or more disks in it and some other box had none? You might say that with the Slide Puzzle we have a box or two with no disks, thinking of the blank as no disk. However, we are going to think of a blank as a specific but especially marked disk.

5.2 Two Notations for Permutations

As we have said, we will be using lower case Greek letters to denote permutations. Such a notation will be convenient for talking about properties that apply to permutations in general. However, we need something more when we want to think or talk about particular permutations. Think back to high school algebra. When you started using letters like "x" and "y" to write and talk about general equations, this did not eliminate the need of a notation for dealing with particular numbers such as 1 or 2/5 or π.

There are two notations that are regularly used for denoting particular permutations and we will introduce both of them. The first of these is *array notation*, a notation which represents a permutation from the set $\{1, 2, \ldots, n\}$ to itself by a 2-by-n matrix. In this notation, we have the integers from 1 through n in order in the first row. Then below each of them, we write the number to which the first-row number maps under the permutation.

A general way to express this definition for a permutation α is with the equation

$$\alpha = \begin{pmatrix} 1 & 2 & 3 & \cdots & n \\ 1\alpha & 2\alpha & 3\alpha & \cdots & n\alpha \end{pmatrix}.$$

For the permutation α with $n = 8$ that is represented in Figure 5.1, we have that

$$\alpha = \begin{pmatrix} 1 & 2 & 3 & 4 & 5 & 6 & 7 & 8 \\ 6 & 4 & 5 & 8 & 3 & 7 & 2 & 1 \end{pmatrix}.$$

Perhaps when looking at this notation for the permutation α and at the representation of it on the puzzle in Figure 5.1, the thought has occurred to you that it might be easier to represent α differently in array notation. One might interpret the numbers in the first row as the numbers of the boxes. If we did this, the number in the second row below a given one in the first row would be the number on the disk in that box. If we were to do this, then the permutation in Figure 5.1 would be represented by the 2-by-8 array

$$\begin{pmatrix} 1 & 2 & 3 & 4 & 5 & 6 & 7 & 8 \\ 8 & 7 & 5 & 2 & 3 & 1 & 6 & 4 \end{pmatrix}.$$

This notation would be quicker to write down. You simply read the numbers on the disks in boxes 1 through 8 in order and write them down as the second row.

This alternative notational system would work, but the ease of writing the permutations down is offset by slightly more cumbersome processing in other ways. After considering both of the options, we have settled on the system that we first indicated. A major reason for doing this is that the system reinforces the idea that a function maps elements from a domain set to a range set. Sometimes we think about this concretely by visualizing arrows going from elements in the domain to elements in the range. Figure 5.2 shows such a representation for our permutation α. It shows by means of an arrow from dot 1 in the upper row, the domain, to dot 6 in the lower row, the range, that 1 maps to 6 under α, or $1\alpha = 6$. The remaining seven arrows make it clear what each of the other elements maps to.

These arrows convey the feeling of 1 physically going to 6, and 2 to 4, and 3 to 5, etc. This corresponds well to what we see on the puzzle in Figure 5.1, where disk 1 has physically gone from box 1 to box 6, and disk 2 has physically gone to box 4, etc.

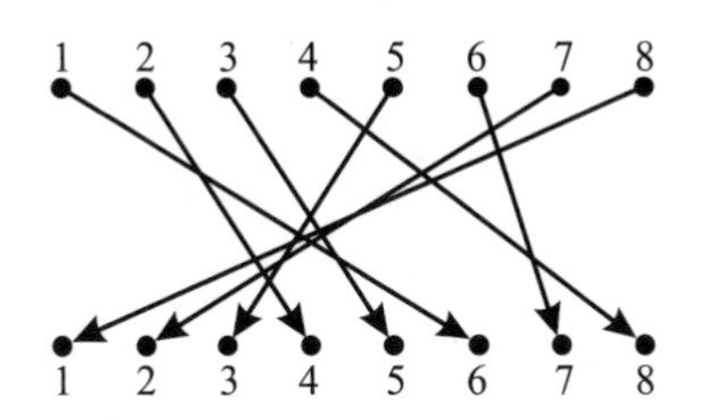

Figure 5.2. An arrow diagram for α.

Another observation about how these representations relate to each other is that if you rearranged the dots in the second row of Figure 5.2 so that all of the arrows are vertical, carrying along with each dot its number identifier, and then removed the dots, you would have the numbers arranged as they are in the array notation for α.

Now let us look at the second notation, which is called *cycle notation*. This type is more compact and displays, at a glance, other information about the permutation that will be of use. It will be the preferred notation for most of what we do.

In cycle notation, we denote a permutation α by starting with an element of the underlying set, $\{1, 2, \ldots, 8\}$ in our example, and creating a list of successive images under α that we get, continuing until we get back to our starting element. For example, if we look back at Figure 5.1 or Figure 5.2, we observe that α maps 1 to 6. It maps 6 to 7 and 7 to 2 and then maps 2 to 4, 4 to 8, and finally maps 8 back to 1. Having gotten back to 1, we have completed the cycle. We represent this behavior of α by writing these elements in order and separated by commas within a pair of parentheses thus: (1, 6, 7, 2, 4, 8). When we read such a notation, we understand it to mean that each of the numbers in the list maps to the next one, and that the last one maps to the first.

These six elements of the set $\{1, 2, 4, 6, 7, 8\}$ all lie on one of the cycles of the permutation α. However, there are two elements of the set that are moved by α that are not part of this cycle. Let us start with one of them, say 3, and build a second cycle starting with it. We see that α maps 3 to 5, and then maps 5 back to 3. Thus, we see that α also includes the cycle (3, 5).

Now we have all of the eight elements listed in these two cycles. Together, they contain all the information one needs to know α. We can write α in the form $\alpha = (1, 6, 7, 2, 4, 8)\,(3, 5)$. We say that α is composed of a 6-cycle and a 2-cycle. We will also refer to 2-cycles as *transpositions*, formalizing the concept first introduced in Chapter 2 in conjunction with the Transpose Puzzle.

Figure 5.3 is a second diagram that represents α with dots for the eight integers and arrows showing which map to which. This time we use only one set of dots rather than two representing $\{1, 2, \ldots, 8\}$ in its two roles as domain

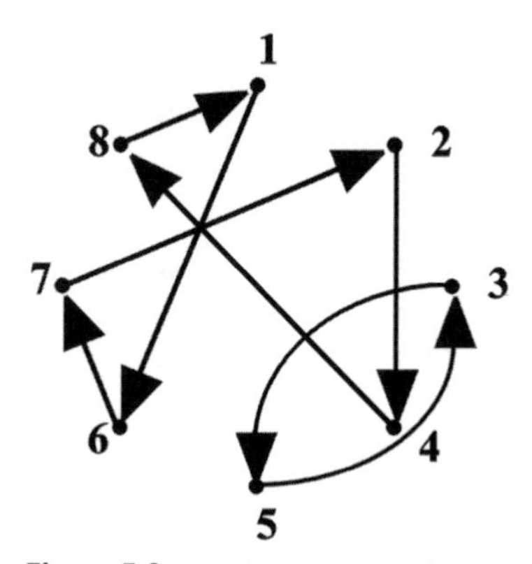

Figure 5.3. Another arrow diagram for α, which displays the cycle structure.

and range. By doing this, we can see the cycle structure. If you start at dot 1, you will see that you follow six arrows before getting back to dot 1, and that the numbers of the dots that you visit on the way are in the order given by the 6-cycle (1, 6, 7, 2, 4, 8). The diagram also shows the transposition (3, 5) by means of the pair of arrows that arc from each of these dots to the other.

The cycle notation we have introduced allows for representing the same permutation in different ways. This is so because cycles can be "cycled" themselves without changing the information they convey. For example, the 6-cycle component of our example α has these alternative representations:

$$(1,6,7,2,4,8) = (6,7,2,4,8,1) = (7,2,4,8,1,6) = (2,4,8,1,6,7) = (4,8,1,6,7,2) = (8,1,6,7,2,4)$$

Each of these six 6-cycles has the underlying six numbers mapping the same way. You see this from Figure 5.3. Start at any of the dots for the numbers in the set $\{1, 2, 4, 6, 7, 8\}$ and follow the arrows until you return to the place where you started. The dots you have passed through are as given by the 6-tuple above that starts with the number of the dot where you jumped on the cycle.

Let's take one more example, this time interpreting as a permutation the scrambling of the Oval Track Puzzle number 2 with 10 active disks shown in Figure 5.4. We will call this one β.

In both array and cycle notation, β is represented as:

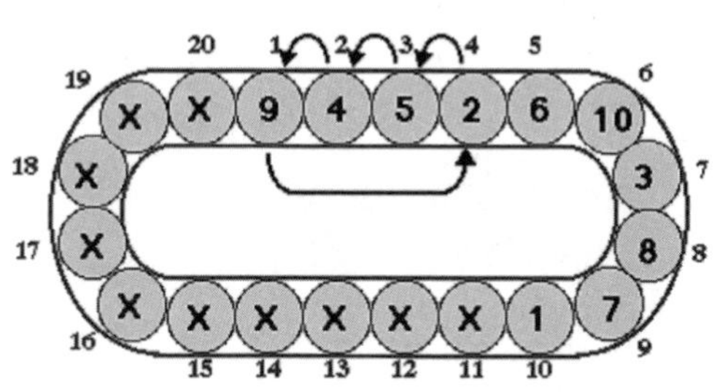

Figure 5.4. A permutation on $\{1, 2, \ldots, 10\}$ shown on Oval Track Puzzle 2 with 10 active disks.

$$\beta = \begin{pmatrix} 1 & 2 & 3 & 4 & 5 & 6 & 7 & 8 & 9 & 10 \\ 10 & 4 & 7 & 2 & 3 & 5 & 9 & 8 & 1 & 6 \end{pmatrix} = (1,10,6,5,3,7,9)(2,4).$$

This permutation is made up of a 7-cycle and a 2-cycle. You will notice that $7 + 2 = 9$, but there are ten numbers in the set to which the permutation applies. Also, note that 8 maps to itself under β, which is seen visually in the puzzle as disk 8 being in spot 8. A conventional practice is to omit "one-cycles" when writing a permutation using cycle notation. If some number does not appear in any of the cycles given, it is understood that it maps to itself. Thus, the fact that 8 is not present in either of the cycles of β means that 8 maps to itself under β. In spite of this convention, there is no prohibition on writing out 1-cycles explicitly. We can write our example permutation as $\beta = (1, 10, 6, 5, 3, 7, 9)(2, 4)(8)$. There might be times when you will want to give the 1-cycles to lay out explicitly which numbers map to themselves.

One situation in which you have no choice but to use a 1-cycle is when you want to express the *identity permutation* in cycle notation, that is, the permutation, denoted by ε, that maps every element to itself. To be specific, let us consider the identity permutation ε on the set $\{1, 2, 3, 4, 5, 6\}$. Expressed in full detail in cycle notation, $\varepsilon = (1)(2)(3)(4)(5)(6)$. We can be implicit on all but the first 1-cycle and write $\varepsilon = (1)$. We could equally well write $\varepsilon = (i)$ for $i = 2, 3, \ldots, 6$. But can you see a way to write ε in cycle notation with all of the 1-cycles being implicit? No. It cannot be done.

5.3 Composition of Permutations

Now that we have some basic notations for permutations, it is time to introduce a method for combining two permutations in order to produce a third one. This method is called *composition*. Composition gives us exactly the tool we need to describe and analyze the results of doing multiple moves on our puzzles.

If α and β are two permutations on $\{1, 2, \ldots, n\}$, the set of positive integers from 1 to n, then we will use them to define a new permutation on this set. We will denote this new one by the symbol $\alpha \circ \beta$ and call this permutation the composition of α and β. We define $\alpha \circ \beta$ by telling how it maps every element of the set. The defining equation for $\alpha \circ \beta$ is $x\,\alpha \circ \beta = (x\alpha)\beta$. This equation says that the result of applying $\alpha \circ \beta$ to a given element x is the element you get by first applying α to x and then applying β to $x\alpha$. That is, $\alpha \circ \beta$ is the mapping you get by applying both α and β in that order.

An important matter related to this notation is that

(5.1) We read expressions involving compositions from left to right.

This may seem like a no-brainer. Unless one grew up reading Hebrew or Arabic or Chinese, don't we expect to read from left to right? Not always in mathematics, where the more common practice for evaluating composition of functions is to read from right to left.

You may be familiar with compositions of functions from some mathematics course, and it is likely that you did them in the right-to-left order. The reason is that, following hundreds of years of tradition, it is common practice to put the variable symbol representing the input to a function on the right of the function symbol contained by a pair of parentheses. Thus, if f denotes a function and x denotes an element from its domain, then $f(x)$ denotes the element of the range set to which x is mapped.

If one is writing functions this way, it is most natural to interpret a composition $f \circ g$ by regarding g, the function closest to the variable, as the one that gets performed first. The composition is defined by the equation $f \circ g(x) = f(g(x))$.

This right-to-left order goes against the grain of our standard practice with our natural language, English, and the habits of eye and mind that it has ingrained within us. When the expressions we are using are short, we adjust, but when they get long, which the expressions we will have to use for studying the puzzles do, it gets more cumbersome to be switching directions as we switch from reading English to reading mathematical notations.

Some mathematicians, particularly algebraists, have tried to break with tradition and switch sides on the composition of functions. If one does this, one puts the input variable on the left of the composition $f \circ g$ and uses the rule $x f \circ g = (x f)\, g$.

The bold few who adopt this system are by far in the minority. Most give in to hundreds of years of tradition, even though it is awkward. Since the expressions we need to process in this book are so long, the awkwardness of reading right to left becomes overwhelming. For this reason, we will join the minority and read from left to right.

If you have invested years of sweat and tears into remembering to go from right to left like this, you may be frustrated for a while with the left-to-right notation. But trust me. You will get used to the new system, and will see its value as we process lengthy expressions representing manipulations of the puzzles.

You will be able to switch back and forth between the systems with no difficulty. This notational switching is at about the level of switching between two different word processing programs on your computer that do things slightly differently. If you have to make a switch, you are annoyed for a while as you break deeply ingrained habits. But once you do, it is no big deal, especially if the new word processor has features that make having switched worthwhile.

Something that we need to think about is whether or not you always get back a permutation when you compose two permutations. This is not something that we should just take for granted. We should convince ourselves that this is always the case.

We will give an argument for this using Figure 5.5 to guide our thinking. In this one diagram, we show two permutations by means of arrows. We can tell them apart because one of them, α, is denoted with dashed arrows, and the other one, β, is denoted with solid arrows. Both sets of arrows represent permutations because for each of them, there is exactly one arrow going out from and exactly one coming into each dot.

How can we picture the composition $\alpha \circ \beta$ using this diagram? Easy. Start at any dot x. Follow the one and only dashed arrow going out from that dot to another one. Then follow the one and only solid arrow out of that second dot to another one. The dot where you have ended up is $x\alpha \circ \beta$. Composing α and β can be thought of as merging together a dashed and a solid arrow whenever the destination end of the dashed one (the pointy end) meets the source end of a solid one. Use six inch pieces of virtual duct tape to hold them together. This ought to hold long enough for the concepts to sink in.

Does this scheme of duct-taping arrows together lead us to a set of taped double-arrows that corresponds to a permutation? Isn't it clear that it does? At each point, we have exactly one dashed arrow coming in and exactly one solid arrow going out, so there is no ambiguity about which ones we should tape together. After the joining is complete, at each point there will be one double-arrow going out. It is the double-arrow you made with the unique dashed arrow that goes out from that point. Similarly at each point there will be a unique double-arrow coming in. It will be the double-arrow you taped together using the solid arrow that comes into the point you are considering. These observations convince us that this taped-together double-arrow mapping, the composition of α and β, is again a permutation.

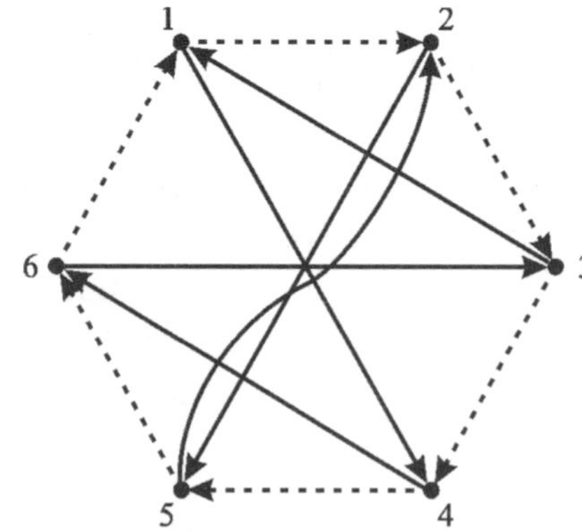

Figure 5.5. Two permutations on one diagram: α denoted by dashed arrows and β denoted by solid arrows.

Now let us start to get some notational practice with compositions of permutations, using the two that are shown in Figure 5.5. We begin by noting that

$$\alpha = (1, 2, 3, 4, 5, 6) \quad \text{and} \quad \beta = (1, 4, 6, 3)\,(2, 5).$$

Converting the expression for $\alpha \circ \beta$ to cycle notation, we get

$$\begin{aligned}\alpha \circ \beta &= (1, 2, 3, 4, 5, 6) \circ (1, 4, 6, 3)(2, 5) \\ &= (1, 2, 3, 4, 5, 6)\,(1, 4, 6, 3)\,(2, 5).\end{aligned}$$

In the second rewrite, we have dropped the symbol " $\circ$ " which we use for composition. This follows the practice that we use with multiplication of dropping the operation symbol and interpreting juxtaposition as indicating application of the operation.

This expression for $\alpha \circ \beta$ is less than efficient because it is not a "disjoint" representation. The first two cycles have numbers in common. We can read the expression to rewrite it in "disjoint cycle notation" in which none of the cycles have anything in common. We see that

$$\begin{aligned}\alpha \circ \beta &= (1, 2, 3, 4, 5, 6)\,(1, 4, 6, 3)\,(2, 5) \\ &= (1, 5, 3, 6, 4, 2).\end{aligned}$$

5

This shows that α, a 6-cycle, and β, a permutation consisting of a 4-cycle and a transposition produce another 6-cycle when they are composed. To write $\alpha \circ \beta$ in this way, start with 1 and apply the first cycle to it, the 6-cycle. The result is 2. This value becomes the input for the 4-cycle. Since 2 is not moved by the 4-cycle, 2 maps to itself under the 4-cycle and becomes the input for the 2-cycle. The 2-cycle maps 2 to 5. The result is that after applying this sequence of steps, 1 has been mapped to 5, as is recorded in the second representation. Now let us track what the sequence of cycles does to 5. The 6-cycle maps 5 to 6, which the 4-cycle then maps to 3, which the 2-cycle leaves fixed. The result is that 5 maps to 3. We will do one more step. We see that 3 maps to 4 under the 6-cycle, which maps to 6 under the 4-cycle, which remains fixed under the 2-cycle. The result is that 3 maps to 6. We will leave it to you to follow through the remaining steps to complete the analysis leading to the final expression for $\alpha \circ \beta$.

Now let us give an example of calculating compositions of permutations in array notation. Consider this one:

$$\begin{pmatrix}1 & 2 & 3 & 4 & 5 & 6 & 7 & 8\\ 5 & 6 & 1 & 7 & 2 & 3 & 8 & 4\end{pmatrix} \circ \begin{pmatrix}1 & 2 & 3 & 4 & 5 & 6 & 7 & 8\\ 2 & 1 & 6 & 5 & 4 & 3 & 8 & 7\end{pmatrix} = \begin{pmatrix}1 & 2 & 3 & 4 & 5 & 6 & 7 & 8\\ 4 & 3 & 2 & 8 & 1 & 6 & 7 & 5\end{pmatrix}.$$

Here is how you interpret this. The first (left) permutation maps 1 to 5 and the second maps 5 to 4, so the composition maps 1 to 4. The first maps 2 to 6 and the second maps 6 to 3, so the composition maps 2 to 3. The first maps 3 to 1 and the second maps 1 to 2, so the composition maps 3 to 2.

Once again, we emphasize that we are performing permutations from the left to right. If we were using the more common right-to-left system, we would have done calculations just as we are here, except for the fact that we would move between permutations from right to left. For example, in cycle notation calculated right-to-left, the reduction to disjoint cycles of (1, 2, 3) (2, 4) (3, 5, 2) results in (1, 2) (3, 5, 4). Start with 1. It maps to itself under the rightmost cycle and also under the next one to the left. Finally 1 maps to 2 under the leftmost cycle, so 1 maps to 2 under the composition. Now let us see what 2 maps to. Under the rightmost cycle, 2 maps to 3, which the next cycle leaves fixed. The last cycle maps 3 to 1, so the composition maps 2 to 1. Calculating in our left-to-right order, we get the result (1, 4, 3) (2, 5), which is clearly different.

5.4 Inverses of Permutations

Imagine that you are working with the Oval Track Puzzle, and you set for yourself the challenge of solving a scrambling without ever using the Undo buttons. Suppose at one point you mistakenly do a rotation of the disks three positions clockwise when you should have done a "Do Arrows." What next?

This is a no-brainer! Just rotate the disks three positions counterclockwise, and this will take the disks back to where they were before the mistaken move. If this seems totally obvious to you, as it ought to if you have played with the puzzles sufficiently, then you already have a working understanding of the concept of inverses of permutations. An intuitive definition is that for a permutation α, another permutation β is an inverse of α if β undoes what α does. More formally, for a permutation α, an *inverse* of α is another permutation β such that $\alpha \circ \beta = \varepsilon$ and $\beta \circ \alpha = \varepsilon$, where ε denotes the identity permutation. Recall that the identity permutation is the one that maps every element to itself.

An important fact is that

(5.2) Every permutation α on any set S has a unique inverse permutation, which is denoted α^{-1}.

If we think visually of a permutation as a set of arrows between dots as in Figure 5.3, it is easy to see why this claim is true. Suppose we have a set of arrows representing a permutation α. Let us assume that they are all red in color. Let us now draw in arrows that will give us an inverse permutation for α, coloring them all blue to distinguish them from the arrows representing α. The system is simply to put in a blue arrow from a point y to the point x whenever there is a red arrow from x to y. Let us use β to denote the function defined by the blue arrows.

Do the blue arrows indeed define a permutation on the set S? The criterion, remember, is that at each dot, there must be exactly one arrow coming in and one going out. Consider what happens at a fixed dot x. Since α is a permutation, the red arrows satisfy the criterion we have just stated. Thus, there is exactly one red arrow coming into dot x. Given the way we constructed the blue arrows, this means that there is exactly one blue arrow going out from x. Similarly, since there is exactly one red arrow going out of x, it follows that we constructed exactly one blue arrow coming into x. We see that the blue arrows define a permutation.

Now let's think about why the blue-arrow permutation β is an inverse for α. The criterion is that each undoes the other. Notationally, this means that $\alpha \circ \beta = \varepsilon$ and $\beta \circ \alpha = \varepsilon$, but on a concrete level in the context of our visualization, it means that no matter where you start, if you follow a red arrow and then a blue arrow, you get back to where you started, and similarly, if you follow a blue arrow and then a red arrow, you also get back to where you started. Again, we have a no-brainer. We constructed the blue arrows in a way that makes this happen. The blue arrows define an inverse for α.

Still, there is that word "unique" in (5.2) that we have not yet addressed. We want to be sure that, for a given permutation, α, there is only one inverse permutation. We will use the traditional way to make an argument for the uniqueness of something, which is to introduce two symbols for things that meet the specified conditions and then show that they must in fact be equal. We have already defined an inverse for α that we have called β. Let us suppose that γ also represents a permutation that satisfies the criteria for being an inverse for α. We wish to show that in fact, γ really is β. Thinking visually again, let us represent γ by means of a set of green arrows.

By saying that γ is an inverse for α, we are saying that γ undoes what α does and vice-versa. Let us see what follows from this. Pick any dot x, and let y denote the dot at the source end of the red α-arrow which comes into x. Since we are assuming that γ undoes what α does, and since α maps y to x, then γ must map x to y. That is, the green arrow going out from x must point to y. But doesn't the blue β-arrow going out from x point to y also? Yes. Since we were thinking of just any dot x, what this means is that in all cases, the blue β-arrow and green γ-arrow pointing out from a dot x both point to the same dot. This is the criterion for two permutations to be equal, so we have just shown that $\gamma = \beta$. That is, in spite of our attempt to have two inverses for α just by using different names for them, a brief investigation shows that the differently named inverses for α are really the same.

The final thing we need to do in getting acquainted with inverse permutations is to look at how they behave notationally. This works out quite simply in either the array notation or in the disjoint cycle notation.

First, let's look at array notation. Consider a permutation α on the set {1, 2, …, n}:

$$\alpha = \begin{pmatrix} 1 & 2 & 3 & \cdots & n \\ 1\alpha & 2\alpha & 3\alpha & \cdots & n\alpha \end{pmatrix}.$$

The listing of numbers in order in the first row represent the set {1, 2, …, n} in its role as the domain and the listing in the second row in a scrambled order represent the same set in its role as the range set. When we consider α^{-1}, these roles are reversed, but the pairings reflected by the columns of the array remain the same. The arrangement of the numbers in the array tells us that for each integer i between 1 and n, α maps i to $i\alpha$. The arrangement also tells us that α^{-1} maps $i\alpha$ to i.

5

If we switch the two rows of this array for α, we now have the first-row entries mapping to the second-row entries below each of them under α^{-1}. If we then keep these columns intact but rearrange them so that the first row entries are in increasing order, we have the array notation for α^{-1}.

To illustrate this, if we have

$$\alpha = \begin{pmatrix} 1 & 2 & 3 & 4 & 5 & 6 & 7 & 8 \\ 4 & 1 & 5 & 7 & 3 & 8 & 2 & 6 \end{pmatrix}.$$

Then by switching the rows and sorting the columns of this array we get the following two expressions for α^{-1}:

$$\alpha^{-1} = \begin{pmatrix} 4 & 1 & 5 & 7 & 3 & 8 & 2 & 6 \\ 1 & 2 & 3 & 4 & 5 & 6 & 7 & 8 \end{pmatrix} = \begin{pmatrix} 1 & 2 & 3 & 4 & 5 & 6 & 7 & 8 \\ 2 & 7 & 5 & 1 & 3 & 8 & 4 & 6 \end{pmatrix}.$$

To summarize,

(5.3) To get from the array notation for a permutation α to the array notation for its inverse α^{-1}, just switch the two rows of the array for α and then resort the columns to put the first row entries into increasing order.

For cycle notation, the inversion rule is even more straightforward. It is:

(5.4) To get from a cycle notation representation for a permutation α to a cycle notation representation for α^{-1}, just write the representation for α down in the reverse order.

To illustrate, if $\alpha = (1, 5, 3, 6)(2,4)(7, 10, 8, 9)$, then $\alpha^{-1} = (9, 8, 10, 7)(4, 2)(6, 3, 5, 1)$. The reason is that in a cycle notation representation for α, each number listed in a cycle maps under α to the next one to the right in the cycle, with the last wrapping around and mapping to the first. The inverse of α, in undoing what α does, takes things the opposite way, including wrapping the first (leftmost) entry to map it to the last one. Thus, to write α^{-1} so that you can read what maps to what in the standard left-to-right order within the k-tuples, you just need to reverse the k-tuples of α.

5

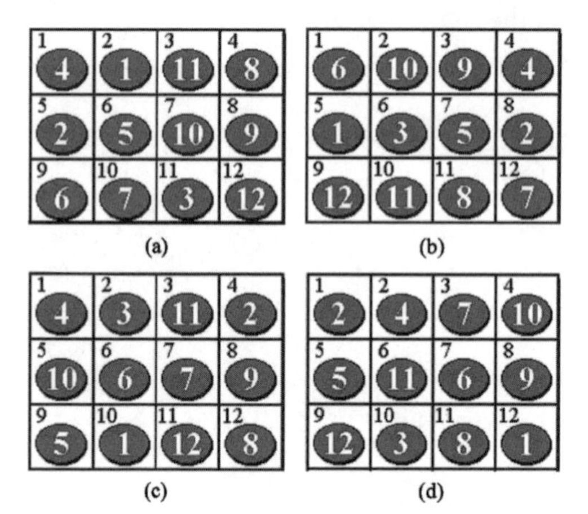

Figure 5.6

Exercises

Note: *In these exercises, we introduce informally the notation S_n for the set of all permutations on the set $\{1, 2,\ldots, n\}$ of integers between 1 and n. Thus, for example, S_{12} denotes the set of all permutations on the set $\{1, 2, \ldots, 12\}$. This notation will be formally introduced in Chapter 6.*

5.1 Transpose Puzzle arrangements into permutation notations. Figure 5.6 shows four arrangements of numbered disks in numbered boxes on the Transpose Puzzle. In each case, express the arrangement shown as a permutation in S_{12}. Use both the 2-by-12 array notation and the cycle notation to represent them. Recall that we identify an arrangement with a permutation α such that $i\alpha = j$ means that disk number i has gone to box number j.

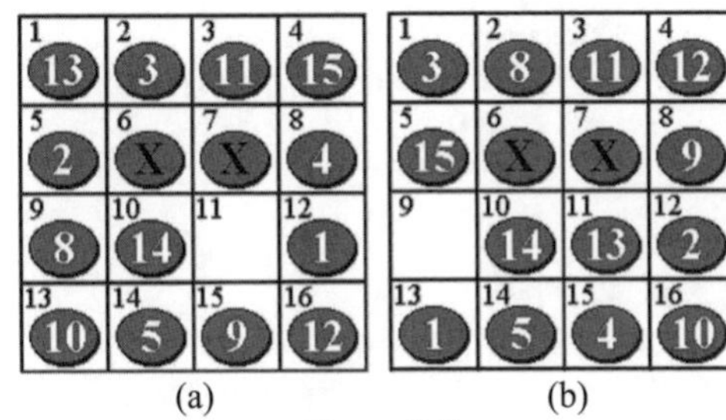

Figure 5.7

5.2 Slide Puzzle arrangements into cycle notation. Figure 5.7 shows two scramblings of the disks on a configuration of the Slide Puzzle that is like the basic "15" configuration except that spots 6 and 7 are out of action. Express these as permutations in S_{16} using cycle notation. Recall that in the initial configuration the blank is in spot 16 so we identify the blank with the number 16. Treat the numbers that are out of action as mapping to themselves, which means that they need not appear in the cycle notation.

5.3 Oval Track Puzzle arrangements into cycle notation. Figure 5.8 shows two scramblings of the disks on the Oval Track Puzzle. Express them in cycle notation.

5.4 Hungarian Rings Puzzle arrangements into cycle notation. Figure 5.9 shows two scramblings of the Hungarian Rings Puzzle in numbers mode. Express these in cycle notation. Make note of the lengths of the cycles that are involved.

5.5 Converting between array and cycle notations. Given here are several permutations in S_{10} expressed in either two by ten array notation or cycle notation. Convert each of them from its current notation to the other one.

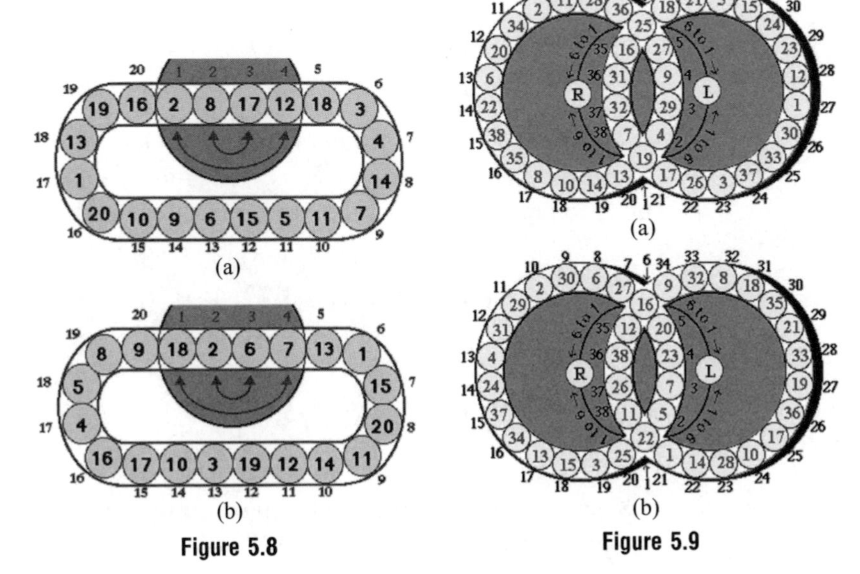

Figure 5.8

Figure 5.9

(a) $\alpha = \begin{pmatrix} 1 & 2 & 3 & 4 & 5 & 6 & 7 & 8 & 9 & 10 \\ 5 & 7 & 4 & 3 & 2 & 6 & 1 & 10 & 8 & 9 \end{pmatrix}$

(b) $\beta = \begin{pmatrix} 1 & 2 & 3 & 4 & 5 & 6 & 7 & 8 & 9 & 10 \\ 10 & 4 & 3 & 8 & 2 & 5 & 6 & 1 & 7 & 9 \end{pmatrix}$

(c) $\gamma = \begin{pmatrix} 1 & 2 & 3 & 4 & 5 & 6 & 7 & 8 & 9 & 10 \\ 2 & 3 & 4 & 1 & 5 & 10 & 9 & 7 & 6 & 8 \end{pmatrix}$

(d) $\delta = (1, 5, 8, 4)\,(2, 9, 10, 7)\,(3, 6)$

(e) $\mu = (2, 6, 8, 3, 10, 9, 7, 4)$

(f) $\nu = (1, 2, 3, 4)\,(5, 6)\,(7, 8, 10)$

5.6 Compositions of permutations. Using the permutations in S_{10} that are defined above in Exercise 5.5, give expressions for the compositions given below using both of the notational systems.

(a) $\alpha \circ \beta$ (b) $\beta \circ \alpha$ (c) $\alpha \circ \alpha$ (d) $\beta \circ \beta$
(e) $\gamma \circ \delta$ (f) $\delta \circ \gamma$ (g) $(\mu \circ \nu) \circ \delta$ (h) $\mu \circ (\nu \circ \delta)$

5.7 Reducing cycle notation to disjoint cycles. When permutations expressed in cycle notation are composed, the cycles are not necessarily disjoint. When this happens, we want to put the compositions into disjoint cycle notation. To practice this, write these permutations in disjoint cycle notation.

(a) $\alpha = (1, 4, 3, 5)\,(3, 7, 6)\,(2, 5, 7, 3, 1)\,(6, 4)\,(2, 3, 5, 4)\,(4, 5, 3)$
(b) $\beta = (1, 2, 3)\,(1, 4, 5)\,(1, 6, 7)\,(1, 8, 9)$
(c) $\gamma = (1, 2, 3, 4)\,(1, 4, 2, 5, 3)\,(3, 6, 4, 2)\,(3, 6)$
(d) $\delta = (1, 2, 3, 4)\,(1, 5, 6)\,(1, 7, 8, 9)$
(e) $\mu = (9, 3, 5, 6)\,(4, 5, 2, 3, 7)\,(3, 7, 8, 2)\,(1, 4)\,(7, 4)$
(f) $\nu = (5, 3, 1)\,(5, 7, 9)\,(2, 4, 6, 1, 3, 5)\,(1, 8)$

5.8 Expressing the inverse of a permutation. For each of the permutations in Exercise 5.5, express its inverse in both notational systems.

5.9 Is composition commutative? A *binary operation* on a set is a rule for combining two elements of the set to get a new element of the set. We usually use an operation symbol, such as "Δ", to denote a binary operation, and if x and y denote two elements of the set, we denote the result of combining them using the operation by the symbol $x \Delta y$. We say that an operation is *commutative* if it is the case that for any two elements x and y in the set, $x \Delta y = y \Delta x$. That is, the operation is commutative if, when you reverse the order of the combination of any two elements, you get the same result. Review your results from Exercise 5.6 and decide whether the operation of composition on S_{10} is commutative.

5.10 Everything doesn't commute with something (except the identity). Prove that if $n \geq 3$, then for every element α in S_n, if α is not the identity permutation ε, then there is some other permutation β in S_n with which α does not commute.

5.11 Is composition associative? We defined what we mean by an operation "Δ" on a set in Exercise 5.9. We say that an operation is *associative* if it is the case that for any three elements x, y, and z in the set, $(x \Delta y) \Delta z = x \Delta (y \Delta z)$. That is, the operation is associative if, when you combining any three elements in a specific order, the final result is the same no matter which two of the three you combine first. Review your results from Exercise 5.6 and decide whether the evidence suggests that the operation of composition on S_{10} is associative. (We will show that composition is indeed associative for S_n in the next chapter.)

Permutation Groups: Just Enough Theory

Now that you have learned the basic concepts and notations of permutation groups in Chapter 5, you will see just enough of the theory of these structures in this chapter to guide our study of the puzzles. We will continue to include informal justifications for the properties set forth, using concrete visualization as the vehicle whenever possible. In some cases where a notation-based argument is more efficient, we will give symbolic justifications.

The context for the development of the theory in this chapter will be the set $\{1, 2, \ldots, n\}$. Let's denote this set by $\mathbb{Z}_n$. Following the standard notation found in nearly all books that discuss permutation groups, S_n will denote the set of all of the permutations on $\mathbb{Z}_n$.

6.1 Associativity

Our first claim is that

(6.1) Composition of permutations is an *associative* operation on S_n. That is, if α, β, and γ are any three permutations in S_n, then $\alpha \circ (\beta \circ \gamma) = (\alpha \circ \beta) \circ \gamma$.

To see why this is true, let us imagine that we have three permutations on $\mathbb{Z}_n$ represented by three sets of arrows in different colors, red for α, blue for β, and green for γ. In Chapter 5, we visualized composing two transpositions as creating a set of double-arrows by putting pairs of arrows together with duct tape. Let's think about how that idea plays out when we put together triples of arrows.

Let's consider first duct-taping to produce $\alpha \circ (\beta \circ \gamma)$. The pair of parentheses is indicating to us that we need to first deal with the composition of β and γ and after that compose α with the result of the first step. Composing β with γ means to make the rounds of every dot duct taping the incoming blue β-arrow at that dot to the outgoing green γ-arrow. Once we have completed this process, we need to make another pass through all the dots and this time duct-tape the incoming red α-arrow to the outgoing, earlier-conjoined, blue-to-green $(\beta \circ \gamma)$-arrow.

Now let's think about producing the arrows for $(\alpha \circ \beta) \circ \gamma$. This time the parentheses tell us to first make the rounds of all the dots duct-taping the incoming red α-arrow to the outgoing blue β-arrow. After this is done, we make the rounds again, this time joining at each dot the incoming red-blue double $(\alpha \circ \beta)$-arrow to the outgoing green γ-arrow.

Now let's compare the results of each of these processes. In each case, we have produced a set of red-blue-green triple-arrows with the terminal end of one being taped to the initial end of the next. The only difference is that in the first case we did the taping at the second joint first and at the first joint second, and in the second case, we did the taping at the first joint first and the second joint second. Does this make any difference in what set of red-blue-green triple-arrows we get? Clearly, it does not. This justifies the claim that $\alpha \circ (\beta \circ \gamma) = (\alpha \circ \beta) \circ \gamma$. Figure 6.1 might help you see what we are talking about here.

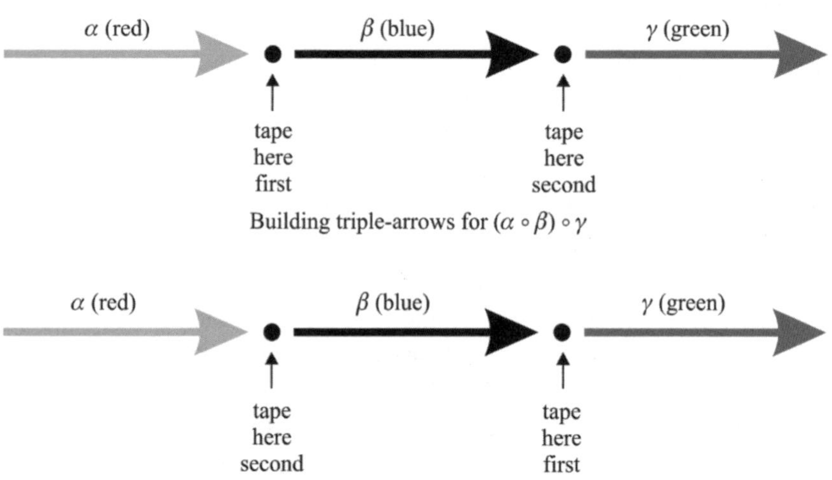

α (red) β (blue) γ (green)

tape here first — tape here second

Building triple-arrows for $(\alpha \circ \beta) \circ \gamma$

α (red) β (blue) γ (green)

tape here second — tape here first

Building triple-arrows for $\alpha \circ (\beta \circ \gamma)$

Figure 6.1. Comparing triple-arrows that represent the compositions $(\alpha \circ \beta) \circ \gamma$ and $\alpha \circ (\beta \circ \gamma)$.

Here is another way to see why (6.1) is true that builds on your experience with the puzzles. Suppose that we identify the permutations α, β, and γ with three procedures that we have programmed into buttons A, B, and C of the Oval Track Puzzle. How can we interpret the two expressions $\alpha \circ (\beta \circ \gamma)$ and $(\alpha \circ \beta) \circ \gamma$ in this context? The first expression tells us to press button A to produce the permutation it gives. Then the grouped expression $(\beta \circ \gamma)$ tells us to press buttons B and C. What does this mean? For now, let's interpret the grouping as pausing after pressing button A, perhaps leaving the puzzle for a moment to check e-mail, but that we are to press buttons B and C without a pause.

Playing along with this sort of game, you will see that other grouping $(\alpha \circ \beta) \circ \gamma$ tells us to regard the pressing of button A and B as a group, and, after doing these steps, take our pause to check our e-mail before returning to press button C. Now think about it. Does it make a difference in the outcome of executing the sequence of macros A, B, and C in that order at which point we break away to check our e-mail? Clearly, the result is the same either way. If you don't believe it, try a few examples.

6.2 Inverses of Products

Our second claim gives a useful rule for taking the inverse of a composition of a permutation. It says that

(6.2) For any two permutations α and β in S_n, $(\alpha \circ \beta)^{-1} = \beta^{-1} \circ \alpha^{-1}$. That is, the inverse of the composition of two permutations is the composition of their inverses in the reverse order.

This property is rather clear if we think of permutations as being represented with sets of arrows. Suppose we have red arrows for α again and blue arrows for β, and that we have taped them together to form the red-blue double arrows for $\alpha \circ \beta$. Now we want to think about undoing the effect of following one of these double-arrows. If you have gone from one dot to another along one of these double-arrows, the way to get back where you came from is to go backwards along the blue arrow that comes into the dot where you are, and then backwards along the red arrow that you find at the dot that you got to on the red arrow. And in doing so, what have you done? You've performed the composition $\beta^{-1} \circ \alpha^{-1}$.

Let's look at (6.2) from the point of view of one of the puzzles as we did in expanding our understanding of (6.1). Suppose you have done two steps on the Hungarian Rings Puzzle that you need to undo. Say the steps are R and L. It ought to be clear from your experience with the puzzles that to back up, you must first undo the last move you made. That is, you must perform the inverse of L first, and then the inverse of R. This means that $(RL)^{-1} = L^{-1}R^{-1}$. The rule is, "last in, first out."

What happens if you need to compose more than two permutations? Well, the "last in, first out" rule continues to hold. Specifically, the rule in (6.2) generalizes to:

(6.3) For any integer m with $m \geq 2$ and for any m permutations $\alpha_1, \alpha_2, \ldots, \alpha_m$ in S_n,

$$\left(\alpha_1 \circ \alpha_2 \circ \cdots \circ \alpha_m\right)^{-1} = \alpha_{m-1}^{-1} \circ \cdots \circ \alpha_1^{-1}.$$

6

6.3 Non-Commutativity and Commutativity

The mathematical ideas that we have been exploring in an informal way are a part of group theory, a subject that mathematics students usually see for the first time in an abstract algebra course. Those readers who are somewhat familiar with group theory will recognize that the properties in (6.1) through (6.3) are ones that are seen not only with permutations. They are true in all groups. (If you do not know what a group is, don't worry. You're picking up concepts of group theory right now.) Our next observations, in contrast, are more specific to permutations.

We will look next at the matter of commutativity. We will say that two permutations α and β in S_n *commute with each other* provided that $\alpha \circ \beta = \beta \circ \alpha$. We observe that

(6.4) Much of the time, permutations in S_n do not commute with each other. Specifically, whenever $n \geq 3$, there are permutations α and β in S_n such that $\alpha \circ \beta \neq \beta \circ \alpha$.

Statements that say fuzzy things like "much of the time" do not lend themselves to the same kind of rigorous justification as ones that say something is "always" or "never" true. We could find a way to be more precise about this, such as trying to come up with a number that gives the probability that two permutations are taken at random will commute with each other, but we do not have need for such precision here. The important point is, as you will learn after enough time with the puzzles, it is far from common that two permutations commute with each other.

Figure 6.2. Sample permutations that do not commute with each other.

The second part of the statement is illustrated in Figure 6.2. We show two permutations by means of arrows. Let us call the one with solid arrows α and the one with shaded arrows β. The arrows show how these two permutations map the three integers in the set $\mathbb{Z}_3$. All of these are available to us if $n \geq 3$. The four unnamed dots with the solid and shaded loops on them are representing the fact that we are defining α and β so that all additional elements in $\mathbb{Z}_n$ map to themselves under both of these permutations.

We claim that the two permutations pictured in Figure 6.2 do not commute with each other. That is, $\alpha \circ \beta \neq \beta \circ \alpha$. In order for it to be true that $\alpha \circ \beta = \beta \circ \alpha$, it would have to be the case that when you start at any dot and follow a solid and then a shaded arrow, you get to the same place as you do if you follow a shaded arrow first and then a solid arrow. This happens sometimes, for example if you start at any of the dots other than those numbered 1, 2, or 3. However, if you start at any of these three, the order in which you follow the arrows makes a difference. For example, from dot 1, the solid then shaded arrow path takes you to dot 3, whereas the shaded then solid path takes you back to where you started at dot 1. This alone is enough to show that in this case, $\alpha \circ \beta \neq \beta \circ \alpha$.

Even though we ought not expect permutations to commute, there is one situation that occurs reasonably often that will assure that they do commute. They will commute when they are disjoint.

For a permutation α on $\mathbb{Z}_n$, let us denote by M_α the set of elements x of $\mathbb{Z}_n$ such that $x\alpha \neq x$. We use the letter "M" to help us remember that these are the elements that are *moved* by α. This set M_α is the set of all elements of $\mathbb{Z}_n$ that are moved by α. In contrast, we will denote by F_α the elements of S_n that are *fixed* by α, that is the elements x such that $x\alpha = x$.

If it is the case that for two permutations α and β in S_n, the sets M_α and M_β have no elements in common, that is, if $M_\alpha \cap M_\beta = \varnothing$, then we will say that α and β are *disjoint*. We claim that

(6.5) Disjoint permutations commute with each other. That is, if there are no numbers in $\mathbb{Z}_n$ that are moved by both α and β, then $\alpha \circ \beta = \beta \circ \alpha$.

As before, we will think through why this is true in terms of a visual model, using the representation of permutations with sets of arrows. The nature of this argument, however, is such that we will have to rely almost as much on the symbolic representations.

Suppose we have two disjoint permutations α and β represented with red and blue arrows respectively. The fact that they are disjoint means that at any dot where the red arrow is not a loop back to the same dot, that is, the red arrow goes to some other point, then the blue arrow at that dot is indeed a loop. Similarly at any point at which the blue arrow goes somewhere else, the red arrow must be a loop.

We need one more observation that is not totally obvious. If x is in M_α, that is, if x is one of the points at which the outgoing red arrow goes somewhere else, then the point $x\alpha$ is also one which has its outgoing red arrow going somewhere else. That is, $x\alpha$ is also in M_α. The alternative would be that the outgoing red arrow at $x\alpha$ is in fact a loop. This would have two red arrows coming into $x\alpha$, namely the one from x and the loop from $x\alpha$. This violates the condition that there is just one arrow coming into each point for a permutation.

Thus, whenever x is in M_α, $x\alpha$ is also in M_α. Clearly the same holds for M_β. Whenever x is in M_β, $x\beta$ is also in M_β.

Now we are ready to say why α and β commute. We can see that α and β will commute with each other if, no matter at what point x we start, the result of applying α to x and β to $x\alpha$ is always the same as applying β to x and α to the result. This translates into saying that, starting at any point x, if you first follow the red arrow and then the blue arrow, you end up at the same point as if you first followed the blue arrow and then the red arrow.

To see that this will always happen, we need to consider different cases according to whether we start at an x that is in M_α, in M_β, or in neither. If we start at an x that is in M_α, then the red arrow going out from x goes to some other point. This point will be another one in M_α, which means, since M_α and M_β are disjoint sets, that this point $x\alpha$ is not in M_β. This means that the blue arrow going out from this point is in fact a loop back to the same point. That is, β must leave it fixed. The result is that $x\,(\alpha \circ \beta) = x\alpha$. On the other hand, returning to x, if we look at the blue arrow going out from x, it is a loop back to x. Why? Because x is in M_α, which is disjoint from $M\beta$, so β has to leave x fixed. Thus $x\beta = x$, and applying α to both sides of this equation, we get that $x(\beta \circ \alpha) = x\alpha$. Since $x(\alpha \circ \beta)$ and $x(\beta \circ \alpha)$ are both equal to $x\alpha$, they are equal to each other as we want. This completes the argument in cases when x is in M_α.

The argument if x is in M_β is similar. It will be a useful exercise for you to think it through, but we will not write out the details.

Finally, what about a point x that is in neither M_α nor M_β? For such an x, both the red and blue arrows going out from x are in fact loops back to x. Clearly if you start at such an x and follow the red and then the blue arrow, or do the same with the order of the colors reversed, you get to the same place, namely right back to the same x. Again we have $x\,(\alpha \circ \beta) = x\,(\beta \circ \alpha)$ as desired.

We have argued why in all cases for x, $x\,(\alpha \circ \beta) = x\,(\beta \circ \alpha)$. This is what it means for $\alpha \circ \beta$ and $\beta \circ \alpha$ to be equal.

As an immediate consequence of this result (6.5), we have that

6

(6.6) Whenever a permutation is written in disjoint cycle notation, the cycles that are used can be commuted.

This is true because we can regard the cycles that we use for expressing a permutation as permutations in their own right. If we are talking about disjoint cycles, this means that they are disjoint permutations in this latest sense also. Thus, by (6.5), they commute with each other.

To illustrate, if we have a permutation α expressed in disjoint cycle notation in the form

$$\alpha = (1, 2, 3)(4, 5)(6, 7, 8, 9)$$

then we are free to reorder these three disjoint cycles in any way that we wish. Thus,

$$\alpha = (4, 5)(1, 2, 3)(6, 7, 8, 9) = (6, 7, 8, 9)(1, 2, 3)(4, 5)$$

6.4 Rules for Exponents

One remaining subject that we need to consider is exponentiation. Recall that when we introduced notation for representing maneuvers on the Oval Track Puzzle in Chapter 1, we began using exponents to denote repetitions of moves. This sort of notation translates naturally to composition of permutations. When α denotes a permutation in S_n, and m denotes a positive integer, we will denote by α^m the permutation we get by composing α with itself m times. That is, $\alpha^m = \alpha \circ \alpha \circ \cdots \circ \alpha$, where the number of copies of α in the repeated composition is m.

We define negative exponents by the rule $\alpha^{-m} = (\alpha^{-1})^m$ where m is any positive integer, and we define the zero exponent by the rule $\alpha^0 = \varepsilon$, where ε denotes the identity permutation which maps all elements to themselves.

An important observation is that some of the familiar "rules of exponents" which students learn in high school algebra apply to this operation of combining permutations. Specifically, for any two integers m and k and for any permutation α in S_n, we have that

(6.7) (a) $\alpha^m\alpha^k = \alpha^{m+k}$

(b) $(\alpha^m)^k = \alpha^{mk}$

6

Note that the little symbol for composition has dropped out. This is a simplification in the notation that is familiar from elementary algebra where it is routinely used with the operation of multiplication.

One of the familiar rules is not true in general, however. Specifically, most of the time if you take two permutations α and β in S_n and an integer m other than 1, the permutations $(\alpha\beta)^m$ and $\alpha^m\beta^m$ are not equal. This is in contrast to the rule $(ab)^m = a^m b^m$ which holds for any numbers a and b and any integer m when the interpretation of the exponent is repeated multiplication. The reason that this "distributivity of exponents" holds for numbers but not for permutations is that the multiplication of numbers is commutative, meaning that every pair of numbers commute with each other. In contrast, not all pairs of permutations commute with each other for composition.

To see why commutativity is relevant, let us consider two permutations that do not commute with each other. Let's take $\alpha = (1, 2, 3)$ and $\beta = (1, 2)$ in S_3. These are the two permutations shown in Figure 6.2, for the case in which $n = 3$, so we have already shown that they do not commute with each other. We see for these two that $[\alpha\,\beta]^2 = [(1, 2, 3)\,(1, 2)]^2 = (2, 3)^2 = \varepsilon$ while $\alpha^2\beta^2 = (1, 2, 3)^2\,(1, 2)^2 = (1, 3, 2)\,\varepsilon = (1, 3, 2)$. We see that $[\alpha\beta]^2 \neq \alpha^2\beta^2$.

In contrast to this, suppose we have two permutations α and β which do commute with each other. Since $(\alpha\beta)^2$ means $(\alpha\beta)(\alpha\beta)$, by our associativity result (6.1) we can change the grouping without changing the order to get the equal expression $\alpha(\beta\alpha)\beta$. Since $\beta\alpha = \alpha\beta$, we could go on to write the string of equalities $(\alpha\beta)^2 = (\alpha\beta)(\alpha\beta) = \alpha(\beta\alpha)\beta = \alpha(\alpha\beta)\beta = (\alpha\alpha)(\beta\beta) = \alpha^2\beta^2$, which gives us the rule for the case $m = 2$. From here, one could extend the rule to all integers. (If you know how to do proofs that use the Principle of Mathematical Induction, you will be able to see how to extend this to all of the positive integers.) In general, we have that

(6.8) If α and β are two permutations in S_n which commute with each other, i.e., if $\alpha \circ \beta = \beta \circ \alpha$, then for all integers m, $(\alpha \circ \beta)^m = \alpha^m \circ \beta^m$.

We emphasize that this rule, which allows for distributing an integer exponent over a composition of two permutations, requires only that these two permutations commute with each other, not that everything commutes with everything else.

Rule (6.8) is not likely to apply if we just take two permutations α and β at random, because the chances that they commute with each other are slim. Rule (6.5) gives us an important criterion for recognizing that (6.8) does hold true. Specifically, we have that

(6.9) If α and β are two disjoint permutations in S_n, then for all integers m, $(\alpha \circ \beta)^m = \alpha^m \circ \beta^m$.

To illustrate how (6.9) helps with computations in S_n, let us consider the permutation shown as a scrambling on the Oval Track Puzzle in Figure 5.4. We see that it is $\beta = (1, 10, 6, 5, 3, 7, 9)(2, 4)$. Thus, β consists of a 7-cycle and a transposition disjoint from it. Since these disjoint cycles commute with each other, we have that for any integer m, $\beta^m = (1, 10, 6, 5, 3, 7, 9)^m(2, 4)^m$. Being able to rearrange the powers of β this way allows us to derive quickly some interesting conclusions about the powers of β. For example, if we take positive powers of β, then the first one that would produce the identity would be 14. Why? We must have both of the mth powers equal to the identity, and $(2, 4)^m = \varepsilon$ when m is even, while $(1, 10, 6, 5, 3, 7, 9)^m = \varepsilon$ when m is a multiple of 7.

Exercises

6.1 Inverses of compositions manifested in the puzzles. You can get a very practical understanding from the puzzles of the principle in (6.3) that the inverse of a composition of permutations is the composition of their inverses in the reverse order. (a) To see this, follow these steps: (i) Pick one of the puzzles, other than Transpose. (ii) Program a two-step procedure into button A. (iii) Set the macro delay to about a second. (iv) Initialize the puzzle so that it represents the identity permutation. (v) Run the macro program A, paying attention to the steps and their order. (vi) Now press the button A^{-1} which inverts what A did. Pay attention to the fact that it runs the reverse of the two steps that A did, and in the reverse order. (b) Repeat this test with a three step, a four step, and a five-step procedure. Again, pay attention to how (6.3) is manifested.

6

6.2 The practicalities of the negative exponent definition. We defined negative exponents with the formula $\alpha^{-m} = (\alpha^{-1})^m$. (a) Setting α equal to R on the Oval Track Puzzle, write a sentence that interprets this equation, starting your sentence with: An m-position counterclockwise rotation of the disks is the same as .... (b) A consequence of (6.7) (a) and the definition that α^0 is the identity permutation is that $\alpha^{-m}\alpha^m = \alpha^{-m+m} = \alpha^0 = \varepsilon$, and $\alpha^m\alpha^{-m} = \alpha^{m+(-m)} = \alpha^0 = \varepsilon$.

This shows that α^{-m} is the inverse of α^m, or $(\alpha^m)^{-1} = \alpha^{-m}$. But since $\alpha^{-m} = (\alpha^{-1})^m$ also, it follows that $(\alpha^{-1})^m = (\alpha^m)^{-1}$. That is, the mth power of the inverse of a permutation α is the same as the inverse of the mth power of that permutation. Write a sentence interpreting this equation in terms of the Oval Track Puzzle if α is again taken to be R.

6.3 The practicalities of exponent distribution. This exercise explores how the rule (6.8) gets manifested in the puzzles. Follow these steps. (i) Go to Oval Track Puzzle version 4, which has $T = (5, 4, 3, 2, 1)$. (ii) Initialize the puzzle with the number of active disks set to at least 10. (iii) Program into button A the short procedure $R^{-5}\,T\,R^5$. (We will use A to denote the permutation you get from this button. (iv) Perform the procedure A and make note of the numbers of the disks moved by A. (v) Reinitialize the puzzle, and perform the "Do Arrows" T, making note of the numbers of the disks that it moves. (vi) Note that A and T are disjoint because there is no disks that is moved by both of them. (vii) Now perform the procedures $(AT)^3$ and make note of the arrangement of the disks that this produces. (viii) Reinitialize and perform A^3T^3. Note that the disks are arranged exactly as they were following $(AT)^3$. Now explain why this is what (6.8) says should happen.

6

6.4 The set M_α of elements moved by α. Listed here are several permutations defined by procedures on some of the puzzles. Perform the procedures and in each case identify the set of elements that are moved by the procedure.

(a) On Oval Track 1 with 20 disks active, $\alpha = (TR^{-1})^{17}$.

(b) On Oval Track 1 with 20 disks active, $\beta = (R^{-1}T)^{17}$.

(c) On the Hungarian Rings in numbers mode, $\gamma = (L\,R)^3\,(L^{-1}R^{-1})^3$.

(d) On Oval Track 4 with at least 8 disks, $\delta = R^{-3}TR^{3}T^{-1}$.

(e) On the Slide Puzzle in the standard "15" configuration, $\mu =$
(16, 12)(12, 11)(11, 15)(15, 14)(14, 13)(13, 9)(9, 10)(10, 11)(11, 12)
(12, 16)(16, 15)(15, 14)(14, 13)(13, 9)(9, 10)(10, 11)(11, 12)(12, 16)
(16, 15)(15, 11)(11, 10)(10, 9)(9, 13)(13, 14)(14, 15)(15, 16)
(16, 12)(12, 11)(11, 10)(10, 9)(9, 13)(13, 14)(14, 15)(15, 16).

Exercises 6.5 through 6.9 are ones that develop the theory a bit more. They provide additional insight into the basics of group theory, and some more theoretical tools for understanding the puzzles.

6.5 Is the converse of (6.8) true? The answer to this question is "no." It is possible to have $(\alpha\beta)^m = \alpha^m\beta^m$ for certain choices of permutations α and β and certain integers m even when α and β do not commute. To see this represented in one of the puzzles, consider Oval Track Puzzle version 1, with only 4 active disks, and let $\alpha = R$ and $\beta = T$. (a) Verify that R and T do not commute with each other. (b) Verify that $(RT)^4 = R^4T^4$. [Hint: Note that every element raised to the fourth power on this puzzle is the identity, so that the claimed equation reduces to $\varepsilon = \varepsilon\,\varepsilon$.] (c) See if you can find additional examples that show that the converse of (6.8) is not true.

6.6. The cancellation laws. (a) Suppose that α, β, and γ are three permutations in S_n such that the equation $\alpha \circ \gamma = \beta \circ \gamma$ is true. Show that it must follow that $\alpha = \beta$. This is called the *right cancellation law.* (b) Suppose this time that $\gamma \circ \alpha = \gamma \circ \beta$. Again show that $\alpha = \beta$. This is called the *left cancellation law.* (c) Give an interpretation of these laws for one of the puzzles and make an argument based upon the puzzle that these claims are reasonable.

6.7 Inverses are unique. In Exercise 6.2, we showed that α^{-m} acts like an inverse for α^m and then concluded that it is indeed the inverse. This is justified because every permutation has a unique inverse. We gave an informal proof of this in Section 5.4. Prove this formally by showing that if β_1 and if β_2 both act as inverses for a permutation α, then in fact $\beta_1 = \beta_2$.

6.8 Inverses of inverses. (a) Prove that for any α in S_n, $(\alpha^{-1})^{-1} = \alpha$. That is, the inverse of the inverse of any permutation is the permutation itself. (b) Give an explanation of how this statement manifests itself on the puzzles, and show from this point of view why it is clearly true.

6.9 Two special cases when the converse of (6.8) *is* true. In Exercise 6.5, we saw that in general the converse of the statement in (6.8) is not true. However it is true for two specific cases, namely for $m = 2$ and for $m = -1$. (a) Prove that for any α and β in S_n, if $(\alpha \circ \beta)^2 = \alpha^2 \circ \beta^2$, then $\alpha \circ \beta = \beta \circ \alpha$. (b) Prove that for any α and β in S_n, if $(\alpha \circ \beta)^{-1} = \alpha^{-1} \circ \beta^{-1}$, then $\alpha \circ \beta = \beta \circ \alpha$.

Cycles and Transpositions

Back in Chapter 5 you learned that cycle notation highlights the "cycle structure" of a permutation. Knowledge of the cycle structure of the permutations manifested in the puzzles will be important for devising strategies to solve them. For this reason we need to consider cycles a bit more. You will see that the simplest nontrivial cycles, namely 2-cycles that we also call transpositions, are important building blocks for all permutations.

7.1 Expressing Permutations with Cycles

Chapter 5 did not go into detail as to why all permutations can be expressed with cycles. Now it is time to consider this matter explicitly.

(7.1) Any permutation α in S_n can be expressed with disjoint cycles.

Visual thinking in terms of dots and arrows works well to see that this is true. Imagine a set of n dots labeled 1 through n, and imagine a set of n arrows between the dots that represents the permutation α. Start at any dot and follow arrows. After traversing some number of arrows that must be at most n, since there are only n arrows, you will have to return to your starting point. The reason is that prior to your return, you are always moving to new points. This is because there is only one arrow coming into each point. If you were to return to a point you had previously visited, other than the first one, this would require that there be two arrows coming into that point.

Now think about it. If you never return to your starting point, you will continue to visit new points indefinitely. This is not possible because after all, there are only n points. As a worst case, you could visit all of them before returning to your starting point, and once this is done, the arrow emanating from the last point visited has nowhere else to go other than the starting point.

Once you return to the first point, you have completed a cycle. Write it down in cycle notation. We will call it σ_1. Let's also trace over this cycle on our dots and arrows diagram with a bright highlighter so that we can see it easily. If this highlighted cycle passes through all of the points of the diagram, then the permutation α is itself an n-cycle and we have expressed α in terms of cycles: $\alpha = \sigma_1$. On the other hand, if there are dots in the diagram that are not on this first highlighted cycle, pick any one of them. Starting at this new dot, trace out a new cycle.

We can say for sure that you will again eventually have to close the loop and return to the starting point. The reason is the same as before. You continue to visit different points until you return and eventually you run out of different points to visit. This "visit different points" principle also assures that this new cycle that you are tracing does not intersect with the first one you identified. If it did, then at some step in tracing the second cycle you would move along an arrow to a point on the first cycle for the first time. This would mean that there would have to be two arrows coming into that point. But that never happens with permutations.

We now have identified a second cycle, which is disjoint from the first. Write it out in cycle notation, name it σ_2, and trace this one out also with your highlighter. Now look around to see if there are any dots that are not on these two cycles. If there are no such dots, we have completed the description of α as being $\sigma_1\sigma_2$. If there are one or more dots not on these two cycles, then you know what to do. Start at one of them and trace out a third cycle disjoint from the first two.

If you continue in this way, eventually you will get all the points identified with cycles. Why? It is because each time you trace out a cycle, you process at least one dot. After at most n such steps of identifying cycles, you will have processed all of the dots, since there are only n of them.

The worst-case scenario in terms of the number of cycles occurs if α is the identity permutation. Then your expression turns out to be $\alpha = (1)\,(2)\,\cdots\,(n)$, and we have expressed α in terms of only 1-cycles. In general, the process we have described has us writing out the 1-cycles, but we can go back and eliminate the 1-cycles, if any, in keeping with our convention that any numbers not listed in a cycle notation are understood to map to themselves. (If α is the identity permutation and came out as n 1-cycles, leave at least one 1-cycle as a placeholder.)

7.2 Expressing Cycles with Transpositions

Let's now explore ways of building up cycles from simpler components, or, looking at it from the opposite perspective, breaking cycles down into simple pieces. Our first observation is

(7.2) Every cycle in S_n can be expressed as a composition of transpositions (2-cycles). Specifically, an m-cycle $(a_1, a_2,\ldots, a_m)$ can be expressed as $(a_1, a_2,\ldots, a_m) = (a_1, a_2)(a_1, a_3)(a_1, a_4) \cdots (a_1, a_m)$.

We remind you once again that we are reading from left to right. Also, recall that we have followed the practice of dropping the symbol "$\circ$" for composition in this expression, relying on the understanding that juxtaposition means composition.

To see why the given expression works, let's use the technique described in Chapter 5 to reduce the non-disjoint expression on the right to an expression in disjoint cycle form. Beginning with a_1, we see that the first transposition maps it to a_2. The remaining transpositions do not include a_2, which means that they keep a_2 fixed. As a result, if we apply all of the transpositions to a_1, the result is that it maps to a_2.

Next, let us see what the sequence of transpositions does to a_2. The first one maps a_2 to a_1. We now are tracking what happens to a_1. The second transposition maps a_1 to a_3, so we now track a_3. We see that the remaining transpositions leave a_3 fixed, so in the end, the sequence has mapped a_2 to a_3.

Now let's think in general about what happens to a_i for any integer i such that $2 \le i \le m-1$. The first transpositions, specifically those of the form (a_1, a_j) where $j < i$, leave a_i fixed, so we track a_i through all of these transpositions. When we get to the transposition (a_1, a_i), we see that it maps a_i to a_1. We are now tracking a_1. The next transposition, (a_1, a_{i+1}), maps a_1 to a_{i+1}, so we switch to tracking a_{i+1}. All of the remaining transpositions are of the form (a_1, a_j) where $j > i+1$, so all of them leave a_{i+1} fixed. The result is that the sequence of transpositions maps a_i to a_{i+1}.

We now know that a_i maps to a_{i+1} for all i such that $1 \le i \le m-1$. The final case to consider is the a_m case. Reading from the left, we see that all of the transpositions leave a_m fixed until we get to the last one. This one maps a_m to a$_1$, so the sequence of transpositions has mapped a_m to a_1.

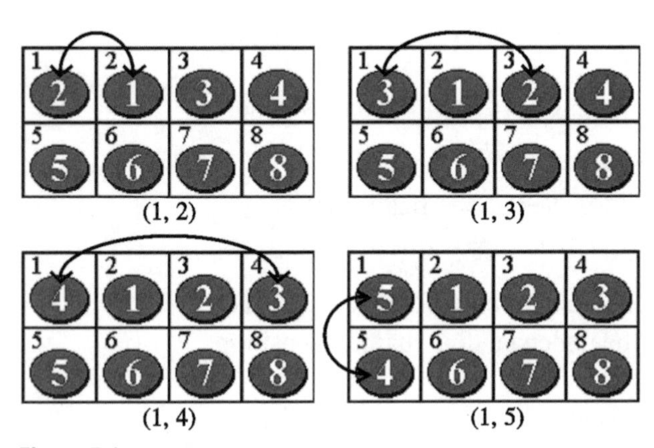

Figure 7.1. Applying swaps from box 1 to boxes 2, 3, 4, and 5 in succession produces the 5-cycle (1, 2, 3, 4, 5).

If we review these observations about how the sequence of transpositions behaves, we see that we have shown that it reduces to the m-cycle $(a_1,a_2,\dots, a_m)$.

This is a symbolically based proof. It proves the point, but you can get a much more concrete and tactile understanding of why (7.2) is true by interpreting it in the context of the Transpose puzzle. To do this, let us drop the a's and consider the cycle $(1, 2,\dots, m)$ with $m = 5$. Go to the puzzle and be sure that the number of active disks is at least five and that the puzzle is initialized into the identity permutation. Now perform swaps between boxes 1 and 2, then 1 and 3, then 1 and 4, and finally 1 and 5, in that order. Figure 7.1 shows the results of doing this. We see that we have produced the 5-cycle $(1,2,3,4,5)$.

The swaps we performed are expressed notationally as $(1,2)$, $(1,3)$, $(1,4)$, and $(1,5)$. Thus, we have verified on the Transpose puzzle that $(1,2)\,(1,3)\,(1,4)\,(1,5) = (1,2,3,4,5)$.

Our rule in (7.2) looks the way it does because of the fact that we are reading compositions from left to right. If you read about these ideas in other books, you will likely find that the rule we have given gets expressed differently because of the more frequent use of right-to-left notation. If you were using this notation, you could use these very same transpositions, but just write them in the reverse order.

7.3 Expressing All Permutations with Transpositions

Now that we know that we can express all permutations using cycles and that we can express all cycles using transpositions, what do you suppose that we are going to observe next? Right. Here is the claim.

(7.3) Any permutation α in S_n can be expressed with transpositions.

To see why, let us take any permutation α in S_n. First express α using disjoint cycles, which we know we can do because of (7.1). Express each of these disjoint cycles using a transposition, which (7.2) assures us we can do. We now have α expressed using transpositions.

If the Transpose puzzle served its intended purpose, claim (7.3) is no surprise at all. If you have played with this puzzle, scrambling and solving it for at least ten minutes, you have a fundamental understanding of the truth of the claim. What we are doing here is simply stating something formally that you have known for a while. But don't dismiss this act of "articulating the known" as being unimportant. Being able to articulate what we know is more important than just knowing it. Surely one of our cave dwelling ancestors had to articulate the knowledge that "round things roll" in order to translate generations of experience with round rocks and chunks of logs into the invention of the wheel. Articulation of knowledge leads to new applications of that knowledge.

Let us illustrate claim (7.3) with an example of a scrambling of the "15" configuration of the Slide puzzle. Figure 7.2 shows us such a scrambling, which we identify with a permutation α on the set $\mathbb{Z}_{16}$. Recall that on the Slide puzzle, the blank is in the highest numbered square when the puzzle is initialized. In this case, the blank started in box 16, so when we write out our expression, we will denote the blank disk as disk 16.

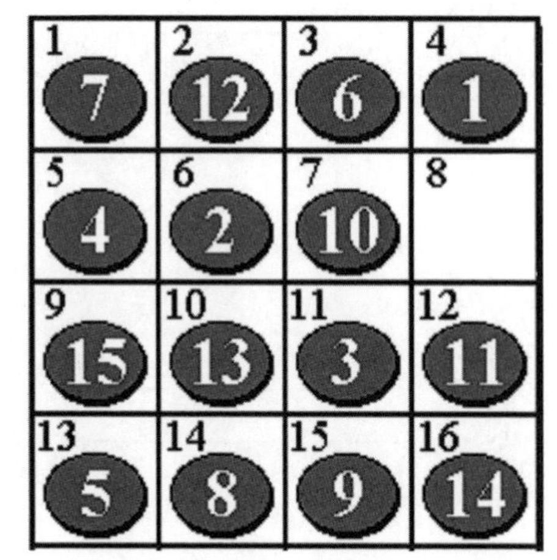

Figure 7.2. This permutation can be expressed with transpositions.

Let's start our analysis by tracing out the cycle involving box 1. We see that disk 1, the one from box 1, has moved to box 4. The disk from there has gone to box 5, the one from there to box 13, the one from there to box 10, the one from there to box 7, and finally, disk 7 moved from box 7 to box 1, completing a 6-cycle (1, 4, 5, 13, 10, 7). We still have boxes not addressed, so let us look next at the box with the smallest number not on our 6-cycle. This would be box 2. Starting there, we see that we get a 5-cycle (2, 6, 3, 11, 12).

By now we see how α maps eleven of the numbers. There are still five of them remaining. Staying with the system of starting at the smallest number not yet addressed, we identify the cycle that involves 8. We see that this is the 3-cycle (8, 14, 16). The number 16 is identified with the blank, which started out in box 16. There are still two numbers to investigate. The smallest is 9. We see that there is a transposition (9, 15) as well. We have now accounted for all sixteen numbers. We have that

$$\alpha = (1, 4, 5, 13, 10, 7)(2, 6, 3, 11, 12)(8, 14, 16)(9, 15)$$

The system given in (7.2) for expressing cycles in terms of transpositions now allows us to go from this cycle expression for α to

$$\alpha = (1, 4)(1, 5)(1, 13)(1, 10)(1, 7)(2, 6)(2, 3)(2, 11)(2, 12)(8, 14)(8, 16)(9, 15)$$

Are you starting to see the potential for applying this knowledge to solving problems that arise in the puzzles? Claim (7.3) tells us that transpositions can play a fundamental role in solving the puzzles. We can infer from (7.3) that if we can figure out how to do any transposition on one of the puzzles, then we will be able to solve it no matter how it is scrambled.

How about that? We have an idea for a method to solve any of the puzzles. We'll follow up on how to make this idea work in later chapters.

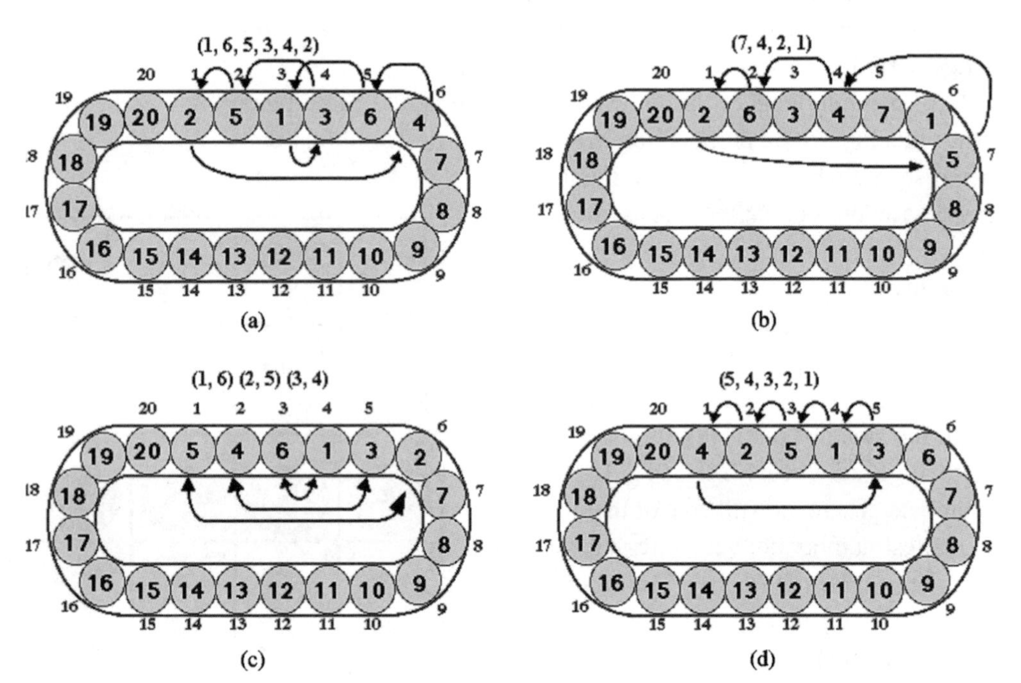

Figure 7.3

Exercises

7.1 Puzzle scrambles with transpositions. Shown in Figure 7.3 are four puzzles with end game scrambles. In each case, express the permutation represented in cycle notation, and then in terms of transpositions.

7.2 From disjoint cycle notation to transpositions. Express these permutations given in disjoint cycle notation as transpositions.

(a) $\alpha = (1, 5, 3, 8)(2, 9, 4, 7, 6)$ (b) $\beta = (1, 4, 2)(3, 7, 6)(5, 8, 9)$

(c) $\gamma = (1, 2, 3, 4, 5, 6)(7, 8)$ (d) $\delta = (1, 2)(3, 4)(5, 6, 7)(8, 9)$

7.3 Exploring orders using the Oval Track Puzzles. We can use the Oval Track Puzzles to explore the order of permutations. (The *order* of a permutation α is the least positive integer m such that $\alpha^m = \varepsilon$.) Specifically, you can go to any one of the versions of Oval Track and see the order of its "Do Arrows" permutation by checking how many times you must press the "Do Arrows" button to return to the identity permutation, ε, assuming that you began with ε. Determine the order of the "Do Arrows" permutation for each of the first 18 versions of the puzzle. Do you see any relationship between the cycle structure of the permutation and its order?

7.4 More exploration of order with the "Build Your Own" OT 19. You can continue your exploration of the relationship between the order of a permutation and its cycle structure by using the "Build Your Own Do Arrows" version of the Oval Track Puzzles. Go to Oval Track 19 and reconfigure it with one of the permutations below as the "Do Arrows." See how many presses of the "Do Arrows" button it takes to get back to the identity. Carry out this exploration for each of these permutations and make note of the order of each. Can you see how the order relates to the cycle structure of the permutation?

(a) $\alpha = (1, 2, 3)(4, 5)$ (b) $\beta = (1, 2, 3)(4, 5, 6)$

(c) $\gamma = (1, 2, 3)(4, 5, 6, 7)$ (d) $\delta = (1, 2, 3)(4, 5, 6, 7, 8)$

(e) $\mu = (1, 2, 3)(4, 5, 6, 7, 8, 9)$ (f) $\nu = (1, 2, 3, 4)(5, 6, 7, 8, 9, 10)$

(g) $\rho = (1, 2)(3, 4, 5, 6)(7, 8, 9)$ (h) $\sigma = (1, 2)(3, 4, 5)(6, 7, 8, 9, 10)$

7.5 The order of a cycle. If you did the exploration proposed in the last two exercises, perhaps you know the order of a cycle of length k. Now prove that the order is k.

7.6 The order of any permutation. Prove that if α is any permutation in S_n, and if the disjoint cycles that make up α have lengths $k_1, k_2, \ldots, k_m$, then the order of α is $\operatorname{lcm}(k_1, k_2, \ldots, k_m)$, the least common multiple of the lengths of its cycles.

7.7 Non-uniqueness of disjoint cycle notation. Permutations can be expressed in terms of disjoint cycles in multiple ways. This is because you can express a cycle starting with any of its elements and because disjoint cycles commute with each other. Thus, for example, $(1, 2)(3, 4, 5)$, $(1, 2)(4, 5, 3)$, and $(5, 3, 4)(2, 1)$ are three of the twelve different representations of one permutation in S_5. How many different ways are there to express a given permutation in terms of disjoint cycles?

7.8 A certain kind of uniqueness for disjoint cycle notation. One can get a unique representation of a permutation α in S_n in the form $\sigma_1\sigma_2 \cdots \sigma_m$, where the σ_i's are disjoint cycles and σ_1 starts with the smallest number moved by α, σ_2 starts with the smallest number moved by α that is not moved by σ_1, σ_3 starts with the smallest element moved by α but not by either σ_1 or σ_2, etc. Furthermore, assume that each of these cycles σ_i is expressed with the smallest element that it moves as the first one listed. Prove that the representation of any permutation α that fits this description is unique.

7.9 The cycle structure of permutations and their inverses. We define the *cycle structure* of a permutation α other than the identity to be the sequence $k_1, k_2, \ldots, k_m$ of integers, all 2 or greater, such that $k_1 \leq k_2 \leq \cdots \leq k_m$ and such that α can be express as disjoint cycles of these lengths. (Define the cycle structure of the identity to be 1.) (a) Prove that every permutation in S_n has a uniquely determined cycle structure. (b) Prove that for any permutation α, its inverse, α^{-1}, has the same cycle structure as α.

The Parity Theorem

8.1 The Parity of a Permutation

Now that you know that any permutation can be expressed using transpositions, let's explore the question of how many transpositions we need for this. We begin by observing that this number is not fixed. To illustrate this, let's start with an inefficient expression for a permutation α in terms of non-disjoint cycles. Specifically, let $\alpha = (1, 3, 2, 5)\,(3, 4, 1)\,(2, 4)$. If we rewrite α in terms of disjoint cycles, we get that

$$\alpha = (1, 3, 2, 5)\,(3, 4, 1)\,(2, 4) = (1, 2, 5, 3, 4)$$

If we reexpress all of the cycles above using transpositions, we get that

$$\alpha = (1, 3)\,(1, 2)\,(1, 5)\,(3, 4)\,(3, 1)\,(2, 4) = (1, 2)\,(1, 5)\,(1, 3)\,(1, 4)$$

We have one expression for α that uses six transpositions and another that uses four of them.

Although these two numbers of transpositions used to express α are different, both of them are even. This is not merely a coincidence. It is a manifestation of a very important fact called the Parity Theorem, which we will state in a moment.

8

Before stating this theorem, we need to tell you what we mean by the *parity* of an integer in case this term is unfamiliar to you. We say for an integer m that its *parity is even* if m is a multiple of 2. If m is not a multiple of 2, then its *parity is odd*. Perhaps the term "parity" was not familiar but certainly the distinction between even and odd numbers is. The positive integers of even parity are the so-called even integers 2, 4, 6, 8,… and those of odd parity are the odd integers 1, 3, 5, 7,…. Note that $0 = 0 \cdot 2$, so is a multiple of 2 and therefore is even.

We can now state the theorem.

(8.1) **The Parity Theorem:** For any permutation α in S_n, all possible ways of expressing α using transpositions have the number of transpositions used all sharing the same parity.

To illustrate what this is saying, let us suppose that we have some permutation α in S_n and that we have expressed α using transpositions in two different ways. Specifically, suppose that we know that $\alpha = \sigma_1\sigma_2 \cdots \sigma_r$ and that also $\alpha = \tau_1\tau_2 \cdots \tau_s$, where all of the σ_i's and all of the τ_j's are transpositions. The Parity Theorem tells us that either the integers r and s are both odd integers or they are both even integers.

One of the ways to see why the Parity Theorem is true involves making use of this fact:

(8.2) Any expression for the identity permutation ε in terms of transpositions uses an even number of them. That is, if

$$\varepsilon = \tau_1\tau_2 \cdots \tau_m$$

where the τ_i's are all transpositions, then m is an even integer.

Before considering why this is true, let's see how knowing it leads to the Parity Theorem. Suppose that we have a permutation α expressed using transpositions in the two ways given in the paragraph above. Let's do some manipulations that follow from having these two expressions for α, and from the simple fact that $\alpha\alpha^{-1} = \varepsilon$. Since α is expressible in the two ways above, we may replace the α in this simple equation by one of these expressions and replace α^{-1} by the inverse of the other one. Thus, we have that

$$(\sigma_1\sigma_2 \cdots \sigma_r)\,(\tau_1\tau_2 \cdots \tau_s)^{-1} = \varepsilon.$$

Two additional facts allow us to do some useful simplification. First, the extended "inverse of the a product" property (6.3) is applicable to the inverse of the product of the τ_j's. Second, we note that if we consider the inverse of a transposition (a, b), we have, because of our observations about inverses of cycles, that $(a, b)^{-1} = (b, a)$. Clearly, because of the way the entries in a cycle can be cycled, $(b, a) = (a, b)$. Thus, $(a, b)^{-1} = (a, b)$, which says that any transposition is its own inverse. (See Exercise 8.3 for more insight into this point.)

These two observations now allow us to rewrite our last displayed equation in the form

$$(\sigma_1\sigma_2 \cdots \sigma_r)\,(\tau_s\tau_{s-1} \cdots \tau_1) = \varepsilon.$$

8

This gives us an expression for the identity permutation ε that uses $r + s$ transpositions. If (8.2) is true, then $r + s$ must be even. This will clinch the Parity Theorem, since r and s must have the same parity to have $r + s$ be even.

We now know that the Parity Theorem is true provided that the claim in (8.2) is true. So how could we justify (8.2)? One way is to prove that:

(8.3) If there is an expression $\tau_1\tau_2 \cdots \tau_m$ for the identity permutation ε that has m transpositions, then there is an expression for ε that uses $m - 2$ transpositions.

Before thinking about why this claim is true, let us see how the claim (8.2) that ε can only be expressed with an even number of transpositions follows from it.

Let us assume to the contrary that it was possible to have an expression $\tau_1\tau_2 \cdots \tau_m$ for the identity permutation involving an odd number of transpositions. We will reason from this assumption to get a contradiction.

If we have an expression $\varepsilon = \tau_1\tau_2 \cdots \tau_m$ with m being odd, according to (8.3), we can get a new expression for ε that uses $m - 2$ transpositions. If m was 3, then $m - 2$ is 1, but if $m > 3$, we can cite (8.3) again to reduce this expression using $m - 2$ transpositions to a new one that uses $m - 4$ transpositions.

Clearly by repeated applications of (8.3), we could whittle down our original expression for ε using an odd number m of transpositions to an expression for ε using just one transposition. But this is ridiculous. Think of performing a single swap on the Transpose puzzle. Have you left the puzzle in the state it was before you did the transposition? Certainly not. The impossibility of a single swap leaving Transpose unchanged, translates into the impossibility of a single transposition being the identity permutation.

The fact that we can deduce something impossible from the assumption that an expression for ε exists that uses an odd number of transpositions forces us to conclude that (8.2) is true.

At this point, we know that once we have shown why (8.3) is true, we will have a complete argument for the truth of the Parity Theorem. The steps to this point have involved a limited amount of work.

In order to convince ourselves that the "can reduce by two" claim in (8.3) is true, let's reinterpret what it is saying into a familiar context, namely the context of the Transpose Puzzle. Suppose that you started out with Transpose initialize so that it was displaying the identity permutation. Now imagine that you did some swaps at random, not paying particular attention to what you were doing. After doing this for a while, imagine that you decided to put everything back into its home box so that you were displaying the identity again. Clearly, you could do this, and when you were done, you would have produced a sequence of transpositions that equals ε. Press the eye-icon "View Moves" button to see the list in the standard ordered pair notation.

If you do this several times, you will notice that the number of transpositions you have used, as displayed on the "Number of Moves" register, is always even. This gives empirical evidence of the truth of (8.2), but that is not good enough for us. We would like to get a rigorous demonstration of (8.2), and thus the Parity Theorem, (8.1), by *proving* (8.3).

So let us think about what we need to do to show that (8.3) is true. We want to take any sequence of transpositions generated on Transpose that starts and ends with the identity permutation being displayed, and find a way to generate another such sequence that uses two fewer transpositions.

We can do this. Not only can we do it, but we can also do it in real-time. By that we mean that if we asked a friend, let's call her Inga, to produce a sequence of the desired sort on one copy of Transpose, we could work on a second copy, match her transposition for transposition—except for using two fewer—and get back to the identity at the same time that she does.

The secret for doing this is simple. Since Inga starts out with every disk in its home box, when she does her first swap, the first disk she touches gets moved out of its home box. Since she is going to produce the identity permutation, at some step in her sequence, this first disk is going to have to come back to its home box. There are two specific transpositions involved here that we are going to target for elimination from our shorter derived sequence: the very first transposition, and the one that brings the first disk moved back home.

Our strategy will be to shorten our sequence by never moving this first disk that she moves. The trick will be to figure out how to follow what she does step for step, while adjusting for the fact that she is moving a disk that we are on purpose keeping at home.

There is an easy strategy for modifying her sequence so that ours is shorter by two, but before we describe it, let's modify the visualization a bit for simplification. We have been imagining two copies of Transpose one that Inga uses, and the other that we use. Let's instead imagine one copy of Transpose with double-wide boxes and two sets of numbered disks—a green set for Inga and a blue set for us. Having both sets of markers in one set of boxes will save us from having to look back and forth between the two displays.

With this new doublewide, two-marker Transpose, let us return to our task of bettering Inga by two. When she does her first transposition, we will not do one. Instead, we will make a note of the two boxes from which she took the disks, making a distinction between them.

Let us make it on the basis of which disk she touched first. Let's call the box that it came from the *First Box*. We will call this disk that Inga takes from the First Box *Inga's First Green Disk*, and call the box to which it went the *Tagged Box*. To aid our memory, let us put markers into these two boxes. The marker we place in the First Box will stay there throughout this process, but the one in the Tagged Box may move, depending on what Inga does.

This is our only response to Inga's first transposition. From here on, until she returns the First Disk home to the First Box, we will respond to each transposition that she does with one of our own which will match hers or will be a small modification of it.

Here are the rules we will use.

(8.4)
(a) If Inga does a transposition on her green markers between boxes other than the First Box or the Tagged Box, then we do a transposition on our blue markers between the same two boxes.
(b) If Inga does a transposition on her green markers between the First Box and a box other than the Tagged Box, then we respond with a transposition on our blue markers between the Tagged Box and the other box, not the First, that she used.
(c) If Inga does a transposition on her green markers between the Tagged Box and a box other than the First Box, then we respond with a transposition on our blue markers between the same two boxes. However, we also move the tag so that this other box now becomes the Tagged Box.
(d) If Inga does a transposition on her green markers between the Tagged Box and the First Box, then we do not do one.
(e) Once Inga has done a transposition of the type described in (d), which she must, then for every transposition of hers thereafter, we do the same transposition.

Figure 8.1 shows a step-by-step example of the application on these sequence-shortening rules. It shows a doublewide Transpose-like grid with two sets of markers. Inga uses the shaded set on the left, and we respond with the plain set on the right. In response to her first transposition, we mark the boxes involved. The next three steps illustrate the use of rules (c), (b), and (a) in that order.

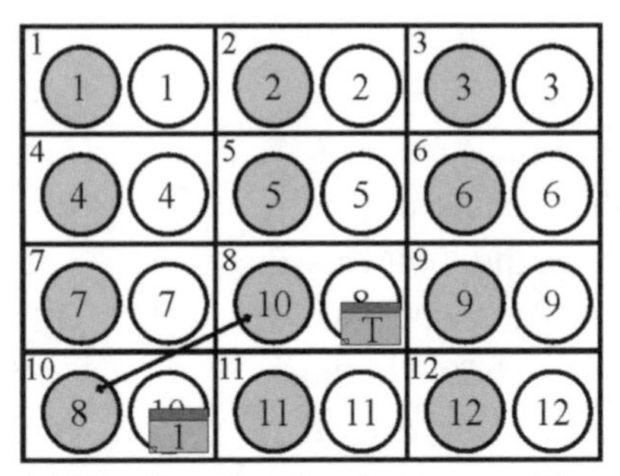

1. When Inga does her first transposition on the shaded left disks, (10, 8) here, we do not do one, but instead mark the First and Tagged boxes.

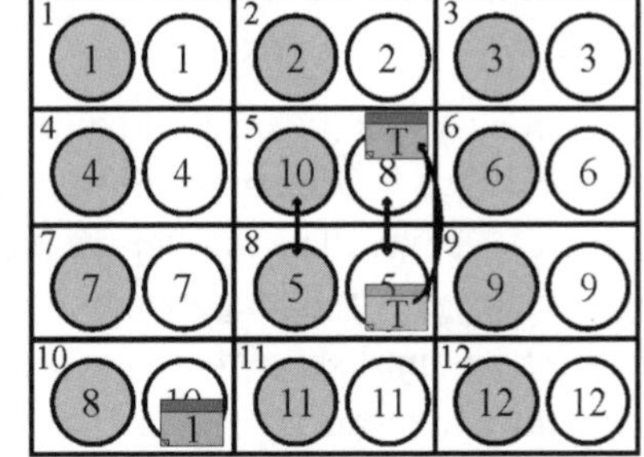

2. If Inga's swap involves the Tagged Box but not the First, we do the same one, and also move the tag.

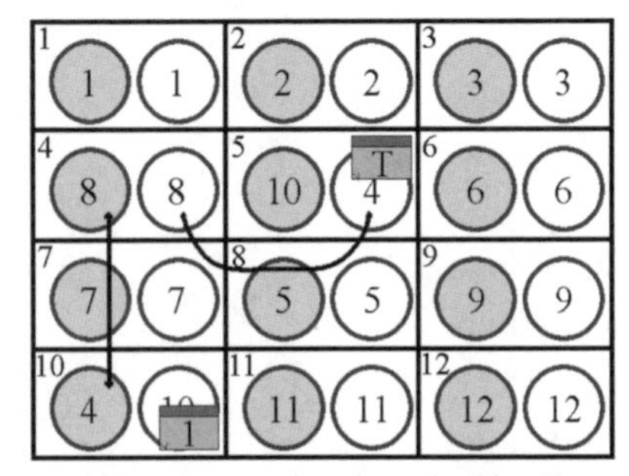

3. If Inga's swap involves the First Box but not the Tagged, we do one with the Tagged Box and her other one.

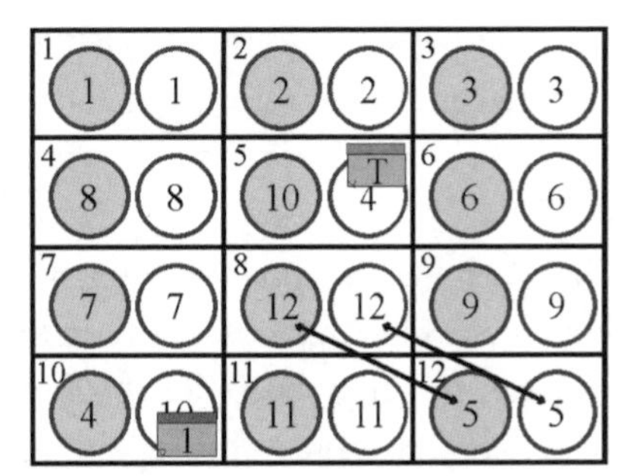

4. If Inga's swap involves neither the First Box nor the Tagged Box, we do the same as she.

Figure 8.1 An example illustrating the shortening algorithm.

Here are some important consequences of following these rules.

(8.5) After each of our responses to Inga's doing a transposition, and up until she returns Inga's First Green Disk to the First Box:

- (i) Inga's First Green Disk is always in the Tagged Box.
- (ii) Our First Blue Disk is always at home in the First Box.
- (iii) Whatever green disk number that Inga has in the First Box, we have the blue disk with that number in the Tagged Box.
- (iv) All boxes other than the Tagged Box and the First Box contain green disks and blue disks with the same number.

Let's think about why these conditions in (8.5) will always be true.

After Inga does her first transposition, we do not do one. The placement of her green and our blue disks differ only in the First Box and the Tagged Box that she changed and we did not, so clearly (iv) is true. We set the Tagged Box in a way that makes (i) true, and (ii) is true because we did not move our blue disk from the First Box. Also, the disk that Inga put into the First Box had on it the number equal to that of the box that we made the Tagged Box. Thus, after Inga does her first transposition followed by our non-transposition tagging ploy, the conditions of (8.5) are true.

Now let us suppose they are true at some point in the process and see why they remain true if we follow the rules for the next step. Let us take the rules in the order they are given. If Inga has done a transposition between boxes other than the First Box and Tagged Box, then following rule (a), we do the same transposition. Since condition (iv) was true prior to this step, namely that we had the same numbered disks in each box other than the First Box and the Tagged Box, then since we have made the same change, this condition remains true. The other conditions (i) through (iii) relate to the contents of the First Box and the Tagged Box. Since we made no changes in these, and since the conditions were true prior to this step, they remain true following this step.

If rule (b) applies because Inga's transposition involved the First Box but not the Tagged Box, then we need to focus our attention on three boxes: the First, the Tagged, and the other one that we used, which we will call Box O for now. Whatever green disk Inga had in the First Box is now in Box O. Because of the way we responded, the green disk that we had in the Tagged Box is now in Box O. But because (iii) was true prior to this transaction, both the green and blue disks in Box O have the same number after the transaction. Also, whatever green disk was in Box O, Inga put into the First Box. Our blue disk with the same number was in Box O because (iv) was true prior to the transaction, and it is in the Tagged Box after the transaction. This means that (iii) is true following the transaction.

Since we both have the same numbered disks in Box O after the transaction as noted above, and since none of the other non-First, non-Tagged boxes were touched, and since (iv) was true prior to the transaction, we have that (iv) is true after the transaction. Condition (i) remains true after this transaction since Inga did not move a green disk out of the Tagged Box. Since Inga's First Green Disk was in the Tagged Box prior to the transaction, it is still in the Tagged Box after it. Condition (ii) remains true because we did not touch our blue disk in the First Box. That puppy's staying put, man! We see that all of the conditions (i) through (iv) remain true after a transaction of type (b) assuming that they were before it.

For a transaction according to rule (c), the argument would be similar. The important difference is that we move the tag, thereby changing the Tagged Box to follow Inga's placement of Inga's First Green Disk. This insures that (i) remains true. You can think through why the other conditions remain true also.

We have now analyzed rules (a) through (c) and have seen that while we continue to follow these rules, all four conditions of (8.5) will remain true. Specifically, this means that Inga's First Green Disk is always the Tagged Box. Since she is duty-bound to eventually end this process by bringing every disk home, eventually she will bring that First Green Disk back home to the First Box. That is, eventually she will have to do a transposition that fits with case (d) of the set of rules. When she does this, we do not do a transposition. As a result, her transposition count is two greater than ours, because we did not do one in response to her first, and we are not doing one in response to this case (d) transposition.

So where does that leave us with respect to placement on the disks? Exactly the same! Up to this point, as the conditions of (8.5) show, the only difference between our placements have been in the First Box and the Tagged Box, with Inga and us having the contents of these two switched with respect to each other. When Inga does her First Box-Tagged Box switch, she brings her arrangement into conformity with ours. But she has used two more transpositions than we have, the two we avoided by leaving our disk in the First Box at home throughout.

Since we agree at this point, and since we will duplicate Inga's transpositions from this point on according to rule (e), when she finally gets every disk home and thus has produced the identity permutation, we will have done the same, but with two fewer transpositions!

We now have an argument for why (8.3) is true. To reiterate, this is the heavy lifting that underlies the Parity Theorem.

Do you have to go through all that to prove the Parity Theorem? Well, not necessarily. There are many proofs known for this theorem, which you can verify by looking at a few introductory texts in abstract or modern algebra. Most of them are shorter, made so by clever uses of mathematical notation. The approach taken here is designed to build on the tactile experience that you have developed from playing with the Transpose Puzzle.

Our approach, motivated by a hands-on way of dealing with the ideas, really lends itself to understanding through hands-on experience rather than the written presentation that is necessary in a book. To this end, we have prepared a computer program that sets up the very doublewide, two-marker representation of the problem that we have been describing.

A Parity Demo program is included as a bonus on the CD ROM with this book. It gives experience working with the shortening algorithm that the rules of (8.4) produce. Figure 8.2 shows the program's screen after several steps of the process have been done. You see that it shows a doublewide Transpose-like grid with two sets of numbered disks. (In color, there is a green set on the left and a blue set on the right.)

The program demonstrates the shortening algorithm of (8.5) in two ways. In the first way, you play the role of Inga and do transpositions on the left set of green disks while the computer plays the "us" role from the above discussion and responds to Inga's moves on the right according to the rules. In the second way, the roles are reversed.

The screen image shows the state of the demonstration after you (in the Inga role) have done six transpositions. The computer, in the role of the responder, has done five transpositions in response. Note that the first four steps in this illustration coincide with the example shown in Figure 8.1.

The First Box is box 10 in this example. The computer shows this by shading the background in box 10. The pattern in the background of box 3 indicates that this is the current Tagged Box. (A pattern is used in Figure 8.2 because of lack of color. The program really uses a different color.) Can you tell from the record which other boxes have been the Tagged Box at some point? Note too that all of the boxes, except for the First Box and the Tagged Box have the same numbers on each disk, and that these two special boxes have opposite things in them with the computer's First Box disk being in its home spot.

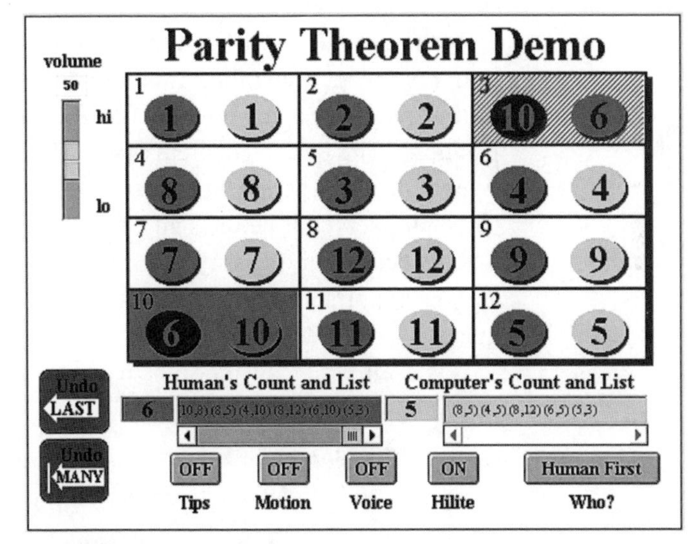

Figure 8.2. The Parity Demo program screen after several steps of the shortening algorithm have been run. Note that the only differences between the placements of the two sets of disks are in box 10, which is the First Box, and box 3, the Tagged Box.

8

If you do the role reversal, you can demonstrate that you understand the shortening algorithm rules by responding to the computer's transpositions. The program gives you the option of having the highlighting which shows the First Box and Tagged Box turned off so that you can demonstrate that your understanding of the algorithm does not depend on the visual cues.

Try the program out, and see if you don't come away understanding the ideas of the proof, even if the written account above seemed murky or impenetrable.

8.2 Even and Odd Permutations

Now that we have justified the Parity Theorem, it is possible to define what we mean by the parity of a permutation. For any permutation α, we will define the *parity* of α to be *even* if α can be expressed using an even number of transpositions and we will define the parity of α to be *odd* if α can be expressed using an odd number of transpositions. The Parity Theorem (8.1) tells us that there is no ambiguity in defining parity of permutations this way. Even if different people arrive at expressions for α using transpositions that have differing numbers, the parities of those numbers will be the same, so that all will arrive at the same parity.

This definition of parity may not stir your passion right now, but just wait. We will see that it will allow us to answer important questions about the puzzles, and sometimes gives us the insight needed to abandon impossible quests.

It is useful to be aware of the parity of cycles. For a cycle $(a_1, a_2,\ldots, a_m)$ with m entries in it, we will call m the *length* of the cycle. We note that

8

(8.6) If the length of a cycle $(a_1, a_2,\ldots, a_m)$ is even then the cycle is an odd permutation and if its length is odd, then the cycle is an even permutation.

This claim follows from the formula

$$(a_1, a_2,\ldots, a_m) = (a_1, a_2)\,(a_1, a_3)\,(a_1, a_4) \cdots (a_1, a_m),$$

which we introduced in (7.2). The transpositions on the right side all share the entry a_1, and we can use the subscript on the other entry to count them out. The counter runs from 2 through m inclusively, which is a total of $m - 1$ numbers. Thus, there are $m - 1$ transpositions here. If m is even, then $m - 1$ is odd and vice versa. This is why the parity of an m-cycle is opposite from the parity of its length m.

With experience we will become increasingly aware of the difference between even and odd permutations. In recognition of their importance as two different classes, let us give them names. We will denote the set of all even permutations in S_n by A_n and the set of all odd permutations by O_n. Since every permutation in S_n is either even or odd and no permutation is both even and odd, it follows that $A_n \cup O_n = S_n$ and $A_n \cap O_n = \varnothing$.

Why do we use the letter "A" for the even permutations, you might ask, rather than "E" which would suggest "even"? The reason is that the set of even permutations is called the *alternating group (on n symbols)*, reflecting some properties that were important in the first context in which they were studied. We will not go into such matters, but we will adopt the conventional notation.

One difference between these two sets is that

(8.7) The set A_n of even permutations is closed under composition and the taking of inverses. The set O_n of odd permutations is closed under the taking of inverses but distinctly not closed under composition. In fact, the composition of two permutations in O_n is always in A_n and thus never in O_n.

When we say that A_n (or any subset X of S_n) is *closed under composition*, we mean that whenever one composes two permutations in A_n, (or in X) the result is again in A_n (in X). We say that a subset X of S_n is *closed under the taking of inverses* if, whenever a permutation α is in X, then α^{-1} is also in X.

The claims regarding composition in (8.7) follow from the fact that if we compose a permutation α that can be expressed with m_1 transpositions with a permutation β that can be expressed with m_2 transpositions, the composition $\alpha \circ \beta$ can be expressed with transpositions by simply stringing together the separate expressions we had for α and β. This gives us an expression using $m_1 + m_2$ transpositions. If m_1 and m_2 are both even or both odd, then their sum is even. Thus, if α and β are both in A_n, or are both in O_n, then $\alpha \circ \beta$ is in A_n.

To see that the claims regarding the taking of inverses are true, recall the observation we made earlier about the inverse of a sequence of transpositions, specifically that $(\tau_1\tau_2 \cdots \tau_m)^{-1} = \tau_m\tau_{m-1} \cdots \tau_1$ if all of the τ_i's are transpositions. This shows that if a permutation can be expressed with a given number of transpositions, then its inverse can be expressed with the same number of transpositions. It follows from this that any permutation and its inverse have the same parity.

As long as we are comparing and contrasting even and odd permutations, we might as well add that there are equal numbers of each type. Since it is easy to see (Exercise 8.5) that the total number of permutations in S_n is $n!$ equal number of even and odd ones means that

(8.8) For $n \geq 2$, there are $n!/2$ even permutations and $n!/2$ odd permutations in S_n.

To see this, pair up with each even permutation α the permutation $\alpha \circ (1, 2)$. By extending an even permutation by one transposition, we get an odd permutation, so we have paired together one of each type. One can see that different even permutations will always pair up with different odd ones in this way and that every odd permutation gets paired with some even one. (See Exercise 8.6 for the details.)

This pairing of even with odd permutations means that there are the same numbers of each. Since together they yield all $n!$ permutations in S_n, there must be $n!/2$ of each type.

8.3 Expressing Even Permutations with 3-Cycles

There is one additional fact about the even permutations that will be important to us for solving puzzles. It has to do with how we can build all even permutations from the *simplest possible* even components.

What is the simplest possible even permutation? The identity permutation is even because it can be expressed using two transpositions, $\varepsilon = (1, 2)(1, 2)$, so clearly it is the simplest. But beyond this, what is the simplest? A permutation that is not the identity cannot move just one number, so it has to move at least two. A permutation that moves exactly two numbers has to be a transposition, which clearly is odd since it is expressible with an odd number of transpositions, namely one transposition. Thus, an even permutation not the identity is going to have to move at least three numbers.

The simplest thing it could do is to move exactly three numbers, say represented by a, b, and c. The only way to map these three and only these three to something other than themselves is to have them form a 3-cycle, either

8

(a, b, c) or (a, c, b). A 3-cycle is an even permutation by (8.6), and is thus the simplest possible nontrivial (the identity is trivial) even permutation.

Note that in order to be talking about 3-cycles at all, we must be thinking in terms of being in S_n where $n \geq 3$. An important fact is that:

(8.9) If $n \geq 3$, then every even permutation in S_n can be expressed using 3-cycles.

To see why this is true, let us take any even permutation α. Since α is even, we can express it in the form $\alpha = \tau_1\tau_2 \cdots \tau_m$, where the τ_i's are transpositions and m is an even number. Since there is an even number of transpositions, we can group them into adjacent pairs without any being left over. We will group them like this: $\alpha = (\tau_1\tau_2)(\tau_3\tau_4) \cdots (\tau_{m-1}\tau_m)$. Each of the pairs of transpositions can either be dropped from the sequence or expressed using one or two 3-cycles without changing the value of the expression.

To see this, let us consider one of the pairs $(\tau_{k-1}\tau_k)$. We will consider three cases based upon how many numbers the two transpositions move in common. Since transpositions move two numbers, the possibilities are that they both move the same two numbers, that they move one number in common, or that they move no numbers in common, i.e., they are disjoint.

If the transpositions τ_{k-1} and τ_k move the same two numbers, then in fact they are the same transposition. Since composing a transposition with itself yields the identity, we have in this case that $\tau_{k-1}\tau_k = \varepsilon$. We can drop such a pair from the sequence without changing its value, while expressing mild pique that whoever gave us the sequence of transpositions had not taken advantage of this reduction option before giving it to us.

If the transpositions move one thing in common, then they are of the form $\tau_{k-1}\tau_k = (a, b)(a, c)$, where $b \neq c$. Since $(a, b)(a, c) = (a, b, c)$, we can replace this pair of transpositions with a single 3-cycle.

If the transpositions move nothing in common, then we have a pair that can be expressed in the form $\tau_{k-1}\tau_k = (a, b)(c, d)$, where a, b, c, and d represent four distinct numbers. This pair of disjoint transpositions can be expressed using two 3-cycles. There are several ways to do this, one of which is this: $(a, b)(c, d) = (a, b, c)(a, d, c)$.

This last expression might seem more mysterious than it really is if you only deal with it in terms of notation. However, if you think about representing the job by moving labeled disks between labeled boxes as shown in

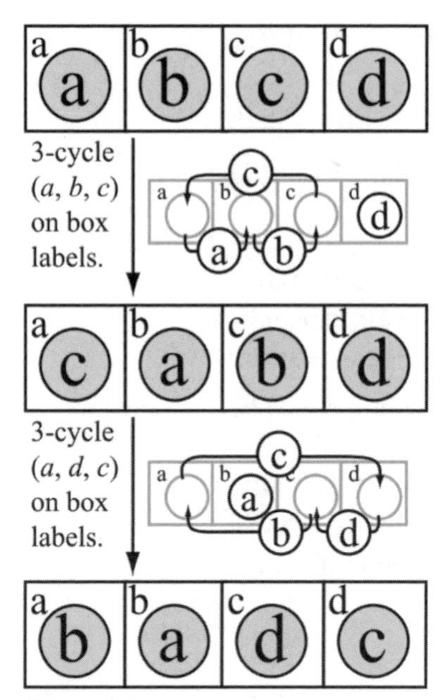

Figure 8.3. Performing a pair of disjoint transpositions with two 3-cycles.

Figure 8.3, it is not that big a deal. Remember you want to perform the permutation $(a, b)(c, d)$ by doing 3-cycles. Just start with any 3-cycle that gets one disk where you want it. Let's move disk a to box b, disk b to box c, and disk c to box a, producing the outcome shown in the second set of boxes in Figure 8.3. This gets disk a into box b where we want it so we'll leave it there. We still have to move disk b, disk c, and disk d to get them where we want them. A piece of cake! Just move the disk from box a (disk c) to box d, the one in box d (disk d) to box c, and the one in box c (disk b) to box a. We've got it. Notationally this comes out as we show it in the last paragraph.

We have seen that under all circumstances, zero, one, or two 3-cycles can replace each of the pairs of transpositions. As a result, we have shown that any even permutation can be expressed using the simplest even permutations, namely 3-cycles.

At the end of Chapter 7 we noted that being able to express all permutations using transpositions is going to be very helpful for solving end game problems with the puzzles. This result on expressing even permutations with 3-cycles will also be of importance for the same reasons.

There are a few other properties of permutations that will be extremely useful to us, but we will postpone these until later chapters when we will have developed additional "need to know."

Exercises

8.1 The identity really is even. The Transpose Puzzle allows you to generate abundant evidence that the identity permutation is even. Start the Transpose Puzzle and initialize it so that it is displaying the identity permutation. Now perform transpositions for a while. Then return every disk to its home box. You will note that when all are returned home, the transposition counter will be even. Repeat this process at least ten times. It works out to be even every time.

8.2 The parity of several end game scrambles. Look again at the end game scrambles for Oval Track shown in Figure 7.3. Determine the parities of each of these.

8.3 Inverses of sequences of transpositions. To see why display (8.2) is true, we needed to show that the inverse of a sequence of transpositions is expressed by taking the same transpositions in the reverse order. This exercise guides you through an exploration of the practical evidence for this that the Transpose Puzzle offers. (a) First, do any transposition on Transpose, and then do it again. Did you return to the state of the puzzle to what it was before the first transposition? Of course. This is the practical interpretation of a transposition being its own inverse. (b) Now initialize Transpose, and enter a sequence of several transpositions, say at least five of them. Make note of which ones you did. Now run the same transpositions in reverse order. You have restored the puzzle to the identity state. Do this several times to make it absolutely obvious.

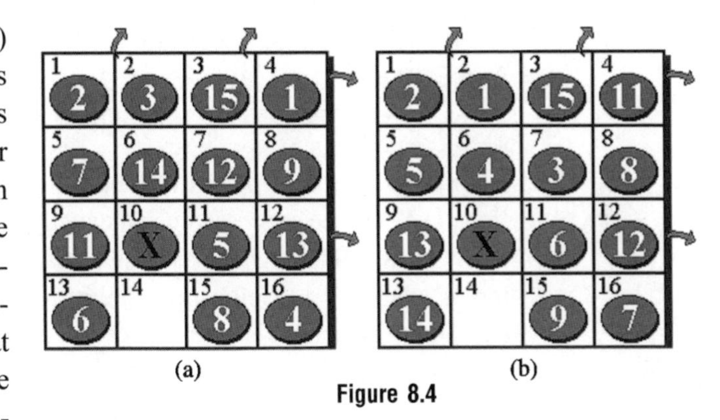

Figure 8.4

8.4 Some parity checks in the Slide Puzzle. Figure 8.4 shows two scramblings on a configuration of the Slide Puzzle. Determine the parities for these two.

8.5 The total number of permutations in S_n. Show that there are $n!$ permutations in S_n. [Hint: When defining a permutation α, think about how many ways there are to define, in order, the image of the number 1, then 2, then 3, etc., all the way up to n.]

8.6 The total number of even and odd permutations in S_n. Fill in the details of the proof of (8.8), namely that if $n \geq 2$, then half of the $n!$ permutations in S_n are even and half are odd. [Hint: Show that the mapping which sends an even permutation α to the permutation $\alpha(1, 2)$ pairs up the even and the odd permutations, thereby showing that the set of even permutations and the set of odd permutations are the same size.]

8.7 Expressing even permutations in terms of 3-cycles. Demonstrate the truth of statement (8.9) by expressing these even permutations with 3-cycles. [Read the discussion following (8.9) for a description of how to do this.]

(a) $\alpha = (1, 2, 3, 4, 6, 5)\,(3, 4, 5)\,(2, 5)\,(1, 4)\,(5, 2)$
(b) $\beta = (1, 2)\,(2, 3)\,(4, 5)\,(1, 3)\,(6, 7)\,(6, 8)\,(9, 10)\,(11, 12)$
(c) $\gamma = (1, 2, 3, 4)\,(2, 3, 4, 5)\,(4, 5, 6, 7)\,(8, 9)$

8.8 Expressing odd permutations in terms of 3-cycles and one transposition. (a) Show that all odd permutations in S_n can be expressed using exactly one transposition together with zero or more 3-cycles. [Hint: Modify the proof of (8.9).] (We will make use of this property later to support claims about the level of solvability of some of the puzzles.) (b) Demonstrate the truth of this claim by expressing these odd permutations with a single transposition and 3-cycles.

(a) $\alpha = (1, 2, 3, 4, 5, 6)$
(b) $\beta = (1, 2, 3, 4)\,(5, 6, 7)\,(8, 9, 10)$
(c) $\gamma = (2, 5, 3, 7, 6)\,(3, 5, 8, 4)\,(6, 8, 2, 1, 9)$

8.9 Practice with the reduction algorithm of claim (8.3). The claim in display (8.3) underlies our proof of the Parity Theorem (8.1), and a reduction algorithm underlies (8.3). Follow these steps to explore the algorithm. (a) Run the Parity Demo program that came on your CD ROM to practice using the algorithm. Run it several times in the "Human First" mode with you producing a sequence of transpositions for the identity permutation while the computer applies the algorithm to reduce your expression by two transpositions. (b) Then switch to the "Computer First" mode and reduce the sequence of transpositions that the computer presents to you. Run the demonstration in this mode several times until you are sure you understand the algorithm.

8.10 Notational practice with the reduction algorithm. Demonstrate your understanding of the reduction algorithm behind (8.3) in a notational way by following the algorithm to reduce these expressions by two transpositions.

(a) $\varepsilon = (6,12)\,(11,7)\,(12,10)\,(9,6)\,(5,6)\,(7,6)\,(7,5)\,(10,6)\,(9,12)\,(10,11)\,(10,7)\,(10,9)$
(b) $\varepsilon = (4,5)\,(5,6)\,(10,6)\,(8,6)\,(11,6)\,(8,4)\,(5,6)\,(8,5)\,(11,8)\,(10,4)$

8.11 What do you get with 4-cycles? With 5-cycles? We saw in (7.3) that all permutations in S_n are expressible using transpositions and in (8.9) that all even permutations in S_n are expressible using 3-cycles provided that $n \geq 3$. Stated another way, this says that you get all permutations by taking all possible products of sequences of transpositions, and similarly, you get all of the even permutations by taking all possible products of sequences of 3-cycles. What do you get when you combine all possible sequences of 4-cycles? Or 5-cycles? Or k-cycles? Explore these questions and see what you can discover. Note of course that we must assume that $n \geq k$ before we can talk about k-cycles within S_n.

8.12 Splicing and dicing cycles. What happens to the cycle structure of a permutation α when you follow α by a transposition? The answer is that you either splice two of the cycles of α into one bigger cycle, you cut one of the cycles of α into two smaller cycles, you extend one cycle by one element, or you add a new transposition to the cycle structure. Verify the special cases of this statement that are set forth below, and then make an argument that the claim follows in general from these special cases.

(a) If $\alpha = (a_1, a_2, \ldots, a_r)(b_1, b_2, \ldots, b_s)$ where these two cycles are disjoint, then
$\alpha(a_1, b_1) = (a_1, \ldots, a_r, b_1, \ldots, b_s)$

(b) If $\beta = (a_1, a_2, \ldots, a_r)$ and $1 \leq i < j \leq r$, then
$\beta(a_i, a_j) = (a_1, \ldots, a_{i-1}, a_j, a_{j+1}, \ldots, a_r)(a_i, a_{i+1}, \ldots, a_{j-1})$

(c) If $\gamma = (a_1, a_2, \ldots, a_r)$ and $b \neq a_i$ for all i, then
$\gamma(a_1, b) = (a_1, a_2, \ldots, a_r, b)$

(d) If $\delta = (a_1, a_2, \ldots, a_r)$ and if (b_1, b_2) is disjoint from δ, then
$\delta(b_1, b_2) = (a_1, a_2, \ldots, a_r)(b_1, b_2)$

8.13 The least number of transpositions to express a permutation. Suppose α is a permutation in S_n, that has cycle structure $(k_1, k_2, \ldots, k_r)$ with $k_1 \leq k_2 \leq \cdots \leq k_r$. If one expresses α in terms of disjoint cycles, and then uses

(7.2) to express each of the cycles in terms of transpositions, then we get an expression for α that uses $k - r$ transpositions, where $k = \sum_{i=1}^{r} k_i$. Show that this is the fewest transpositions that there can be in any expression for α in terms of transpositions. (See [3] in the list of journal articles in Appendix C for one possible proof of this fact.)

8.14 An obvious algorithm for using a minimal number of transpositions. (a) Expressing a permutation with a minimal number of transpositions is equivalent to solving a scramble on the Transpose Puzzle with a minimal number of transpositions. Show why. (b) Here is a rather obvious algorithm for solving scrambles on the Transpose Puzzle: (i) Find the lowest numbered box that does not contain the right disk. Call it box i. Keep dragging disks from box i to their home boxes until disk i gets returned to box i. (ii) Repeat step (i) until there are no more disks out of place. If the original scramble is denoted by α, show why this algorithm produces the sequence of transpositions for the essentially unique representation for α^{-1} that is described in Exercise 7.8.

8.15 A surprising algorithm for using a minimal number of transpositions. Here is another, far more general algorithm for solving any scramble α on the Transpose Puzzle. (i) Go to any box, say box i, which does not contain its home disk. Note the cycle of the scramble that involves i. Swap the disk in box i with any other box on this cycle. (ii) Repeat (i) until you can find no more disks out of place. Prove that this algorithm also leads to a minimal number of transpositions for solving the puzzle. [Hint: Each application of step (i) leads to an application of part (b) of Exercise 8.12. Follow up on the consequences of this.]

8.16 An alternative approach to the Parity Theorem. There are many ways to prove the Parity Theorem. (See [4] in the list of journal articles in Appendix C for a bibliography and a review of the types of proof that have been used in textbooks over the past forty years.) Following this outline, work out an alternative proof, based upon one by Edwin L. Gray that appeared in the *American Mathematical Monthly* in 1963. (i) Verify the result for S_2, using the fact that $(1, 2)$ is the only transposition. (ii) Suppose that the Parity Theorem is already verified for S_{n-1}, where $n \geq 3$. Let α, be any permutation in S_n, and suppose that $\alpha = \tau_1\tau_2 \cdots \tau_r$ and $\alpha = \sigma_1\sigma_2 \cdots \sigma_s$ are two expressions for α in terms of transpositions. If none of the transpositions from either expression moves n, then α is in S_{n-1} and the result follows by the induction hypothesis. Thus, assume that at least one of the transpositions, say one of the τ_i's,

8

moves n. (iii) Starting with the first occurrence of n in one of the τ_i's, repeatedly use the identities $(n, k)(r, s) = (r, s)(n, k)$ (if $\{n, k\} \cap \{r, s\} = \varnothing$), $(n, k)(n, j) = (j, k)(n, k)$, $(n, k)(j, k) = (j, k)(n, j)$, and $(n, k)(n, k) = \varepsilon$ to rewrite the sequence of transpositions $\tau_1\tau_2 \cdots \tau_r$ as $\tau_1^* \tau_2^* \cdots \tau_r^*$ so that either n appears only in the final transposition τ_r^* or not at all. Do the same for the sequence $\sigma_1 \sigma_2 \cdots \sigma_s$ yielding a sequence $\sigma_1^* \sigma_2^* \cdots \sigma_s^*$. (iv) Show that if n drops out of one of the sequences, then it must drop out of the other also, and that, because of the induction hypothesis, $r - 2i$ and $s - 2j$ share the same parity for some integers i and j. (v) Show that if n does not drop out of either expression, then $\tau_r^* = \sigma_s^* = (n, k)$, where $k = n\alpha$. From this, $\tau_1^* \tau_2^* \cdots \tau_{r-1}^* = \sigma_1^*\sigma_2^* \cdots \sigma_{s-1}^*$, and the induction hypothesis tells us that $r - 1$ and $s - 1$ share the same parity, and thus, so do r and s.

The Role of Conjugates

A relationship between permutations called conjugacy will be of great importance for solving puzzles. For two permutations β and γ in the group S_n, γ is a *conjugate* of β if there is some permutation α such that $\gamma = \alpha\beta\alpha^{-1}$. When this happens, γ is said to be a conjugate of β by α. Notice that if α and β are elements that commute with each other, then the expression $\alpha\beta\alpha^{-1}$ will reduce simply to β. Thus, the only element that is conjugate to β by an element α that commutes with β is β itself.

However if two elements α and β do not commute, then the conjugate $\alpha\beta\alpha^{-1}$ of β will not equal β. It is an easy exercise to show that $\alpha\beta\alpha^{-1} = \beta$ if and only if α and β commute with each other. For a given element β, the set of *all* elements, which can be expressed in the form $\alpha\beta\alpha^{-1}$ for some element α of S_n is called the *conjugacy class* of β.

This idea of conjugacy applies to any group, not just to the permutation group S_n that is of interest to us here. A careful study of the conjugacy classes of all elements in a finite group leads to some of the interesting theorems about finite groups, such as the Sylow theorems. These can be found in most introductory textbooks.

If a conjugate $\alpha\beta\alpha^{-1}$ of a permutation β is not equal to β; ought it not be like β in some ways? Yes. The conjugates of β share some important characteristics with β. In particular, it is the case that

(9.1) For any positive integer n and for any two permutations α and β in S_n, β and $\alpha\beta\alpha^{-1}$, its conjugate by α, have the same cycle structure.

To see that this is true, let us suppose that we have expressed β as $\sigma_1 \circ \sigma_2 \circ \cdots \circ \sigma_k$ where the σ_i's denote cycles that are pairwise disjoint. That is, this expression for β displays its cycle structure. It follows that

$$\alpha\beta\alpha^{-1} = \alpha(\sigma_1 \circ \sigma_2 \circ \cdots \circ \sigma_k)\alpha^{-1} = (\alpha\, \sigma_1\alpha^{-1}) \circ (\alpha\sigma_2\alpha^{-1}) \circ \cdots \circ (\alpha\sigma_k\alpha^{-1}).$$

[Note that we have mixed use of the composition symbol "∘" and of juxtaposition to denote composition in this expression. We will completely drop use of the symbol "∘" shortly.]

It will suffice to show that for each of the cycles σ_j, the permutation, $\alpha\sigma_j\alpha^{-1}$, its conjugate by α, is a cycle of the same length as σ_j. We observe that if σ_j is the m-cycle $(c_1, c_2,\dots, c_m)$, and if $d_i = c_i\alpha^{-1}$ for each i with $1 \le i \le m$, then

$$\alpha\sigma_j\alpha^{-1} = \alpha(c_1, c_2,\dots, c_m)\alpha^{-1} = (d_1, d_2,\dots, d_m).$$

We see that this is true by observing that for any i,

$$d_i(\alpha\sigma_j\alpha^{-1}) = (c_i\alpha^{-1})(\alpha\sigma_j\alpha^{-1}) = (c_i\alpha^{-1}\alpha)(\sigma_j\alpha^{-1}) = c_i(\sigma_j\alpha^{-1}) = (c_i\sigma_j)\alpha^{-1} = c_{i+1}\alpha^{-1} = d_{i+1}.$$

The understanding when we write $c_i\sigma_j = c_{i+1}$ in the second line of this last display is that this is "addition modulo m," so that $m+1 = 1$. Using addition modulo m produces the wrap-around that we need at the end of the cycle notation. The display shows that $\alpha\sigma_j\alpha^{-1}$ maps d_i to d_{i+1} (where $m+1 = 1$) for each i with $1 \le i \le m$. In addition, if x is an element that is moved by $\alpha\sigma_j\alpha^{-1}$, that is, $x \ne x\alpha\sigma_j\alpha^{-1}$, then we have $x\alpha \ne (x\alpha)\sigma_j$, which means that σ_j moves $x\alpha$. Since the only elements that σ_j moves are the c_i's, this means that $x\alpha = c_i$ for some i, which means that $x = c_i\alpha^{-1} = d_i$ for this i. That is, the only elements that $\alpha\sigma_j\alpha^{-1}$ moves are the d_i's. With the earlier observation, this shows that $\alpha\sigma_j\alpha^{-1}$ is the m-cycle $(d_1, d_2,\dots, d_m)$ as claimed.

Claim (9.1) tells us in particular that if β is a cycle, say a 3-cycle, then no matter what permutation you take for α, the conjugate $\alpha\beta\alpha^{-1}$ will also be a 3-cycle. If β is a permutation that moves only three items, then $\alpha\beta\alpha^{-1}$ will also be a permutation that moves only three items.

This insight is a very important one when it comes to solving permutation puzzles, since in the end, the difficult thing is to get the last few pieces into their right places. This theorem tells us that if we can find a way to perform one small change, such as a 3-cycle, we have the potential of using it to get other small changes, such as other 3-cycles.

To illustrate the principle, let us use it to prove a broad general statement about the Slide puzzle.

9

(9.2) On the Slide puzzle with 16 active squares and one blank, every 3-cycle that places the blank in its initial spot (spot 16) can be achieved.

Let's begin by showing that a particular 3-cycle can be achieved. We will start the puzzle with every piece in its initial place as in Figure 9.1. If we now click in spots 12, 11, 15, and 16, we perform the sequence of transpositions (16, 12) (12, 11) (11, 15) (15, 16). [Remember that, contrary to the more common practice, we are performing the transpositions in the order one reads them, that is, from left to right. Remember also that the blank spot is denoted in this representation by the number of its initial location, namely 16.] The result of these four clicks is the arrangement of the pieces shown in Figure 9.2. We see that the permutation we have performed is the 3-cycle (11, 12, 15). That is, the disk that was initially in spot 11 is now in spot 12, the one from spot 12 is now in spot 15, and the one from spot 15 is now in spot 11. Every other disk, including the blank, is in its original spot. We have demonstrated that the one particular 3-cycle (11, 12, 15) is achievable. This will be our β.

Suppose we want to perform some arbitrarily chosen 3-cycle (a, b, c). Starting from the identity state, we will first perform a permutation, which we will denote by α, which puts disk number a into spot 11, puts disk number b into spot 12, and puts disk number c into spot 15. It does not matter at all how much additional scrambling of pieces that α causes; this will all get put right later.

Having done this rearrangement α, let us follow this by performing the four-step 3-cycle β = (11, 12, 15). The effect of doing β is to change the places of disks a, b, and c, which were, respectively, in spots 11, 12, and 15. Doing β causes no other changes.

Now follow β with α^{-1}, the inverse of α. What will be the effect of this? Well, since α^{-1} undoes

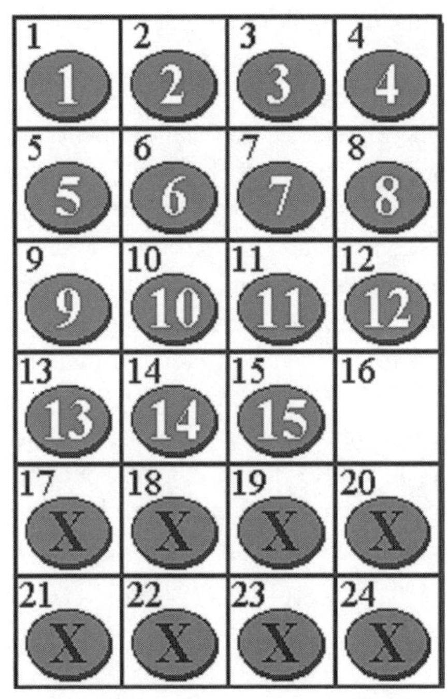

Figure 9.1. The identity permutation.

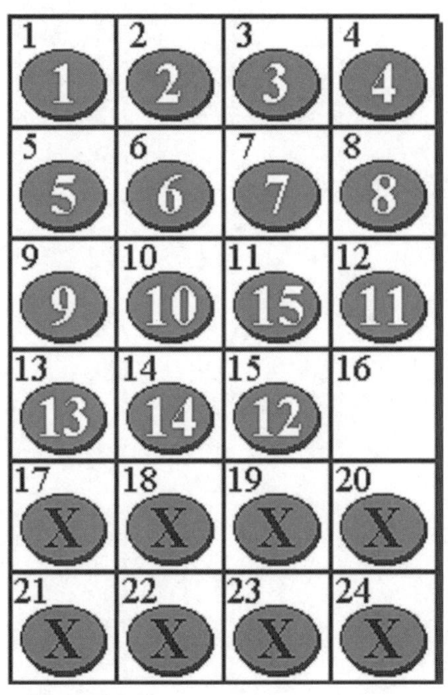

Figure 9.2. The 3-cycle β = (11, 12, 15).

everything that α has done, it will return things to where they came from. The exception to this will be that disks a, b, and c will not return to their original spots because of the fact that between having moved them elsewhere with α and having returned them using α^{-1}, we also did some switching of them in the middle with β. But of course that is what we wanted to do. To summarize, the composition of permutations $\alpha\beta\alpha^{-1}$ has produced the 3-cycle (a, b, c).

Let's illustrate this with the puzzle. We will pick three numbers at random, say 7, 2, and 12, and decide that we would like to produce the 3-cycle (7, 2, 12) on the Slide puzzle. We first perform a permutation α which moves the disks numbered 7, 2, and 12 into spots 11, 12, and 15, respectively, and which also returns the blank to its original spot 16. As we mentioned earlier, what happens to all the other disks during this process is not of any concern as all of them will move back to where they came from later.

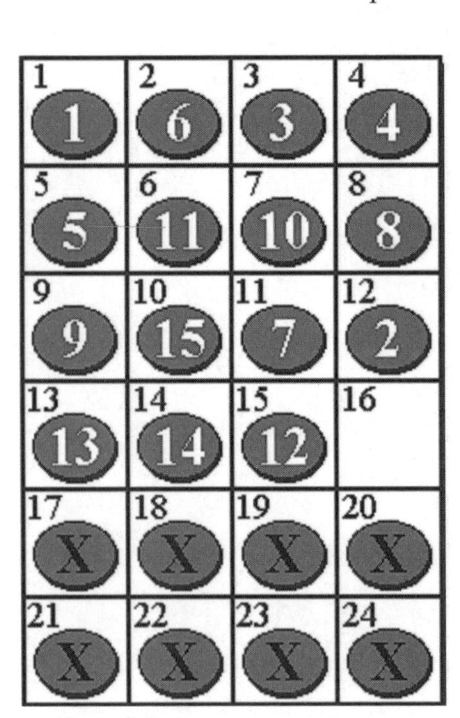

Figure 9.3. A setup permutation α …

There is one more detail to mention. Since we will have to perform the inverse of α in a moment, and since this will involve too many steps for most mere mortals to remember, let us have the computer do the remembering for us. Before you begin to perform the permutation α, start the "Record a Macro" process to record the series of steps that produce the permutation α into one of the buttons, say button A. When you have gotten the three disks of interest where you want them, conclude the recording process.

There are many different permutations that could do what we want α to do by getting the disks 7, 2, and 12 into spots 11, 12, and 15, while returning the blank to spot 16. In fact, since the remaining twelve disks can go into any of the remaining twelve spots, there are 12! possibilities for α. A goal would be to pick an α that uses the fewest possible transpositions. One of the possible α's is shown in Figure 9.3. This permutation requires 32 transpositions, and is given by

(9.3) $\alpha = (16, 12)(12, 8)(8, 4)(4, 3)(3, 2)(2, 6)(6, 10)(10, 11)$
$(11, 7)(7, 3)(3, 4)(4, 8)(8, 12)(12, 16)(16, 15)(15, 11)$
$(11, 7)(7, 6)(6, 10)(10, 14)(14, 15)(15, 16)(16, 12)(12, 11)$
$(11, 10)(10, 14)(14, 15)(15, 11)(11, 10)(10, 14)(14, 15)(15, 16).$

It is, of course, not important to produce the desired result in this same way, but the interested reader can reproduce the exact result shown in Figure 9.3 by starting with the Slide puzzle in identity mode and performing the transpositions shown above in the indicated order. When you get to each pair, the blank will be in the box having the first number. Click on the disk in the box with the second number to make the switch. Exercise 9.2 deals with this example, and gives the details about loading α into one of your macro buttons from Puzzle Resources.

Now let us apply the 3-cycle permutation β, which acts on the disks in spots 11, 12, and 15. The result is as shown in Figure 9.4. Disk 7 has moved to where disk 2 was before, disk 2 has moved to where disk 12 was, and finally, disk 12 has moved to where disk 7 was. Finally, we perform α^{-1}. This means performing the 32 transpositions that produced α in exactly the reverse order. By reading the listing above in reverse, we could do this manually, but since we have recorded α into button A, we have its inverse available to us at the push of a single button, namely the button A^{-1}. When we push this button, the computer performs the 32 steps that produce α in the reverse order, producing Figure 9.5. We examine that figure and see that the only changes from the identity arrangement are that disk 7 has gone to spot 2, disk 2 has gone to spot 12, and disk 12 has gone to spot 7. This gives us the 3-cycle (7, 2, 12) that we were after.

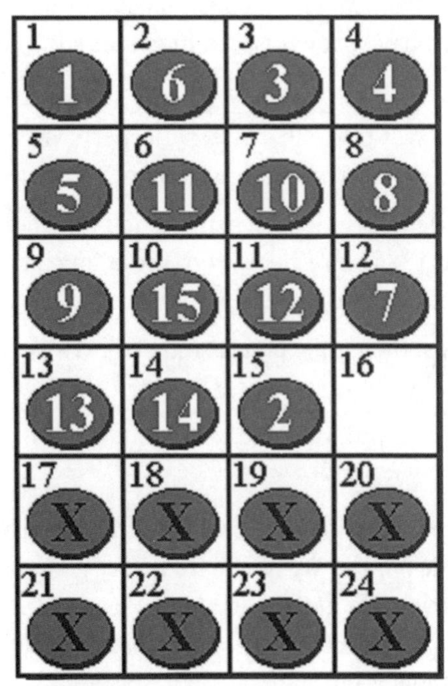

Figure 9.4. ... followed by a 3-cycle β ...

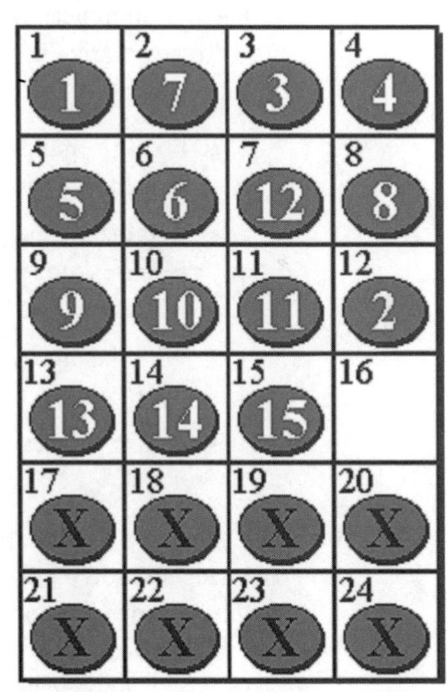

Figure 9.5. ... followed by using α^{-1} to undo the setup

We make the claim that the method we used here to produce the 3-cycle (7, 2, 12) could be used to produce *any* 3-cycle whatever. But it takes more than one illustration to have a fully rigorous proof of (9.2). Before sketching out a general proof, try several more examples modeled after this one, to become convinced that this claim is true, and then to think of how you might prove this.

Basically, you must be able to show that this puzzle gives us enough freedom to bring any three disks together with each other and the blank in a two-by-two subsquare of the grid. A little experience playing with the puzzle convinces you that this is the case, and suggests how to determine an algorithm for doing this. Verifying that the algorithm performs as claimed would constitute as formal a proof as you might expect to see here.

To summarize, what we have shown is that to get an arbitrary 3-cycle (a, b, c) on the Slide Puzzle, you do the following:

- Fix on some particular 3-cycle that is easy to perform; call it $\beta = (x, y, z)$. (To make it easy to repeat this sequence, you might want to program it into one of the macro buttons, say B.)
- Reinitialize the puzzle to the identity permutation and then perform any permutation α which moves disks a, b, and c into spots x, y, and z respectively. (Program α into a macro button, say button A, in order to be able to reproduce it and its inverse later.)
- With the puzzle still showing the permutation α, perform the permutation β. (If you programmed it into button B, do this by clicking on button B.)
- Follow β by α^{-1}. (If you programmed α into button A, then you can perform α^{-1} by a single click on button A^{-1}.)

Not only have we proven (9.2), albeit somewhat informally, but we have also developed a powerful tool for analyzing and solving the Slide puzzle. We now have a very general method for making small changes, which is something we need to know how to do in the end game phase of the solution process.

The principles we used to justify (9.2) for the Slide puzzle can be applied to the other puzzles as well. This method is quite general. In later chapters, we will develop more insight into the other puzzles using conjugates.

Exercises

9.1 Evidence for (9.1) developed notationally. For each of the pairs of permutations, α and β, in S_{12} listed below, calculate the conjugate $\alpha\beta\alpha^{-1}$. Note that it has the same cycle structure as β has.

(a) $\alpha = (1, 2, 3, 4, 5, 6, 7)\,(8, 9, 10)$ $\beta = (1, 5, 8)\,(2, 6)\,(3, 7, 4)$

(b) $\alpha = (1, 5, 8)\,(2, 6)\,(3, 7, 4)$ $\beta = (1, 2, 3, 4, 5, 6, 7)\,(8, 9, 10)$

(c) $\alpha = (5, 7, 3, 6, 12, 10, 11, 8)$ $\beta = (1, 2)\,(5, 11, 7, 9, 12, 4, 10)$

9.2 The 3-cycle (2, 12, 7) in Slide. Carry out for yourself the steps described in this chapter for getting the 3-cycle (2, 12, 7) on the Slide Puzzle in the traditional "15" configuration. To do so, follow these steps. (i) Load the 32-step procedure in (9.3) into button A of your puzzle. (You can either carry out the steps yourself to program the button, or you can load the procedure from Puzzle Resources, where you will find it as the first entry in the Locally Resident Macros collection.) (ii) Program macro button B with the 4-step 3-cycle (11, 12, 15) = (16,12)(12,11)(11,15)(15,16). (iii) Reinitialize your puzzle to the identity arrangement of the disks. (iv) Perform the conjugate permutation $A\,B\,A^{-1}$ by pressing the buttons, A, B, and A^{-1}.

9.3 A few more 3-cycles on Slide. To strengthen your conviction that it is possible to do any 3-cycle on the Slide Puzzle in its basic "15" configuration, get these 3-cycles on Slide: (a) (5, 11, 14); (b) (6, 1, 12); (c) (3, 9, 13) Rather than following the strategy used in Exercise 9.2 of bringing the three disks together in the lower right corner, bring them together with the blank in any two-by-two subsquare that is convenient. [Note: You will find solutions for these three 3-cycles as the second through the fourth entries in the Locally Resident Macros collection for the Slide Puzzle if you want to see them demonstrated.]

9.4 All 3-cycles are possible on the "15" puzzle. Building on the experience of the last two exercises, develop an argument for why the claim in (9.2) is true. That is, prove that every 3-cycle on the "15" puzzle which returns the blank to its initial spot can be achieved.

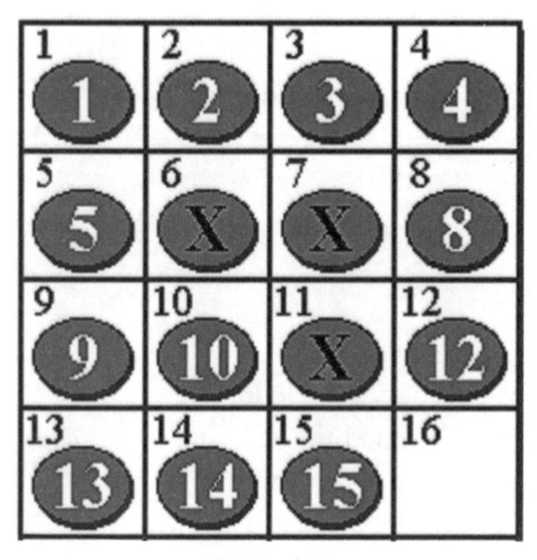

Figure 9.6

9.5 An "L" obstacle does not block 3-cycles. Reconfigure your Slide Puzzle with 16 active spots, one blank, the boxes 6, 7, and 11 out of action, and no wrapping. The screen will look like Figure 9.6. Show that all 3-cycles which return the blank to spot 16 are achievable on this puzzle. That is, the L-shaped obstacle shown still allows considerable freedom of movement. [Hint: Note that there is just one two-by-two square of active boxes, namely in spots 9, 10, 13, and 14. Show that you can move any three disks, along with the blank, into this square. Then draw the relevant conclusions from this. A demonstration of performing the 3-cycle (2, 10, 12) is included as the fifth entry in the Slide Puzzle Resources Locally Resident Macros.]

9.6 Conjugates of "Do Arrows" on Oval Track 9. The Oval Track Puzzle version 9 has the 3-cycle (4, 2, 1) as its "Do Arrows" permutation. Produce five other 3-cycles on this puzzle, say (2, 8, 14), (9, 4, 15), (11, 3, 7), (6, 1, 15), and (7, 3, 9). Based upon this experience, give an argument for the claim that every 3-cycle can be obtained on this puzzle assuming that the number of active disks is at least 5.

9.7 All 3-cycles need not be conjugates of a "Do Arrows" 3-cycle. Consider Oval Track Puzzle 7, with (5, 3, 1) as its "Do Arrows." (a) Show that on this puzzle, if the number of active disks is even, then any permutation that is possible on the puzzle has the disks alternating in parity. (b) Show that on this puzzle, if the number of active disks is even, any 3-cycle involving both odd and even disks is not achievable. Explore which 3-cycles involving disks of only one parity can be achieved. (c) Show that on this puzzle, if the number of active disks is odd and at least 7, then every 3-cycle is achievable as a conjugate of the "Do Arrows" 3-cycle.

The Role of Commutators

In the last chapter we looked at certain permutations, called conjugates, and set forth some facts about them that will be of use to us in solving puzzles. In this chapter, we will look at a class of closely related permutations called commutators and see what they can do for us.

As you read about commutators, you might find it useful to have your puzzles up and running so that you can try things out for yourself.

If α and β are any two permutations in S_n, the *commutator* of this pair is the permutation $\alpha\beta\alpha^{-1}\beta^{-1}$. We will denote this commutator by the shorthand symbol $[\alpha, \beta]$.

A first observation about the commutator $\alpha\beta\alpha^{-1}\beta^{-1}$ is that if α and β commute with each other, then

$$\alpha\beta\alpha^{-1}\beta^{-1} = (\alpha\beta)\,(\alpha^{-1}\beta^{-1}) = (\beta\alpha)(\alpha^{-1}\beta^{-1}) = \beta(\alpha\alpha^{-1})\beta^{-1} = \beta\varepsilon\beta^{-1} = \beta\beta^{-1} = \varepsilon.$$

That is, the commutator of a pair of elements that commute with each other is simply the identity permutation, and therefore not of much interest. If you look at Exercise 10.6, you will see that the converse of this observation is also true; if $[\alpha, \beta] = \varepsilon$, then α and β commute with each other.

When two permutations α and β fail to commute with each other "just a little bit," $[\alpha, \beta]$ will not be the identity, but will be close to it in the sense that it moves just a few things. This is why commutators will be of interest for solving puzzles.

To start exploring commutators, consider a given permutation α in S_n. Recall that in Chapter 6 we defined M_α to be the subset of $\{1, 2,\ldots, n\}$ consisting of those numbers m such that $m\alpha \neq m$. That is, M_α is the set of those numbers moved by α. When we think about α in terms of moving numbered disks between numbered locations, we blur the distinction between the numbers in M_α and the locations identified with these numbers. Thinking in this way, we will talk in terms of a location being in M_α, meaning that the permutation α moves a disk in this location to somewhere else as opposed to leaving it there.

Here are two consequences of a number being moved by a commutator $[\alpha, \beta]$ that will be of use to us.

(10.1) (a) If a number k is in $M_{[\alpha,\beta]}$, then $k\alpha$ is in M_β or k is in M_β.
(b) If a number k is in $M_{[\alpha,\beta]}$, then $k\beta$ is in M_α or k is in M_α.

A way to see that part (a) is true is to show that if both k and $k\alpha$ are *not* in M_β, then k is not in $M_{[\alpha,\beta]}$. That is, the commutator $[\alpha, \beta]$ leaves k fixed. Let's see why. Since $k\alpha \notin M_\beta$, this means that $(k\alpha)\beta = k\alpha$. Since $k \notin M_\beta$, this means that $k\beta = k$, or the disk in location k is left fixed by β. It follows from this that k must also be left fixed by β^{-1}. Putting these facts together, we see that $k\,(\alpha\beta\alpha^{-1}\beta^{-1}) = (((k\alpha)\beta)\alpha^{-1})\beta^{-1} = ((k\alpha)\alpha^{-1})\beta^{-1} = k\beta^{-1} = k$. This shows that the commutator $[\alpha, \beta]$ leaves k fixed as claimed, and we have verified (a). You can show part (b) similarly.

You can translate the statements in (10.1) completely into set notation. If A is any subset of $\{1, 2,\ldots, n\}$ and if γ is any permutation in S_n, then we can denote the set of all images under γ of elements of A by $A\gamma$. That is, $A\gamma = \{k\gamma \mid k \in A\}$.

One of the conditions considered in (10.1a) is the condition that $k\alpha$ is in M_β. This means that $k\alpha = j$ for some j in M_β, which is equivalent to saying that $k = j\alpha^{-1}$ for some j in M_β. In terms of this new notation, this is equivalent to saying that $k \in (M_\beta)\alpha^{-1}$.

Using this notation, we can combine parts (a) and (b) together and translate (10.1) into

(10.2) $$M_{[\alpha,\beta]} \subseteq (M_\beta \cup (M_\beta)\alpha^{-1}) \cap (M_\alpha \cup (M_\alpha)\beta^{-1}).$$

The significance of (10.2) is that it tells us that a commutator $[\alpha, \beta]$ can move no more than twice the number of elements in the smaller of the sets M_α and M_β. This is true because the set $(M_\alpha)\beta^{-1}$ is the same size as M_α. It is just a "translation" of M_α produced by the permutation β^{-1}, which takes each of the separate elements of M_α to separate elements. This means that when $M_\alpha \cup (M_\alpha)\beta^{-1}$ is as big as it possibly can be, that is, when the sets M_α and $(M_\alpha)\beta^{-1}$ share nothing in common, it is twice the size of M_α. Similarly, the largest that $M_\beta \cup (M_\beta)\alpha^{-1}$ can be is twice the size of M_β. Since the intersection of two sets is no larger than the smaller of them, this means that the largest the intersection of these two sets $M_\beta \cup (M_\beta)\alpha^{-1}$ and $M_\alpha \cup (M_\alpha)\beta^{-1}$ can be is the size of the smaller of them, which is at most twice the size of the smaller of M_α and M_β.

These observations, together with (10.2), lead to the conclusion that a commutator $[\alpha, \beta]$ can move no more than twice the smaller of the number of elements moved by either α or β. The right notation makes it possible for us to say things like this extremely compactly. If we introduce a standard convention that putting vertical bars on each side of the symbol for a set means the number of elements in that set, and use "min" followed by a set of numbers to denote the minimum number in the set, then that last sentence can be stated this way:

$$| M_{[\alpha,\beta]} | = 2 \times \min\{|M_\alpha|, |M_\beta|\}. \tag{10.3}$$

These last few paragraphs have gone on with only notational reasoning and abstraction for a tad longer than we like to do. Let's think about what this means in the context of the concrete models that our puzzles provide for us.

A natural type of commutator $[\alpha, \beta]$ to consider for the Oval Track Puzzle is one where α is a rotation R^k of the disks by k positions and β is the "Do Arrows" permutation T. In this case, the set M_α contains the numbers of all the active disks (equivalently all active positions) since a rotation moves all the disks. Call this number n. For the first Oval Track Puzzle, the set M_β is $\{1, 2, 3, 4\}$ since the "Do Arrows" button causes only the disks in locations 1 through 4, which are on the turntable, to move.

Claim (10.3) tells us that a commutator of this basic form will move at most 8 disks since $2 \times \min\{n, 4\} = 2 \times 4 = 8$. We can see that this maximum will sometimes be realized. For example, let's consider the commutator $[R^{-4}, T] = R^{-4}TR^4T^{-1}$. By starting out with a rotation R^{-4} of four positions counterclockwise, we move the four disks 1 through 4 off the turntable and the four disks 5 through 8 onto the turntable. The application of T reverses disks 5 through 8, and then the R^4 rotation moves these four disks off of the turntable and disks 1 through 4 back on. The final T^{-1} moves disks 1 through 4. Figure 10.1 shows the results of applying this commutator, after starting with the puzzle initialized with the identity permutation. We have moved 8 disks.

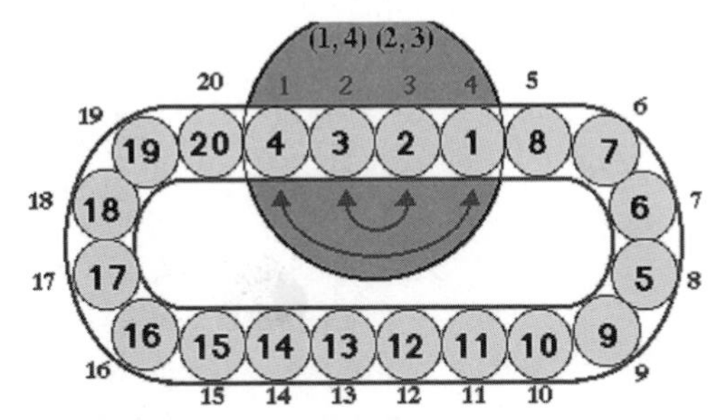

Figure 10.1. The commutator $R^{-4}TR^4T^{-1}$ moves 8 disks, which is the maximum possible given that T moves 4 disks.

For the commutator $[R^{-1}, T] = R^{-1}TR^1T^{-1}$, the number of disks moved is less. Figure 10.2 shows the result of this one. We see that $[R^{-1}, T] = (1, 4, 2, 5, 3)$. This commutator moved only 5 disks. In contrast to R^{-4}, R^{-1} moves three fewer disks onto

10

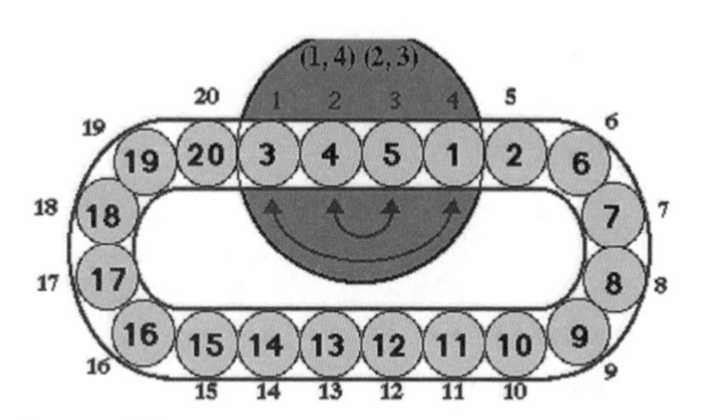

Figure 10.2. The commutator $R^{-1}TRT^{-1}$ moves only 5 disks, giving a 5–cycle.

the turntable. In the notation of (10.2), if $\alpha = R^{-1}$ and $\beta = T$, then $M_\beta = \{1, 2, 3, 4\}$ and $(M_\beta)\alpha^{-1} = \{2, 3, 4, 5\}$. Since M_β and $(M_\beta)\alpha^{-1}$ have three elements in common, their union is smaller, specifically $\{1, 2, 3, 4, 5\}$.

Stated in the physical terms of the puzzle, this commutator moves fewer disks because the smaller rotation brings fewer new disks onto the turntable.

We have identified an important insight here. If a commutator is to move the fewest possible disks, then the first permutation should bring as few new disks as possible into locations where they will be moved by the second one. This insight is especially important when we consider the Hungarian Rings puzzle, which has the feature that every possible permutation that you can perform moves over half of the disks. For this reason, the bound in (10.3) is of no help at all, since it tells us that a commutator moves at most 40 disks, which is more disks than there are. If we find a realistic upper bound, it will be as a result of using this new insight.

The Hungarian Rings has the feature that there are only two locations where the basic permutations L and R both act. These are the intersection points labeled 1 and 6. Specifically, $M_L = \{1, 2,\ldots, 20\}$ and $M_R = \{1, 6\} \cup \{21, 22,\ldots, 38\}$, and the intersection of these two sets is the two-element set $\{1, 6\}$. Consequently, any rotation of the right ring brings exactly two disks into locations that are subject to being moved by a rotation on the left ring. This means that for any integers i and j, since $M_{L^i} = \{1, 2, \ldots, 20\}$ and since R^j leaves all of these elements fixed except for 1 and 6, we have $(M_{L^i})R^j = \{1R^j\} \cup \{2, 3, 4, 5\} \cup \{6R^j\} \cup \{7, 8, \ldots, 20\}$. Note that $1R^j$ and $6R^j$ are numbers from the right ring.

Thus, $(M_{L^i})R^j$ contains the numbers of two locations from the right ring as well as all except 1 and 6 on the left ring, so $M_{L^i} \cup (M_{L^i})R^j$ contains all of the numbers from the left ring and two from the right.

In the same way, $M_{R^j} \cup (M_{R^j})L^i$ contains all of the numbers from the right ring together with two on the left ring. These are two sets with 22 elements in each of them, but given their nature, we can see that their intersection can contain at most 6 elements. For certain, $(M_{L^i} \cup (M_{L^i})R^j) \cap (M_{R^j} \cup (M_{R^j})L^i)$ contains the numbers 1 and 6. In addition, it will contain at most two additional numbers of locations from each of the rings. Bringing (10.2) into the discussion with $\alpha = L^i$ and $\beta = R^j$, and with small adjustments of exponents, we see that

(10.4) On the Hungarian Rings puzzle, any commutator of the form $[L^i, R^j]$ moves at most six disks.

Figure 10.3 illustrates this, showing that the commutator $[L, R^{-1}]$, moves the maximum six disks. Specifically, we get that $[L, R^{-1}]$ is a composition of two disjoint 3-cycles, namely, (7, 6, 34) (2, 1, 38).

It is possible to cut down the number of disks moved by a commutator on the Hungarian Rings puzzle by reducing the size of the unions of sets $M_{L^i} \cup (M_{L^i})R^j$ and $M_{R^j} \cup (M_{R^j})L^i$. Recall that for the first of these, this means arranging so that R^j brings less than two numbers from the left ring onto the right ring. The values of j that can do this are $j = \pm 5$. This is the case since R^5 moves the disk from location 1 to location 6, while R^{-5} moves the disk from location 6 to location 1. Thus, one of the disks that $R^{\pm 5}$ must bring onto the left ring was one that was already there. The disk that was not on the left ring that R^5 brings onto it is the one that was in location 25.

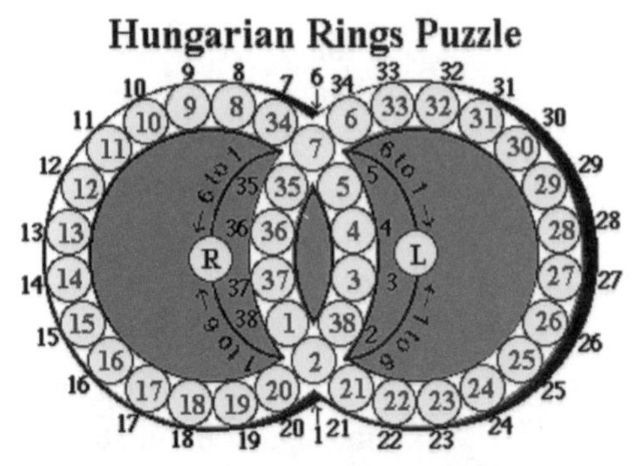

Figure 10.3. The commutator $LR^{-1}L^{-1}R$ moves 6 disks, the maximum possible.

The analogous observation is that the rotations of the left ring that bring only one new disk onto the right ring are L^5 and L^{-5}. If we build a commutator of the form $[L^i, R^j]$ where both i and j are ± 5, then it will move only four disks. Figure 10.4 shows one of these commutators, which produces a pair of transpositions.

We have looked at four examples of commutators so far, and have gotten one composed of four transpositions, a single 5-cycle, one composed of two 3-cycles, and finally a pair of transpositions. Do you notice a pattern here? All of these are even permutations. This is not just a coincidence. In fact,

(10.5) All commutators $[\alpha, \beta]$ in S_n are even permutations.

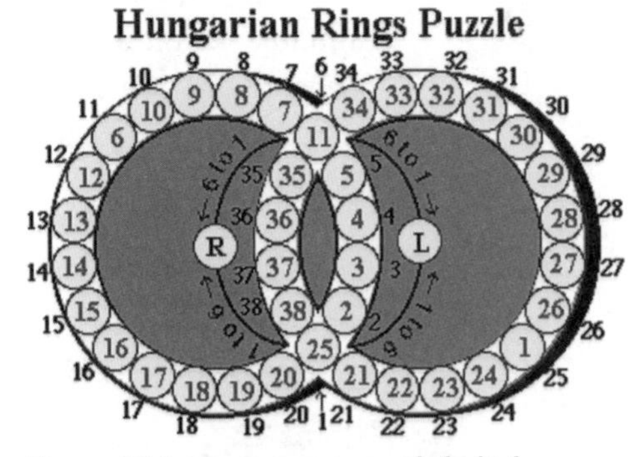

Figure 10.4. The commutator $L^5R^5L^{-5}R^{-5} = (1, 25)(6, 11)$ moves only 4 disks.

The reason for this is that any permutation α and its inverse α^{-1} have the same parity. In fact, if α can be expressed using i transpositions, we have seen that we can express α^{-1} using these same i transpositions in the reverse order. Thus, if we can express α with

i and β with j transpositions, then $[\alpha, \beta] = \alpha\beta\alpha^{-1}\beta^{-1}$ can be expressed with $2(i + j)$ transpositions, which means that $[\alpha, \beta]$ is even.

The example in Figure 10.4 illustrates the principle that we can sometimes take advantage of special features of our puzzles to build commutators that move fewer than the expected number of elements. In our next example from the Oval Track puzzle number 1, we see that not only do we want to explore ways to reduce the number of disks moved but also we want to look at *how* they are moved. That is, we want to go after commutators with advantageous cycle structure.

Consider the commutator $[R^{-3}, T]$ on the Oval Track Puzzle number 1. Figure 10.5 shows the result of applying this commutator to the initialized puzzle. The first rotation of three positions counterclockwise brings three new disks, numbers 5, 6, and 7, onto the turntable while moving the disk from position 4 to position 1. This last fact that we have moved a disk from location 4 to location 1 means that the next permutation called for, which is the turntable permutation T, will move this disk back from location 1 to location 4, while moving disk 7 which had been in location 4 to location 1. When this spinning of the turntable is followed by a clockwise rotation three positions, disk 4, which was back in its home location, now gets shifted to location 7. Meanwhile, disk 7, which was in location 1, has moved those three positions clockwise into location 4, and disk 1, which had been out of the turntable action during the second step, comes back into its home location. The final application of T^{-1} spins the turntable one last time, switching disks 1 and 7 between locations 1 and 4. The result of all of this for the three disks that we have been tracking though locations 1, 4, and 7 is that they have produced a 3-cycle, namely (1, 4, 7).

Meanwhile, two different pairs of adjacent disks have found themselves in the middle of the turntable each time it was spun. The first time disks 5 and 6 were there, and they switched places. The second time, disks 2 and 3 were there, and they switched places. The result is that the commutator $[R^{-3}, T]$ consists of a 3-cycle and two disjoint transpositions. Specifically, $[R^{-3}, T] = (1, 4, 7)(2, 3)(5, 6)$.

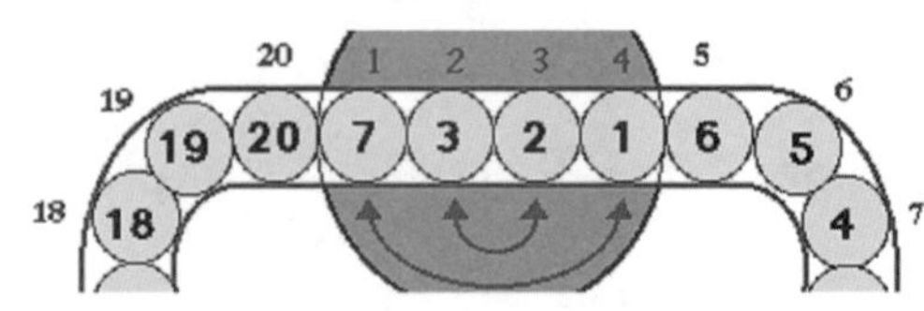

Figure 10.5. The commutator $R^{-3}\,T\,R^{3}\,T^{-1}$ on Oval Track Puzzle 1

This commutator moves a total of seven disks, which is not as small a number as we would like, but the way that it moves them ought to interest us. Since we have broken $[R^{-3}, T]$ down into disjoint cycles, our observations (6.5) and (6.6) about dis-

10

joint cycles commuting with each other and about integral exponents distributing over products of commuting elements allow us to see quickly what happens if we apply this commutator twice. We get that

$$[R^{-3}, T]^2 = ((1, 4, 7)\,(2, 3)\,(5, 6))^2 = (1, 4, 7)^2(2, 3)^2(5, 6)^2 = (7, 4, 1). \tag{10.6}$$

The last step of this sequence follows in part from the fact that transpositions are self–inverse permutations. We have found that by squaring a certain commutator that moves seven disks, we have gotten a permutation that moves just three disks, resulting in a 3-cycle. This should be a useful thing to know for solving end game problems on this puzzle. Also, since a 3-cycle is the simplest possible even permutation and since commutators are even, this is the minimal amount of movement that we will be able to get with a commutator.

We will put this squared commutator to work in the next chapter as we discuss the end game for the Oval Track Puzzle number 1 in detail.

In contrast to this example in which we combine one commutator with itself, we can sometimes find two different commutators that combine to give something useful. The Oval Track Puzzle number 2, where the "Do Arrows" permutation is $T = (4, 3, 2, 1)$, gives us such an example. On this puzzle, you can check that, as long as the number of active disks is at least 5, then $[R^{-1}, T] = (1, 2, 5)$, and that $[T^{-1}, R^{-1}] = (1, 5, 4)$. If you combine these two commutators, you get $[R^{-1}, T][T^{-1}, R^{-1}] = (1, 2, 5)(1, 5, 4) = (1, 2, 4)$. So what's the big deal, you ask? We are just getting a variety of 3-cycles.

Since commutators are even permutations, getting 3-cycles is about what one would expect. However this last 3-cycle meshes with the "Do Arrows" permutation $T = (4, 3, 2, 1)$ to produce the transposition shown in Figure 10.6. Follow the two commutators by a final T and you get

$$[R^{-1}, T][T^{-1}, R^{-1}]T = (1, 2, 4)(4, 3, 2, 1) = (2, 3). \tag{10.7}$$

This is a very significant find for this puzzle when we combine it with our knowledge about the role of transpositions in the overall scheme of things. Once we have found this transposition, we have found a way to build a strategy for fully solving the puzzle. We will return to the details of this in the next chapter.

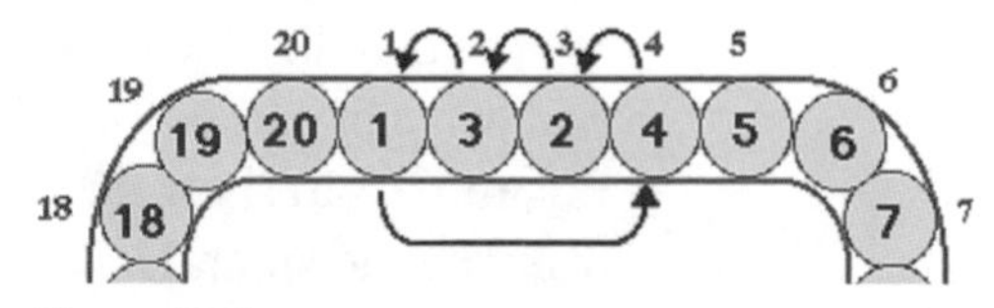

Figure 10.6. On Oval Track 2, two commutators and a "Do Arrows" yields a transposition.

We have now looked at two examples of getting useful results from composing a commutator with itself or by composing two different commutators. As our final example, we will look at a way of getting useful results with "compound commutators." By this we mean a commutator that combines another commutator with some permutation, that is, something of the form $[[\alpha, \beta], \gamma]$. If you expand this expression, you get

$$(\alpha\beta\alpha^{-1}\beta^{-1})\gamma(\beta\alpha\beta^{-1}\alpha^{-1})\gamma^{-1}.$$

This is a rather complex expression. Can we see why an expression this involved might be something that we can plan and control rather than just discover by trial and error?

Well, let's try. Suppose you have found that the commutator $\alpha\beta\alpha^{-1}\beta^{-1}$ does something useful for you. Of course its inverse, $\beta\alpha\beta^{-1}\alpha^{-1}$, would undo this change and restore the puzzle to its original state. Note that the last part of this compound commutator, namely, $\gamma(\beta\alpha\beta^{-1}\alpha^{-1})\gamma^{-1}$, is a conjugate of the inverse of the first commutator. This conjugate of $\beta\alpha\beta^{-1}\alpha^{-1}$ will have the same cycle structure as $\beta\alpha\beta^{-1}\alpha^{-1}$ does.

Suppose we think of γ as being a "setup" permutation that tweaks the arrangement of the disks in a way that modifies part of the change that the original commutator produced but leaves some of it the same. Then when we apply $\beta\alpha\beta^{-1}\alpha^{-1}$, the inverse of the first commutator, we will fully undo part of the original change, but also undo in a modified way the part that we tweaked with γ. After this is done, we reverse the unwanted changes made with the setup procedure when we apply γ^{-1}.

To give a useful example of a compound commutator that implements the strategy described here, we will look at the Hungarian Rings and the commutator $[\alpha, \beta] = [L^5, R^5]$ shown above in Figure 10.4. This commutator produces two transpositions, $(1, 25)$ and $(6, 11)$, as shown again in Figure 10.7(a). One of these transpositions involves the lower intersection point and the right ring and the other involves the upper intersection point and the left ring. We ought to be able to tweak at one of the intersections while leaving the other unchanged. We see that $\gamma = R^{-1}LR$ will leave the disks 1 and 25 in locations 25 and 1 respectively while making some changes at locations 6 and 11. Figure 10.7(b) show the result. We still have disk 11 in location 6 as we did before we applied γ, but now we have disk 12 in location 11.

Now when we apply the inverse of the commutator $[L^5, R^5]$, it does a second pair of swaps between locations 1 and 25 and between locations 6 and 11. This restores disks 1 and 25 to their original locations, but makes for additional change involving locations 6 and 11, bringing the disk from location 12 into the mix. Figure 10.7(c) shows the state of the puzzle following this step.

At this point, we have the majority of the disks on the left ring rotated one position clockwise from their home positions as a result of the L in the middle of the setup permutation γ. We also have a bit of unwanted displacement around locations 1 and 6 that resulted from γ. The application of γ^{-1} takes care of these final matters, yielding the outcome shown in Figure 10.7(d).

What is the final result of this compound commutator? It is a 3-cycle, namely (6, 11, 12). Knowing that we can produce a 3-cycle on this puzzle has profound implications. We will return to these implications in Chapter 15.

This last example is, admittedly, a bit complex. Maybe you are thinking, "I doubt that I can make use of this idea to discover anything on my own." If you think that way, and act on the thought (or fail to act might be more accurate), it will surely be a self-fulfilling prophecy.

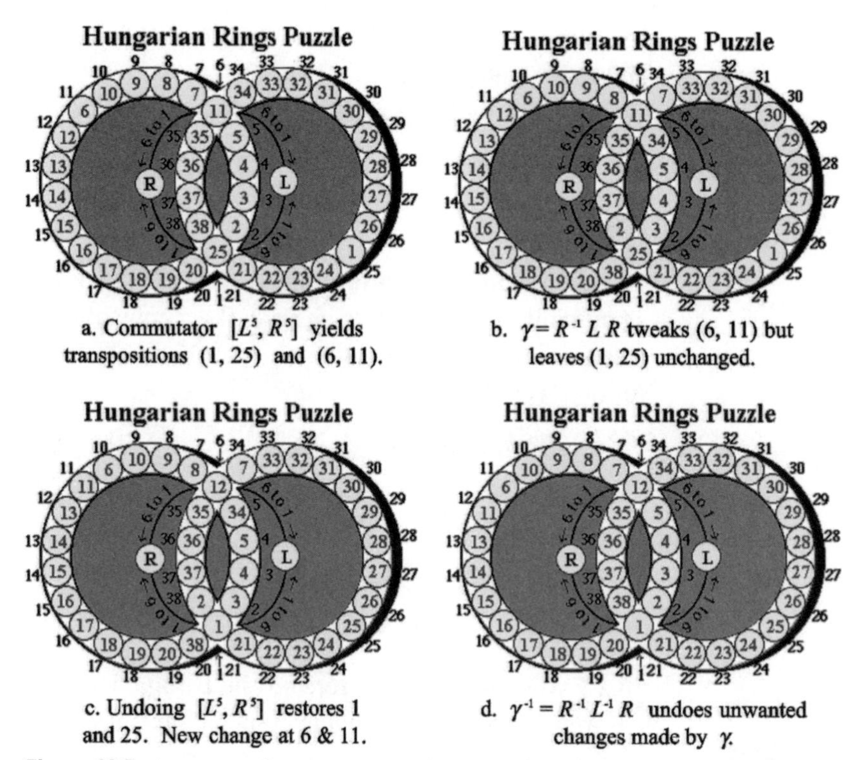

a. Commutator $[L^5, R^5]$ yields transpositions (1, 25) and (6, 11).

b. $\gamma = R^{-1} L R$ tweaks (6, 11) but leaves (1, 25) unchanged.

c. Undoing $[L^5, R^5]$ restores 1 and 25. New change at 6 & 11.

d. $\gamma^{-1} = R^{-1} L^{-1} R$ undoes unwanted changes made by γ.

Figure 10.7. A compound commutator on the Hungarian Rings uses a commutator that yields two transpositions to get a 3-cycle (6, 11, 12).

However, if you reject this thought, and combine positive expectations and persistence with strategic guidance from the theory, you too can make useful discoveries. In fact, an equivalent variation of this compound commutator was discovered by one of the author's undergraduate university students, a young man named Brent Sauvé. He was not guided by the compound commutator structure at the time of the discovery. Instead, his discovery was a result of an intrinsic sense of commutators, trial and error, and persistence.

By analyzing Brent's discovery, we can see the structure that underlies it. This knowledge can guide our further investigations. It will not eliminate the need for persistence, however.

Exercises

10.1 Exploring commutators on the Oval Track Puzzles. Here is a list of commutators that you might explore as possible end game tools on several versions of the Oval Track Puzzles. In each case, write out the permutation resulting from the commutator in cycle notation, and determine the smallest number of active disks for which the formula you derive works.

(a) $[R^{-1}, T]$ on Oval Track 2
(b) $[R^{-2}, T]$ on Oval Track 2
(c) $[T^2, R^{-2}]$ on Oval Track 2
(d) $[R^{-1}, T^2]$ on Oval Track 2
(e) $[R^{-1}, T]$ on Oval Track 4
(f) $[R^{-1}, T^2]$ on Oval Track 4
(g) $[R^{-1}, T]$ on Oval Track 5
(h) $[R^{-2}, T]$ on Oval Track 5
(i) $[R^{-3}, T]$ on Oval Track 5
(j) $[R^{-5}, T]$ on Oval Track 17

10.2 Exploring commutators on the Hungarian Rings Puzzle. Express in cycle notation the permutation resulting from each of these commutators on the Hungarian Rings Puzzle.

(a) $[L, R]$
(b) $[L, R^{-1}]$
(c) $[R, L]$
(d) $[R, L^{-1}]$
(e) $[L^5, R]$
(f) $[R, L^5]$
(g) $[L^5, R^{-1}]$
(h) $[L, R^{-5}]$
(i) $[L^5, R^5]$
(j) $[R^5, L^5]$
(k) $[L^{-5}, R^{-5}]$
(l) $[L^5, R^{-5}]$

10.3 Find some commutator 3-cycles. Explore commutators on the Oval Track Puzzles 8, 10, and 11, looking for ones which are 3-cycles, or which yield 3-cycles when squared. If you find any, think about the consequences of this.

10.4 Metaphoric commutators on the Slide Puzzle. Non-identity commutators involving two one-position moves are impossible on the Slide Puzzle. To illustrate, if the puzzle is in the basic "15" configuration with the blank in spot 16, one possible sequence of two moves is (16, 15)(15, 11). To make this into a commutator, we would have to invert the first transposition, and then the second. But inverting the first one calls for the blank to be in spot 15, but it is now in spot 11. (a) Prove that the only *true* commutators built with two one-position moves on the Slide Puzzle yield the identity permutation. (b) If we encode moves on the slide puzzle by tracking movements of the blank with U for "up," D for "down," L for "left," and R for "right," and define that U and D are inverses for each other and similarly for L and R, then we can do commutators in this sense. (i) Explore the commutators $[L, U]$ and $[U, L]$. Show that these are both 3-cycles and are inverses of each other. (ii) Can you find any commutators in this sense that are other than 3-cycles?

10.5 Inverses of commutators. Prove that for any permutations α and β, $[\alpha, \beta]^{-1} = [\beta, \alpha]$.

10.6 If a commutator is trivial, the components commute. Show that for every pair of permutations α and β, the commutator $[\alpha, \beta]$ equals the identity ε if and only if α and β commute with each other.

10.7 Is the building of commutators associative? (a) Explore the equation $[[\alpha, \beta], \gamma] = [\alpha, [\beta, \gamma]]$. By trying out these compound commutators on one of the puzzles, show that this equation is not true for all permutations α, β, and γ. This shows that the operation of commutator building is not an associative operation. (b) Show that for any permutations α, β, and γ such that β commutes with both α and γ, this associativity equation is trivially true. (c) Are there other special cases of α, β, and γ for which this associativity equation is true?

10.8 Exploring the equation $[(\alpha\beta), \gamma] = [\alpha, (\beta\gamma)]$. Another associativity-like equation involves composition and commutators. Consider the equation $[(\alpha\beta), \gamma] = [\alpha, (\beta\gamma)]$. (a) Show that this equation is not true for some permutations α, β, and γ. (b) Show that for any three permutations α, β, and γ, this equation is true if and only if $\gamma\alpha\beta = \beta\gamma\alpha$, that is, if and only if β commutes with $\gamma\alpha$.

10.9 Conjugates of commutators are commutators of conjugates. Prove that every conjugate of a commutator is a commutator by showing that the formula $\gamma[\alpha, \beta]\gamma^{-1} = [\gamma\alpha\gamma^{-1}, \gamma\beta\gamma^{-1}]$ holds for all permutations α, β, and γ.

10.10 Commutators involving products are products of commutators. The expressions $[(\alpha\beta), \gamma]$ and $[\alpha, (\beta\gamma)]$ from Exercise 10.7 were commutators with one of their terms a product. (We are generalizing by saying "product" rather than "composition" here.) Verify these formulas, which show that commutators of products are also products of commutators.

(a) $[\alpha, (\beta\gamma)] = [\alpha, \beta]\,(\beta[\alpha, \gamma]\beta^{-1}) = [\alpha, \beta]\,[\beta\alpha\beta^{-1}, \beta\gamma\beta^{-1}]$.

(b) $[(\alpha\beta), \gamma] = (\alpha[\beta, \gamma]\alpha^{-1})\,[\alpha, \gamma] = [\alpha\beta\alpha^{-1}, \alpha\gamma\alpha^{-1}]\,[\alpha, \gamma]$.

10.11 Commuting with commutators. Prove these facts about one of the terms of a commutator commuting with the commutator. (a) A permutation α commutes with the commutator $[\alpha, \beta]$ if and only if $[\alpha, \beta] = [\beta, \alpha^{-1}]$ if and only if $[\beta, \alpha] = [\alpha^{-1}, \beta]$. (b) Both α and β commute with $[\alpha, \beta]$, if and only if $[\alpha, \beta] = [\beta, \alpha^{-1}] = [\beta^{-1}, \alpha]$.

10.12 Distributivity of exponents with a twist. Back in Chapter 6, display (6.8), we showed that a positive integer exponent will distribute over a composition of two permutations if the permutations commute with each other. We also looked at examples to show that this rule can fail if the elements do not commute with each other. Show now that if α and β satisfy the weaker hypothesis that both commute with $[\alpha, \beta]$, then for every positive integer n, $(\alpha\beta)^n = \alpha^n\beta^n\,[\beta, \alpha]^{n(n-1)/2}$.

Mastering the Oval Track Puzzle

11

Now that we have developed some relevant theory, we are ready to start putting it to work to solve the puzzles. The remainder of this book will focus on solving puzzles, starting with a few of the variations of the Oval Track Puzzle.

As always, you will find reading this book most useful if you do so while having your puzzles up and running on your computer. As you read, try things out on your puzzles.

11.1 The First Oval Track Puzzle

Let's get right down to business with version one of the Oval Track Puzzle which has as its "Do Arrows" the turntable permutation $T = (1, 4)\,(2, 3)$. In Chapter 1, we talked about the opening game on this puzzle. You should have practiced enough with this part of the solution process so that you can consistently get the puzzle to the point where there are just four contiguous disks out of place. You are now ready to address the end game.

The ideas of conjugates and commutators discussed in the last two chapters provide us with most of what we need. To begin, let us recall that in Chapter 10, we introduced the square of a commutator, (10.6), which produces a 3-cycle. Specifically,

(11.1) On Oval Track 1 with $T = (1, 4)(2, 3)$ and with 7 or more active disks
$$[R^{-3}, T]^2 = (R^{-3}TR^3T^{-1})^2 = (7, 4, 1)\,.$$

Why are we using a commutator that starts out with a counterclockwise rotation rather than, say, $[R^3, T]$, which starts with a clockwise rotation? This one yields a 3-cycle also when squared; with 20 active disks, $[R^3, T]^2 = (1, 4, 18)$.

The disadvantage of starting out with a clockwise rotation is that we begin by bringing some high-numbered disks onto the turntable. With 20 disks active, the ones we bring on are disks 18, 19, and 20. In general, if there are n disks active, we bring into play disks numbered $n - 2$, $n - 1$, and n, and $[R^3, T]^2 = (1, 4, n - 2)$. We have a

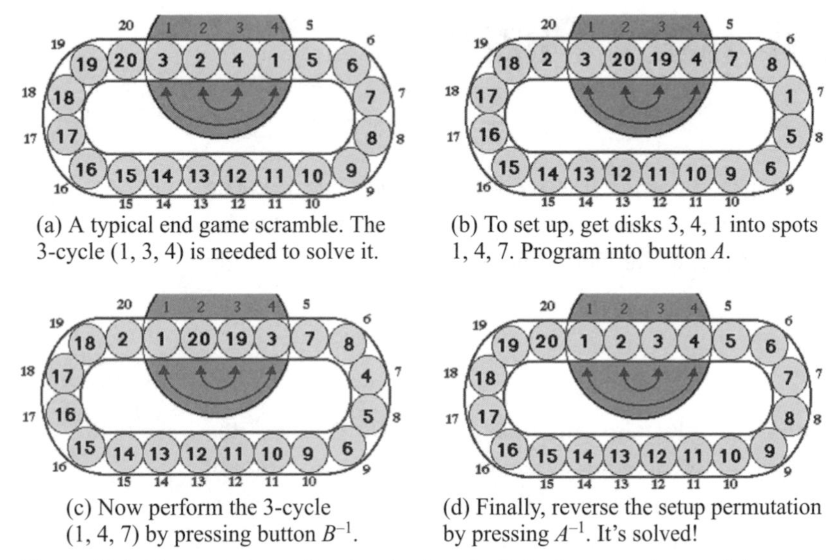

(a) A typical end game scramble. The 3-cycle (1, 3, 4) is needed to solve it.

(b) To set up, get disks 3, 4, 1 into spots 1, 4, 7. Program into button A.

(c) Now perform the 3-cycle (1, 4, 7) by pressing button B^{-1}.

(d) Finally, reverse the setup permutation by pressing A^{-1}. It's solved!

Figure 11.1. The steps for performing the 3-cycle (1, 3, 4) as a conjugate of the fundamental 3-cycle (1, 4, 7)

need for a variable in the formula. In contrast, if we start off with a counterclockwise rotation of three positions, the new disks brought into play are always disks 5, 6, and 7—as long as there are at least 7 active disks. There is no need for a variable.

Maybe you have noticed that when you started the Oval Track Puzzle program for the first time, there was a program already there for macro button B. It was the squared commutator of (11.1) which we are discussing. If you have reprogrammed button B and have saved changes in Oval Track, you no longer have this as the button B program, but it is easy to reprogram it into a button manually or retrieve this macro from Puzzle Resources, where it is part of the permanent collection.

Having this one 3-cycle is valuable to us because we can take conjugates of it to get other 3-cycles that we may need to solve the end game. For example, suppose we have gotten the puzzle down to the end game state shown in Figure 11.1(a). We have here the permutation (1, 4, 3), which is a 3-cycle. To solve the puzzle, we need to apply the inverse 3-cycle (1, 3, 4). That is, the contents of spot 1 (disk 3) needs to go to spot 3, disk 4 in spot 3 needs to move to spot 4, and disk 1 in spot 4 needs to complete the cycle by moving to spot 1.

To accomplish this, we will perform a series of moves that puts disks 3, 4, and 1 into spots 1, 4, and 7. We will record this sequence of steps into button A so as to have the computer remember the sequence for us. But before we start, we look at the current arrangement and record into short-term memory (the one in the head, not in the computer) the phrase, "1 chases 3." By this we mean that disk 1 needs to go where disk 3 is right now. This is a sufficient description of the direction things should move to guide us later.

It does not matter how you go about getting these three disks into spots 1, 4, and 7. You may try for as few ticks on the counter as is possible, but you need not. One sequence of steps that produces the outcome as shown in Figure 11.1(b) is: $R^2TR^{-5}TR^{-1}TR^5$.

Now we review our situation. We have disks 3, 4, and 1 in spots 1, 4, and 7 respectively. We have a procedure programmed into button *B* that makes no changes other than to move disks from spots 7 to 4, from 4 to 1, and from 1 to 7. This is where we need to recall our memory aid regarding what needs to be done: "1 chases 3." If we press button *B*, we will have disk 1 moving from spot 7 to spot 4 where disk 4 is now. That is, we will have "1 chasing 4." This tells us that we do not want to press button *B*, but rather we want to press button B^{-1}. This will give the inverse 3-cycle (1, 4, 7) which will have disk 1 in spot 7 moving to spot 1 where disk 3 is. That is, "1 will chase 3" as we want.

Once we press button B^{-1} and the computer runs the 3-cycle, the result is as shown in Figure 11.1(c). Comparing Figures 11.1(b) and (c), we see that the only difference is in the placement of the three disks 1, 3, and 4. We are now ready to reverse the setup permutation that we programmed into button *A*. We are able to do this simply by pressing the inverse button A^{-1}. The result is shown in Figure 11.1(d). We have solved the puzzle.

Let's look at another example showing how to use our button *B* 3-cycle to solve an end game. Suppose that we arrive at the end game shown in Figure 11.2(a). What we need to do to completely solve the puzzle is to perform a pair of transpositions, (1, 3) (2, 4). This is an even permutation, and we recall that we saw in (8.9) that every even permutation can be expressed using 3-cycles. In showing why this was true, we saw that a pair of disjoint transpositions such as we have here can be expressed using two 3-cycles. Let's solve the puzzle by doing two 3-cycles as conjugates of our fundamental 3-cycle (7, 4, 1) programmed into button *B*.

We have flexibility in what 3-cycle we do first. We just want to involve disks that are currently out of place and make sure that we get one of the disks into its right spot. Let's decide to get disk 4 into its right spot first, and also involve disks 2 and 3 in this first 3-cycle. To keep track of the direction, let's remember "4 chases 2" (meaning that we want disk 4 to go to the spot where disk 2 is now).

We next need to do a setup permutation which puts disks 3, 4, and 2 into spots 1, 4, 7 respectively, programming it into button *A* while we do it. One way to get the result is shown in Figure 11.2(b). This results from the steps $R^3TRTR^{-6}TR^4$.

Note that there is also flexibility in how we do this setup. The three disks that we want to move can go into the target spots 1, 4, and 7 in any order we want. As long as we remember of the "who chases whom" mnemonic, we will be able to decide whether button B or B^{-1} is the right one to use. Even if we get it wrong and do one conjugate, say ABA^{-1}, when we should have done the other, the problem is easily corrected. What we have done is the inverse of the 3-cycle that we wanted. For a 3-cycle, its square is its inverse (see Exercise 11.1) so to get what we want we need to apply ABA^{-1} again by three more presses of the right macro buttons.

Returning to the puzzle, we look at the situation in Figure 11.2(b) and, remembering that "4 chases 2," we see that we want to press button B^{-1} to perform the 3-cycle (1, 4, 7) to have disk 4 chase after disk 2 and move to spot 7. When we run the macros B^{-1} and A^{-1}, the result will be as shown in Figure 11.2(c).

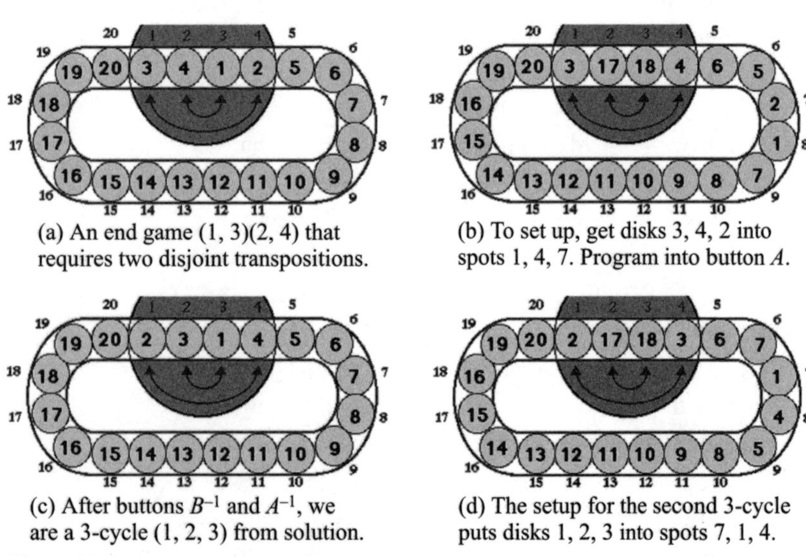

(a) An end game (1, 3)(2, 4) that requires two disjoint transpositions.

(b) To set up, get disks 3, 4, 2 into spots 1, 4, 7. Program into button A.

(c) After buttons B^{-1} and A^{-1}, we are a 3-cycle (1, 2, 3) from solution.

(d) The setup for the second 3-cycle puts disks 1, 2, 3 into spots 7, 1, 4.

Figure 11.2. Performing two disjoint transpositions using two 3-cycles, which are gotten as conjugates of our fundamental 3-cycle.

We now need to perform the 3-cycle (1, 2, 3) to solve the puzzle. This is, of course, what we expected since we are carrying out the process shown in Figure 8.3 for expressing two disjoint transpositions with two 3-cycles. Having applied a first appropriate 3-cycle, the result is the other one.

We perform this second 3-cycle the same way, as a conjugate of our fundamental 3-cycle programmed into button B, or its inverse. Noting that "1 chases 2," we perform and record in button A a setup. Figure 11.2(d) shows one way to do this that gets the disks 1, 2, and 3 into the spots 7, 1, and 4 where they need to be for our fundamental 3-cycle to act on them. The steps that produce this are $R^3TRTR^{-6}TR^{-1}TR^5$. Since "1 chases 2," we again see that button B^{-1} is the one to use rather than button B. Pressing button B^{-1} and following this with button A^{-1}, we will have solved the puzzle.

This is a good point to comment on the way the setup procedure goes for using our button B 3-cycle and its inverse. Typically, the three disks we want to move are going to be adjacent to each other or have one non-involved disk between them. To get disks into spots 1, 4, and 7, we need to have them arrayed so that there are two non-involved disks between the leftmost and the middle one, these to occupy spots 2 and 3, and similarly two non-involved disks to occupy spots 5 and 6 between the center one and the rightmost one. The strategy for getting this spacing between our target disks is to use the turntable to spread them out appropriately.

If the leftmost and center disks are right next to each other, we need to spread them out more than two spaces and then get them closer again. This was the case with disks 2 and 3 when they were in spots 1 and 2 as shown in Figure 11.2(c). To separate them, we did a rotation R^3 that put disk 2 on the right edge of the turntable and disk 3 just off the right edge. Then when we spun the turntable, disk 2 moved to spot 1. This put three non-involved disks between it and disk 3, still in spot 4. Then we needed to cut down the distance between disks 2 and 3 by one disk. By applying R, we moved disk 2 into spot 2, one of the two spots nearest the center of the turntable. When we spin the turntable next, disk 2 moved to spot 3. This puts it one disk closer to disk 3, and gives us the right separation.

The steps just described were the first four steps, namely R^3TRT, of the setup procedure. The remaining five steps of the procedure focused on getting the right spacing between the center disk 3 and the rightmost disk 1, and then finally placing all the three target disks into the right spots.

The procedure that solved the puzzle for the end game scrambles in these two examples can be used to solve all end games that need an even permutation. For the Oval Track Puzzle number 1, the end game will involve at most four disks, so the only type of even permutations on four objects are 3-cycles and pairs of transpositions. Any case that you encounter will be like one or the other of the examples, except that the placement of the disks involved can of course be different.

Does one have to solve the puzzle the way we have described, that is, by making use of one fundamental 3-cycle and performing all other 3-cycles as conjugates of the fundamental one? In a word, no. But the alternative calls for having a larger arsenal of tricks. If, for example, you have found a neat way of performing the 3-cycle $(1, 2, 3)$, but were of a mind to disdain conjugacy, you would need other procedures to perform the 3-cycles $(1, 2, 4)$ or $(1, 3, 4)$.

You might need quite a few different procedures that you have committed to memory or have written down. It may be that you could do many end games with fewer moves than our general procedure involving conjugacy, but at the expense of having to remember many rules rather than just a few.

What we have not yet addressed is the question of how to solve an end game that involves an odd permutation. Since this will be an odd permutation on four disks, the cycle structure possibilities are again only two. The odd permutation is either a 4-cycle or a transposition.

Let us see how we can reduce any end game 4-cycle to a transposition. It may be that a single application of the turntable T will produce such a reduction. For example, if the end game permutation is $(1, 3, 2, 4)$, then following this by T, we get $(1, 3, 2, 4)T = (1, 3, 2, 4)(1, 4)(2, 3) = (1, 2)$.

If T does not reduce the 4-cycle to a transposition, then there will always be a 3-cycle that does. Consider this permutation equation in cycle notation: $(a, b, c, d)(d, c, a) = (a, b)$. This gives a way of following a generic 4-cycle with a 3-cycle involving three of the same numbers that yields a transposition as the composition. We can always identify a 3-cycle that will produce such a reduction, and perform it using our conjugacy method. Just perform any 3-cycle that puts two disks back where they belong.

Given these observations, if we have an end game that is odd, we can always get the puzzle down to a single transposition. Thus, the only thing we need to do to be able to completely solve the puzzle is to find a way to perform a transposition.

The bad news is that this might not even be possible. It will depend on n, the number of active disks. If n is odd, then we will not be able to perform a single transposition.

In fact, if n is odd, then we cannot perform any odd permutation at all. When n is odd, the basic one-position rotation $R = (1, 2, 3,\dots, n)$ is an even permutation, as we noted in (8.6). Since the "Do Arrows" turntable $T = (1, 4)(2, 3)$ for this version of the puzzle is also an even permutation, we will only get even permutations by putting together sequences of these basic moves. Thus, we can never produce an odd permutation, and in particular, cannot perform a transposition.

To have any hope of performing a transposition, we must assume that n, the number of active disks, is even. In this case, we can obtain a transposition. Specifically,

(11.2) On Oval Track 1 with $T = (1, 4)\,(2, 3)$ and with n, the number of active disks, being even and at least 6, $(TR^{-1})^{n-3} = (1, 3)$.

Figure 11.3 illustrates why this is true. It gives a representation of the turntable area of the Oval Track puzzle and the spot on either side of it. The first row shows the contents of these spots with an initialized puzzle, and each successive row shows their contents after successive applications of the moves T and R^{-1}. You might find it helpful to run through a sequence of these moves on your Oval Track Puzzle to confirm that Figure 11.3 represents what really happens.

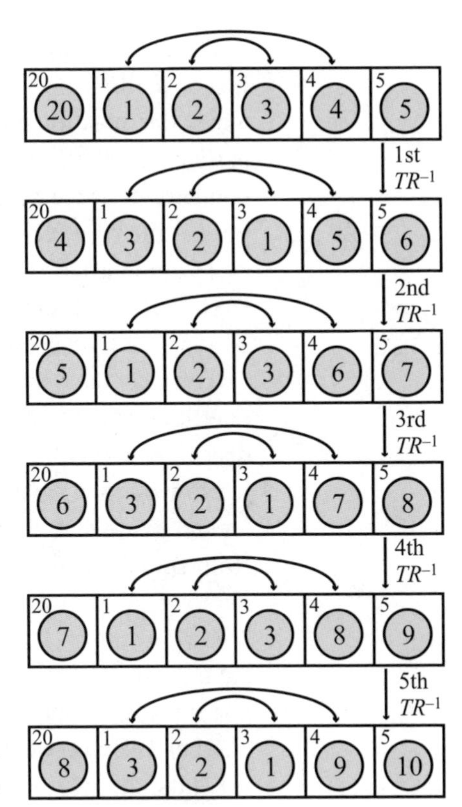

Figure 11.3. The effect of repeated applications of TR^{-1}.

The first T reverses the order of the disks 1 through 4 so that they read 4, 3, 2, 1 from left to right. When we follow this by the R^{-1}, disk 4 gets pushed off of the turntable to the left, and disk 5 is brought into the rightmost turntable location. Disks 1, 2, and 3 remain on the turntable, but in the order 3, 2, 1. If we note the permutation represented just by the spots numbered 1, 2, and 3, it is the transposition (1, 3).

When we apply the second TR^{-1}, disk 5 first moves over to the left edge of the turntable, and then off the turntable on the left. Disk 6 moves onto the right side of the turntable. Meanwhile, disks 1, 2, and 3 continue to stay on the turntable, but this time, since the turntable has spun twice, these disks are back in their normal places. If we consider the permutation on just these three again, this time it is the identity.

Let us continue performing the moves TR^{-1}, focusing attention on the three disks 1, 2, 3. They alternate between the transposition (1, 3) and the identity permutation. After each odd number of these applications, the permutation on the set {1, 2, 3} is (1, 3) and after each even number, it is the identity.

As this alternating behavior is going on, what is happening with the other disks? One by one, they are getting moved counterclockwise around the track. When they get to the right edge of the turntable at the end of one application of TR^{-1}, they move over to the spot just to the left of the turntable on the next application of TR^{-1}.

As a result of the application of TR^{-1} when a disk makes this passage from spot 4 on the turntable to the spot just to the left of the turntable, it moves four positions counterclockwise. Every other application of TR^{-1} moves a disk one position counterclockwise. Other than the disks 1, 2, and 3, there are $n-3$ disks. Thus, after $n-3$ applications of TR^{-1}, each of these other disks will have made one passage through the turntable, and will have moved counterclockwise by one position $n-4$ times. The number of positions of counterclockwise shift it will have undergone is thus $4+(n-4)=n$. Since there are n positions, shifting n positions brings the disk back to its starting point.

To summarize, after $n-3$ applications of TR^{-1}, the disks other than disks 1, 2, and 3 will be back where they started. If n is even, then $n-3$ is odd, so, based upon our earlier analysis, the permutation we have on the set $\{1, 2, 3\}$ is the transposition (1, 3). We have a transposition.

You may have discovered that Oval Track One comes with this transposition (1, 3) programmed into button *C*. (Retrieve the macro from Puzzle Resources if needed.) The program is $(TR^{-1})^{17}$, which yields (1, 3) when all 20 disks are active.

Now that we have one transposition, we can get others as conjugates of this one. For example, if you encounter an end game that requires performing the transposition (1, 4), just perform a setup procedure that moves the disk from spot 4 to spot 3 while keeping the disk in spot 1 where it is. Then run the procedure $(TR^{-1})^{n-3}$ which produces (1, 3), and finally, reverse your setup procedure. The result will be the desired transposition (1, 4).

Our example (1, 4) should make it clear that we can get any transposition whatever. Given this, the theory we have developed tells us that we can totally solve the puzzle. Since all permutations can be expressed using transpositions, we could, if we choose to do it that way, build any permutation of the disks on the Oval Track Puzzle number 1. Since we could build any permutation with a sequence of possible moves, this means that we can also solve the puzzle no matter how it is scrambled by reversing the sequence of moves that produced that scrambling. Remember, all of this assumes an even number of active disks.

In cases where n is odd, which limits us to even permutations, we can say definitely that we can solve all even permutations. The reason is that, as we saw in (8.9), every even permutation can be expressed using 3-cycles, and we can produce all 3-cycles on this puzzle.

Let us summarize these insights for future reference.

(11.3) On Oval Track 1 with $T = (1, 4)(2, 3)$, any scrambling can be solved if the number n of active disks is even and more than 7. If n is odd and at least 7, then every even scrambling can be solved. Under these latter conditions, odd scramblings can be brought down to a single transposition, but cannot be completely solved.

This summary describes in detail what is possible and what is not for all numbers n of active disks with $n \geq 7$. The value 7 is a break point because having at least 7 disks gives us enough room for our fundamental 3-cycle to work and also enough room to build all 3-cycles as conjugates of this one. The situations for $n = 4$, 5, or 6 do not provide enough "wiggle room" to allow our general approach to work. For $n = 6$, the puzzle is fully solvable since our basic transposition formula works: $(TR^{-1})^3 = (1, 3)$. Furthermore, one can see that there is enough wiggle room to set up to get any transposition as a conjugate of this one, and thus we can get all permutations.

For $n = 4$ and $n = 5$, there is not enough wiggle room to allow for full solvability of the puzzle; you can't even solve all of the even permutations. For $n = 4$, eight of the 24 possible scramblings are solvable, but 16 are not. For $n = 5$, there are $5!/2 = 60$ even scramblings, but we can only solve ten of them. For the details of these small cases, see Exercise 13.3.

A vexing question is this: Is it really necessary to have such a long procedure to produce a transposition? Our transposition formula in (11.3) requires $2(n - 3)$ steps. When the number of active disks is the maximum 20, for example, this is 34.

The answer is that we *really do* need to do all this work to get a transposition. A claim that gives insight into why this is true is:

(11.4) If there is a way to get an odd permutation γ on Oval Track Puzzle number 1 that moves only disks from spots in the "turntable set" $\{1, 2, 3, 4\}$, then the sequence of steps that produces γ must move all disks from outside the turntable set through the turntable.

This is saying that, to get a transposition that swaps any two disks with numbers between 1 and 4, it is necessary that all of the disks numbered 5 through n, must pass through spots in the range from 1 to 4.

Why should that be the case? Well, if it were not true, then there would have to be at least one disk with a number 5 or more, call it disk r, that never gets onto the turntable. We will see that the assumption of such a disk r forces the permutation γ to be even, which is a contradiction.

Without any loss of generality, we can assume that the sequence of moves that produces the permutation γ is of the form $R^{i_1}T^{j_1}R^{i_2}T^{j_2} \cdots R^{i_k}T^{j_k}$, where the i's and j's are all integers. This notation is more general than it needs to be for this particular "Do Arrows" permutation T. In this turntable case, all of the j's could be 1. However, the more general notation will be needed when we think about other "Do Arrows" permutations later on.

We can assume that powers of R and T alternate because if we had two of either in succession, we could simply combine them together. Also, it is no loss of generality to have our notation start with a rotation. If in fact we start with a "Do Arrows," we simply set i_1 equal to 0. A similar comment can be made about ending with a power of T.

Returning to our hypothetical disk r, if it never enters the turntable, then clearly when the sequence $R^{i_1}T^{j_1}R^{i_2}T^{j_2} \cdots R^{i_k}T^{j_k}$ is applied, it never gets acted upon by the T^j's but only by the R^i's. Remember also that in the end, disk r returns to spot r from whence it started its journey.

To aid in our analysis, we are going to make some adjustments in the notation that will not change the results. Suppose there is among the R^i's one such that $i < 0$ and $|i| > k$, where k is the number of the box in which disk r is located at the time that this rotation gets performed. This rotation moves all the disks counterclockwise by $|i|$ positions. Given that $|i| > k$, and that disk r is in box k, this would have us passing disk r across the turntable area. We can avoid doing that by replacing this counterclockwise rotation by a clockwise one that does the same thing. This clockwise rotation moves everything by $n - |i|$ places. (We clearly can assume that $|i| < n$, making this difference positive. If $|i| = n$, we are making a complete counterclockwise rotation, and this could be replaced by doing nothing. If $|i| > n$, we are rotating more than a complete rotation, and it is clear how to reduce this.)

Replace all such disk-r-crosses-the-turntable counterclockwise rotations with their clockwise equivalents, and similarly, if there are any clockwise rotations that would move disk r through the turntable area, replace them by their counterclockwise equivalents that do not cause disk r to pass through the turntable area.

We now have our special disk r moving about under the control of the R^i's but never being acted upon by the T^j's. Think of it as rocking back and forth on the bottom and sides of the oval track, but never being run across or landing in the turntable at the top. Also, remember that it ends up back where it started.

Now take a signed count of how many positions disk r moves under this whole process. If you think about it, you will see that this means that we are adding up all the exponents i from the R^i's. What will the sum be? It has to be zero, since disk r ends up where it started. This means that if we separately added up the positive exponents, and the negative exponents, the absolute values of these sums would be the same.

Putting this numerical result into physical terms, if we break the whole process down into one-position rotational components, some clockwise and the rest counterclockwise, *then there have to be equal numbers of each.* A counterclockwise one must offset each clockwise single position rotation. Equivalently if we add up the absolute values of the exponents, we get an even number.

Consequently, even though R and R^{-1}, being n-cycles with n being even, are odd permutations, we are combining an even number of them together, making the combination even.

At the same time, our "Do Arrows" turntable permutation T is also even. This means that $\gamma = R^{i_1}T^{j_1}R^{i_2}T^{j_2} \cdots R^{i_k}T^{j_k}$ is an even permutation. That contradicts our assumption that γ is odd. To resolve this contradiction, we have to abandon our thought that there could be a disk r that never passed through the turntable. The claim (11.4) is now established.

If you think about the consequences of (11.4), it implies that the sort of procedure we used to move through the turntable all 16 disks numbered 5 through 20 in the $n = 20$ case is unavoidable if you want to produce a transposition.

We took them onto the turntable one at a time, and got the desired outcome. One could try to cut down on the number of steps by running them onto the turntable several at a time. However if you do this, you reverse the order of the ones in the group you are processing when you spin the turntable, and this reversal will have to be corrected at some later time. This hardly looks promising, but let's not rule it out entirely as an option for improving on our results.

11.2 The Second Oval Track Puzzle

Once you have figured out how to solve one of the puzzles, the concepts you learn makes solving the next one considerably easier. We will see this as we go on to consider Oval Track Puzzle 2, which has as its "Do Arrows" permutation $T = (4,3,2,1)$.

In Chapter 1, we discussed the opening game for this puzzle. The (4, 3, 2, 1) "Do Arrows" which cycles four consecutive disks at a time is an easy one for reordering scrambled disks. It also lends itself to efficiency-promoting heuristics. By now you should have worked with this puzzle enough so that you can get all but four disks into place without any need for formal mathematical thinking. Let's take a "top down" approach, starting with the highest numbers and working downward to put them in order. If we do so, then the end game always involves disks 1 through 4.

Let us assume that we are down to the last four disks, numbered 1 through 4 that need to be addressed. To guide our thinking, let us consider a specific end game example, the one shown in Figure 11.4(a). This one presents us with a 3-cycle, (1, 4, 2), that we need to solve. It might be that our end game scramble has disk 4 in spot 4, but if not, given the nature of the "Do Arrows," we can apply it or its inverse at most twice to cycle the disks so that disk 4 is in spot 4.

We have very easily reduced our end game problem down to a permutation on the set {1, 2, 3}. If we were not so lucky as to have gotten these last three disks into place along with disk 4, then the remaining scramble has to be either a 3-cycle or a transposition. With our example, the reduced permutation is the 3-cycle (1, 2, 3), as shown in Figure 11.4(b).

The final steps in the solution process can be handled using a result that was included in the previous chapter. Recall from (10.7) that

(11.5) For Oval Track Puzzle 2 with $T = (4, 3, 2, 1)$ and with at least 5 active disks,
$$[R^{-1}, T]\,[T^{-1}, R^{-1}]T = (R^{-1}TRT^{-1})\,(T^{-1}R^{-1}TR)T = (2, 3).$$

That is, we have discovered a nine-step transposition for this puzzle. We can easily put the final three disks in order using this transposition one or more times. You will find this permutation programmed into button B when you first start your Oval Track Puzzle software. If it is not still there, reprogram it into button B.

Various orders of doing things are possible, but let's work first to get disk 3 into spot 3 if it is not already there. If it is not in spot 3, then it may be in spot 2. If so, just perform the transposition (2, 3) by pressing button B and you will have put disk 3 into spot 3. If disk 3 is not in spot 3 or 2, then it must be in spot 1 as shown in Figure

11.4(b). In this case, just shift the disks clockwise one position using R. This puts disk 3 into spot 2. If you now perform the transposition (2, 3) (button B), you will move disk 3 into spot 3. However, we need to undo the changes that resulted from clockwise shift by doing a counterclockwise shift R^{-1}. When we do this, we shift disk 3 back to spot 2. The result of doing this with our example is shown in Figure 11.4(c).

We have not gotten disk 3 into its right place yet, but it is now closer, being in spot 2 rather than in spot 1. One more performance of (2, 3) will put it into spot 3 where we want it. At this stage, we have reduced the problem down to one of placing only disks 1 and 2. Half the time they will fall into place when we place disk 3. If they do not, we need to perform the transposition (1, 2). This transposition is an easy conjugate of the one we have discovered. Clearly, $(1, 2) = R(2, 3)R^{-1}$. That is, shift everything one position clockwise so that disks 2 and 1 move into spots 2 and 3, perform the (2, 3) transposition, and then shift everything one position counterclockwise.

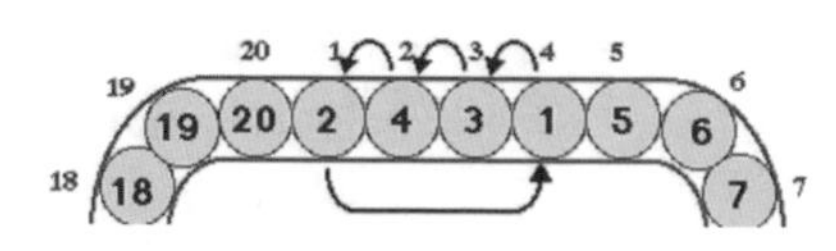

(a) One possible end game permutation, this time a 3-cycle.

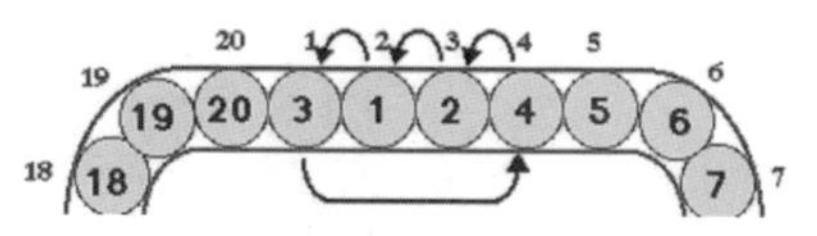

(b) First, cycle disk 4 into spot 4 using "Do Arrows."

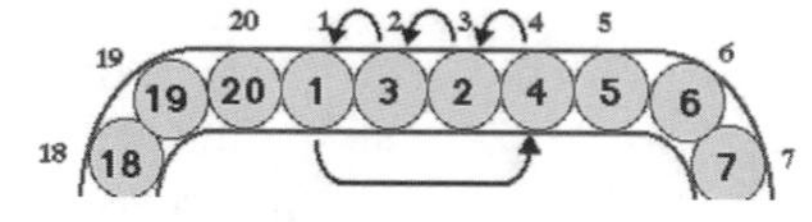

(c) Next, swap adjacent disks with conjugates of (2, 3) to move disk 3 clockwise into its spot.

Figure 11.4. A typical end game for Oval Track Puzzle 2 with the "Do Arrows" permutation (4, 3, 2, 1).

The end game strategy just described is not designed to be the most efficient. As we did with Oval Track Puzzle 1, we have opted for using a limited set of end game tools, in this case just one, and using conjugates of it to get the variation we need. This has the advantage of minimizing the need for remembering many specialized procedures.

The strategy we described used the simplest possible conjugates of (2, 3), namely ones that retain the feature of switching an adjacent pair of disks. In our example, when disk 3 was in spot 1 (Figure 11.4(b)), we first did an adjacent swap between spots 1 and 2 to move it to one position clockwise (Figure 11.4(c)), and then did a second adjacent swap to get it to spot 3. We could have used a more involved setup procedure for our conjugate so that we swapped between spots 1 and 3 the first time. You can experiment with modifications like this to find out what methods you prefer and what tricks give improved efficiency.

One could design a strategy for solving the entire puzzle that uses nothing but adjacent swaps. Such a strategy would be far from efficient, but it would be of interest because it gives a representation of a simple sorting algorithm called a "bubble sort" that students of computer science learn. (See Exercise 11.7.)

A final comment contrasting what we have seen with versions 1 and 2 of the Oval Track Puzzle is in order. It is highly significant that the "Do Arrows" permutation of version 2, a 4-cycle, is odd in contrast to that of version 1 which is even. This makes it possible for us to get a "local" procedure for performing a single transposition, which is also odd. When the "Do Arrows" is even, as with version 1, it becomes necessary to find some sort of "around the track" method to be able to do a transposition, as was made so clear by (11.4).

11.3 The Third Oval Track Puzzle

The third version of the Oval Track puzzle is very similar to the second. The only difference is that the third has as its "Do Arrows" a 3-cycle, (3, 2, 1), on a set of adjacent disks rather than a 4-cycle, (4, 3, 2, 1), on adjacent disks.

As before, we will assume that you have spent time with the opening game for this puzzle and are able to get all but the last three disks into place using common sense and heuristic methods. (Likely your methods for version 3 will be very much like those you use for version 2, given the similarity of their "Do Arrows" permutations.) We will work "top down" again.

There is a very limited end game for this puzzle. After you have gotten all the disks into place from the highest number down to 4, and you only need to get the three disks numbered 1, 2, and 3 into place, you can get disk 3 into place using the "Do Arrows." This will leave you with only having to address disks 1 and 2. Either they will have gone into their right places when you put disk 3 into place, or they will be in each other's place. In this latter case, what we need to be able to do is to perform the transposition (1, 2).

Doing this transposition, since it is odd, may or may not be possible. As with version 1, we have an even "Do Arrows" permutation T. If the number of active disks is odd, then the rotation permutation R is also even, so we can only get even permutations by performing the available moves on the puzzle. If the scramble started out odd, then we will be left with an irresolvable residual transposition.

On the other hand, if the number of active disks is even, then we do have odd permutations available to us because R is odd. However, as was the case with version 1, to get a transposition will require an "around the track" method. The reason for this is that claim (11.4) will apply to this puzzle as it did to version 1 because the "Do Arrows" for version 3 is even like the "Do Arrows" turntable for version 1. You could go through the proof of (11.4) and replace all references to the "Do Arrows" turntable by references to the "Do Arrows" permutation (3, 2, 1) and the proof remains valid.

Knowing that you will need to use around the track methods, why don't you try to find a way to get the transposition (1, 2) on your own before reading farther. If you try, you ought to find this without too much difficulty.

Did you find it? Was it this one?

(11.6) On Oval Track 3 with $T = (3, 2, 1)$ and n active disks, n even and $n \geq 4$, $(TR^{-2})^{(n/2)-1}(TR^{-1}) = (1, 2)$.

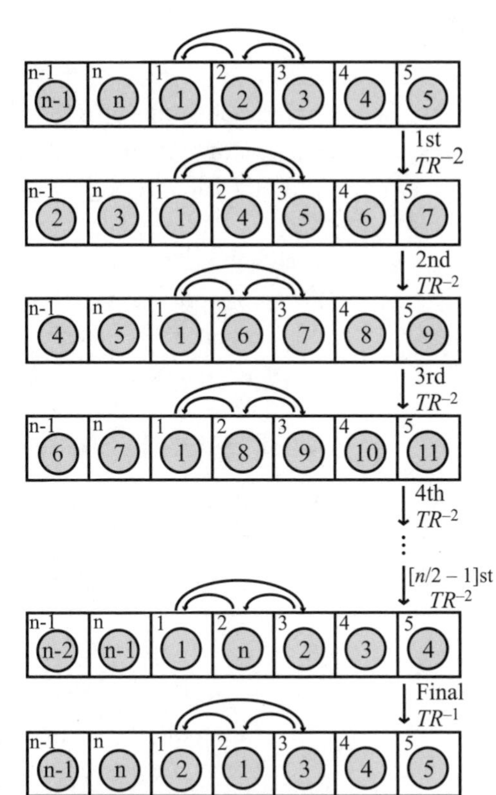

Figure 11.5. The effect of repeating TR^{-2} when n is even and $T = (3, 2, 1)$.

Figure 11.5 illustrates why this works, and we shall talk our way through the process now. Each application of TR^{-2} shifts disk 1 to the right of the next two adjacent disks in a clockwise direction, and then rotates everything counterclockwise two positions, which brings disk 1 back to spot 1. All the other disks that are not in the "Do Arrows" spots numbered 1, 2, and 3 are shifted two positions counterclockwise. The two disks that started out in spots 2 and 3 are shifted three positions counterclockwise. They get a one-position counterclockwise kick from T and then a two-position kick from R^{-2}.

The ith time that TR^{-2} is done, the numbers of the disks that get moved to the other side of disk 1 are $2i$ and $2i + 1$. This is the case as long as $2i + 1 < n$ since the highest number we are using is n. We cannot have $2i + 1 = n$ since n is even.

When $i = (n/2) - 1$, the ith application of TR^{-2} moves disks with numbers $n - 2$ and $n - 1$ from spots 2 and 3 over to the other side of disk 1 and into spots $n - 1$ and n. This application also has at last brought disk n into box 2. Since this is the highest numbered disk, the one that is to its right is disk 2, the one that was put there by the first application of TR^{-2}.

We now have the situation depicted in the second to the last row of Figure 11.5. The next application of T jumps disk 1 to the right of disks n and 2. This puts disk 1 into spot 3 and to the right of disk 2, which is in spot 2. The final application of R^{-1} shifts everything over one more position, finally returning each disk to its original spot, except for disks 1 and 2 which have traded places.

Let's think through why all of the disks other than numbers 1 and 2 really do return to their starting positions. They are all moved counterclockwise two positions for each of the $(n/2) - 1$ applications of R^{-2}. In addition, they each get a one-position counterclockwise kick by the application of T the one time that they are in one of the "Do Arrows" spots. This accounts for a shift of a total of $2[(n/2) - 1] + 1$, or $n - 1$ positions counterclockwise. The final application of R^{-1} adds one to this count, making the total number of positions shifted work out to n. Since there are n active spots on the oval, shifting n positions brings the disk back to where it started.

11.4 General Strategy for the Other Versions

We will not be looking at any more of the versions of Oval Track Puzzles in this chapter. We want to save for you some of the fun of discovering how to solve them. We will, however, be giving some solution ideas for a couple more of them in the next chapters to illustrate a few additional principles.

Another reason to not go into detail about all of the puzzles is that after having looked at how to solve three of the variations of the Oval Track Puzzles, you perhaps have picked up some insights that will help you solve the other ones. To summarize what we have learned, let's set forth a step-by-step description of the method that we have used.

1. Scramble the puzzle. Then put most of the disks into the correct order using a common sense opening game strategy and any heuristics you have developed to improve efficiency. Get all pieces in order this way until what remains out of place is a set of disks of the size of the set of disks spanned by the "Do Arrows" permutation.

2. Make sure in advance that you have identified a specific 3-cycle that works for the version of the puzzle you are using. Use conjugates of this specific 3-cycle to perform other 3-cycles that move you closer to completion of the puzzle. If the remaining end game scramble is even, this approach will be all you need.

3. If the number of active disks is odd and the "Do Arrows" permutation is even, then you can solve the puzzle completely only if the permutation was even. If the permutation was odd, solve the puzzle until only a transposition remains and then quit. You'll not be able to do better than this.
4. Assuming that odd permutations are possible as a result of having an even number of disks, or the "Do Arrows" permutation being odd, be sure in advance that you have identified a specific transposition that you can perform. Use conjugates of this specific transposition to perform other transpositions as needed to finish the solution.

This strategy outline assumes that you have done some research on the puzzle before you get serious about solving a particular end game scramble. Steps 2 and 4 presuppose that you have found some end game tools, that is, some convenient 3-cycles or transpositions or both. Starting with the puzzle initialized with the identity permutation is the best way to explore for these.

As you do this exploration, keep in mind that commutators are going to be good candidates for useful end game tools. As you explore commutators, you will find yourself seeing useful variations without having to think much about it.

Another thing to remember is to pay attention to whether the "Do Arrows" permutation is odd or even. If it is odd, then you ought to be able to find a "local" transposition. By this we mean that the disks that are far away from the "Do Arrows" area ought to just rock back and forth during the procedure rather than travel around the track. The procedure you find ought to work the same for all numbers of active disks, both odd and even, as long as the active number is large enough.

On the other hand, if the "Do Arrows" permutation is even, you know from the start that the puzzle will be at best only half solvable if the number of active disks is odd. You also know in this case that it is a total waste of time to try to find a transposition procedure with an odd number of disks, and if the number of disks is even, you will need to search for some sort of "around the track" procedure as you look for a transposition.

One final comment is that sometimes the theory provides a basis for knowing what is and is not possible for these puzzles based upon only a small portion of the information that you need to solve them.

For example, suppose that after a few minutes of experimentation, you find a way of performing a 3-cycle (a, b, c) on Oval Track Puzzle version j. Let us say that you discover that this same sequence of steps produces this 3-cycle whenever the number of active disks is 6 or more. Let us suppose further that it is clear that there is sufficient maneuverability to the puzzle that any three disks whatever can be put into spots a, b, and c.

What does this knowledge buy for you? Quite a bit. It tells you that whenever you have an even number of active disks, and when that number is at least 6, the puzzle will be completely solvable, and when there are an odd number of disks, the puzzle can be solved from scrambles that are even permutations but never from ones that are odd permutations. Even though you know this much, in the early stages of exploration, you probably do not understand the puzzle well enough to solve it with dispatch, but you know with certainty what can and cannot be done.

The reason is that if you can place any three disks into the spots acted upon by the 3-cycle you have discovered, then you can obtain all the 3-cycles as conjugates of your discovered 3-cycle. Since you can get all the 3-cycles, you can combine them to get all the even permutations. If the two permutations R and T for the version you are using are both even, then this is all that you can get.

However, if at least one of them is odd, then what you have discovered is enough to tell you that the puzzle is completely solvable. Let β denote the one that is odd. Recall how we proved back in Chapter 8 that the sets A_n and O_n of even and odd permutations were the same size? We showed this by pairing elements in these sets by following each even permutation with a transposition. We get the same result if we pair each even permutation α with our performable odd permutation β. Since we can perform any even permutation α and we can perform the odd permutation β on the puzzles, then we can perform $\alpha\beta$. The set of all such $\alpha\beta$'s is the same size as the set A_n of even permutations, so it must be all odd permutations.

We now know that we can start with the identity permutation and by doing moves that the puzzle allows, get to any permutation whatever. This means that we can solve any permutation. A sequence of moves that will solve the permutation is just the sequence of the inverses of the moves that get you to that permutation performed in the reverse order.

It is important after you have found a 3-cycle that you do not simply jump to the conclusion that you will be able to get all others as conjugates of this one. Check this out to be sure that it is true. There are several examples of puzzles that will surprise you in this regard.

The main insight is that after some purposeful experimentation with one of the puzzles, you have the possibility of knowing the whole story on its degree of solvability. You still have plenty of work to do to figure out just how to solve the puzzle. One of the main challenges will be to turn the knowledge that odd permutations are possible to getting a particular transposition. But at least you know that what you are going after is possible, which can motivate putting in the effort.

Exercises

11.1 Powers of cycles. It is frequently handy when seeking end game tools for the Oval Track Puzzles to be aware of certain facts about the positive integral powers of cycles. Prove these facts. (a) Suppose that $\alpha = (a, b, c)$ is a 3-cycle in S_n. Then $\alpha^2 = (c, b, a)$ is also a 3-cycle, and $\alpha^2 = \alpha^{-1}$. More generally, if β is a k-cycle, then β^{k-1} is also a k-cycle, and $\beta^{k-1} = \beta^{-1}$. [Hint: Use Exercise 7.5.] (b) If β is a k-cycle and k is a prime number, then β^i is also a k-cycle if $1 \leq i \leq k-1$. (c) If β is a k-cycle and $k = rs$ where r and s are both greater than 1, then β^r is made up of r s-cycles.

11.2 Practice with some Oval Track 1 end games. Figure 11.6 shows four different end games for the Oval Track Puzzle version 1. (a) Write out each in cycle notation. (b) Plan a strategy for solving the end game, assuming that there are 20 active disks. (c) Using the "Write your own" feature of Puzzle Resources, load each of these end games into your puzzle and implement your strategy to solve the puzzle. (Note: There are solutions for each of these end games included in the Locally Resident Macros collection of Puzzle Resources for Oval Track 1. Load and run these (slowly) to see a solution. Can you see the strategy behind the solution, in particular the use of conjugacy and a basic tool 3-cycle or transposition?)

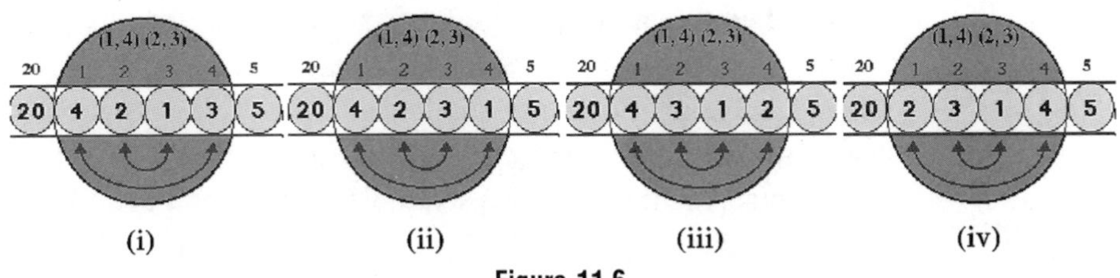

Figure 11.6

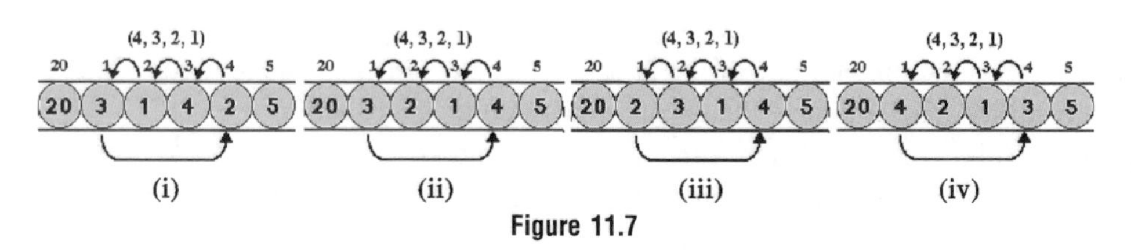

Figure 11.7

11.3 Considering end games with differing numbers of active disks. Suppose the end games shown in Figure 11.6 came up when there are different numbers of active disks. Decide which are solvable and which are not and under what conditions.

11.4 Practice with some Oval Track 2 end games. Figure 11.7 shows four end game scrambles on the Oval Track Puzzle version 2. Plan a strategy, load each of these into your puzzle, and carry out a solution. [Hint. The procedure given in (11.5) which produces the transposition (2, 3) on this version is a sufficient end game tool. It is the first entry in the macro collection in Puzzle Resources for this version. Load it from there, or program it into a button yourself.]

11.5 Major consequences from a single observation. In Exercise 10.1, we considered the commutator $[R^{-3}, T]$ on the Oval Track Puzzle 5, which has its "Do Arrows" $T = (1, 2)\,(3, 4)$. (a) Show that, if n, the number of active disks, is at least 7, then $[R^{-3}, T]^2 = (5, 4, 3)$. That is, by squaring this commutator, one gets a 3-cycle. (b) Give the argument that since this 3-cycle is available for this puzzle, then when the number of active disks is even, the puzzle is totally solvable, and when the number of active disks is odd, then only the even permutations can be solved.

11.6 When unsolvable scrambles can be reduced to a single transposition. Suppose that you are working with a version of the Oval Track Puzzle that has an even "Do Arrows" permutation, and that you have found a way to do a 3-cycle on the puzzle and have verified that you can get all 3-cycles as conjugates of this tool 3-cycle. By generalizing from Exercise 11.5, this tells you that if the number of active disks is odd and sufficiently large, then you can solve all of the even scrambles but none of the odd scrambles. Suppose now that you have scrambled the disks and the permutation α that you got is odd. Prove that you can solve the puzzle to the point of a single transposition, and that you can pick any transposition you want. [Hint: Use Exercise 8.8.]

11.7 Solving Oval Track Puzzles and sorting. When you solve a scramble on the Oval Track Puzzle, you are sorting an unordered list of numbers. Thus, we have the potential of applying things we may know about sorting to solving puzzles. To illustrate, consider the "bubble sort" algorithm for sorting an unordered list with n entries $a_1, a_2, \ldots, a_n$: (i) Set a counter i to 1. (ii) For a second counter j running from 1 to $n-i$, inspect entries a_j and a_{j+1} in the list, and switch them if $a_j > a_{j+1}$. (iii) Increase i by 1 until it reaches $n-1$ and repeat step (ii). (a) Show why this algorithm sorts the list to increasing order. (b) Show that if you can find a way to do a transposition of two adjacent disks, then you can implement a bubble sort as a strategy for solving the puzzle. (c) Do you think that this bubble sort strategy is efficient or not? Why?

Transferring Knowledge Between Puzzles

The last chapter ended with a description of a general strategy for solving any of the Oval Track Puzzles. We showed how the theory provides a basis for analyzing and solving all of the puzzles with a common approach centered on the notations of commutators and conjugates. In this chapter we will look at another more direct connection between some of the puzzles. When two of them are connected in the way we will consider, then it will be possible to transfer your knowledge about what is possible for one of them to knowledge of what is possible for the other.

To illustrate, recall that the Oval Track Puzzle 3, which has as its "Do Arrows" permutation $T = (3, 2, 1)$, is totally solvable if the number of active disks is even, and it is solvable for all even permutations and none of the odd permutations if the number of active disks is odd. This information can give us instant insight into Oval Track Puzzle 4, which has $T = (5, 4, 3, 2, 1)$ as its "Do Arrows."

We observe that:

(12.1) On the Oval Track Puzzle 4 with $T = (5, 4, 3, 2, 1)$ and any number n of active disks with $n \geq 7$,

$$R^{-2}T^{-2}R^{2}T^{-1}R^{-2}T^{-2}R^{2}T^{-1} = (3, 2, 1).$$

Try it out on your copy of Oval Track to confirm that it works. Figure 12.1 shows the result of performing these steps. We have found a way to perform the "Do Arrows" permutation of Oval Track 3 on Oval Track 4.

Suppose we program this procedure into button A of version 4. It now does the same thing as the "Do Arrows" button of ver-

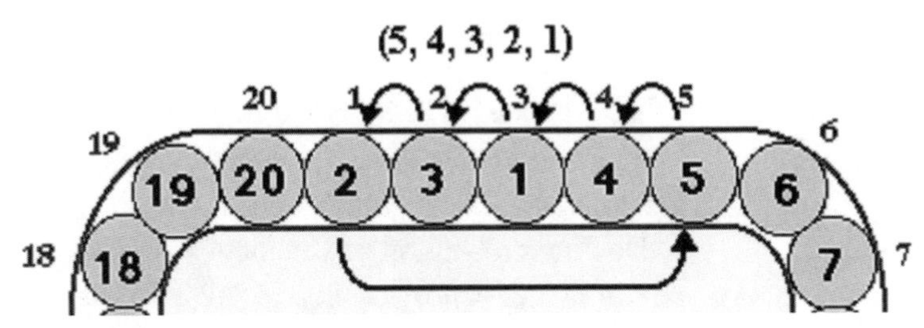

Figure 12.1. When there are 7 or more active disks, the "Do Arrows" permutation (3, 2, 1) of version 3 can be performed on version 4, which has (5, 4, 3, 2, 1) as its "Do Arrows".

sion 3, and button A^{-1} does what the "Back Arrows" button of version 3 does. Of course our button A of version 4 takes ten steps to do what the "Do Arrows" of version 3 does in one step.

We can see now that it is possible to do anything that we did on version 3 of the puzzle on version 4. Just replace a press of button A on version 4 for every press of the "Do Arrows" button on version 3, and replace every press of "Back Arrows" by a press of A^{-1}. The counter will advance more rapidly, and it will take longer, but the final result will be the same.

The consequence of this observation is that since version 3 is totally solvable if the number of active disks is even and greater than 7, then version 4 is totally solvable under these conditions. Note that we know version 3 to be totally solvable if there are four or six active disks as well, but this information does not transfer over to version 4. For one thing, having four active disks on version 4 is not allowed since the "Do Arrows" permutation for this puzzle requires at least five active disks. It is possible to have six active disks with version 4, but the method we found for replicating the "Do Arrows" of version 3 on version 4 requires seven or more active disks. Thus, we will have to investigate this case as a separate one if we want to transfer knowledge for it.

Now let's think about how our knowledge transfers if the number of active disks is odd. We know that on version 3, we can solve all of the even permutations. Since we can solve them all on version 3, we can solve them all on version 4, assuming at least seven active disks.

We also know that on version 3 with an odd number of disks active, we cannot solve any of the odd permutations. Does this tell us that we cannot solve any odd permutations on version 4 when there are an odd number of disks (at least seven)? No, it does not. The inability to solve an odd permutation on version 3 simply means that we cannot solve that same odd permutation on version 4 if we only allow ourselves to do rotations and presses of the buttons A and A^{-1} which replicate the "Do Arrows" and "Back Arrows" of version 3. It is possible that the different "Do Arrows" button of version 4 would allow us to do something on version 4 that we could not do on version 3.

In fact, it is not possible to solve an odd permutation on version 4 with an odd number of disks active. The same obstacle that prevents solving odd permutations on version 3 also applies to version 4 since its "Do Arrows" permutation, being a 5-cycle, is even like the 3-cycle "Do Arrows" of version 3. Thus, we are again restricted to getting even permutations when the number of active disks is odd.

However this impossibility insight is not transferred but is derived separately based upon the characteristics of version 4. Impossibility insights do not transfer. To illustrate this, let us take a look at version 2 which has as its "Do Arrows" $T = (4, 3, 2, 1)$. It is possible to replicate the "Do Arrows" (3, 2, 1) on this puzzle whenever there are at least five active disks. In fact,

(12.2) On the Oval Track Puzzle 2 with $T = (4, 3, 2, 1)$ and any number n of active disks with $n \geq 5$,
$$R^{-1}T^2RT^{-1}R^{-1}T^{-1}R = (3, 2, 1).$$

Try it yourself to see that this is correct.

As we have observed, it is impossible to get a transposition on version 3 of the puzzle with (3, 2, 1) as the "Do Arrows" and an odd number of active disks. We have replicated this 3-cycle "Do Arrows" in version 2, which has a 4-cycle "Do Arrows." Getting a transposition in version 2 is possible. The impossibility of doing something in one version did not translate over when we replicated this version in another.

Our knowledge transfer chains can extend beyond a single link. To illustrate this, we are now going to take our newly acquired transfer-knowledge about version 4 which has the 5-cycle (5, 4, 3, 2, 1) as its "Do Arrows" permutation to version 15 which has the "twisted" 5-cycle (1, 3, 4, 2, 5) as its "Do Arrows."

This later version in the collection performs a 5-cycle on the same five spots as the "Do Arrows" of version 4. However, rather than moving them in an orderly way like version 4, (starting in spot 5, down one, down one, down one, down one, up four), it twists around a lot. Starting in spot 5, it goes down 4, up 2, up 1, down 2, up 3. Chaotic!

Even though this version looks like it ought to be much more complicated than version 4 with its orderly 5-cycle, it really is not. In fact, we can untwist version 15's twisted 5-cycle to replicate version 4's orderly 5-cycle by means of conjugation. It will just take a little planning.

To get the orderly 5-cycle as a conjugate of the disorderly twisted one, let's just preset the disks numbered 1 through 5 in a way that produces the orderly exchange in the hard-wired disorderly way. We can leave disk 5 where it is and then one by one use the twisted "Do Arrows" to arrange the other four to its left in the order that will work for us. Figure 12.2 shows how this will work.

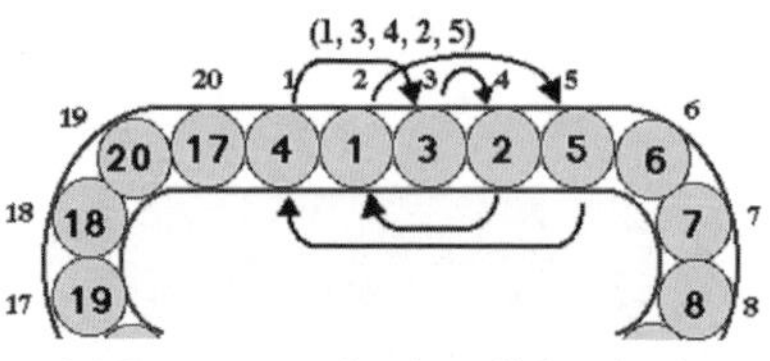

(a) Do a setup that has disks chasing disks in the desired orderly way following the disorderly arrows.

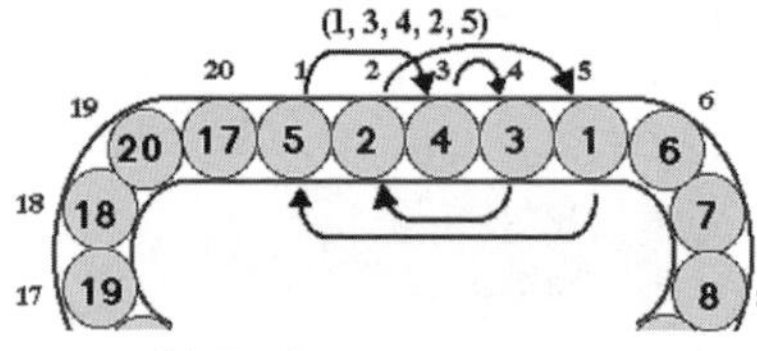

(b) Perform a "Do Arrows."

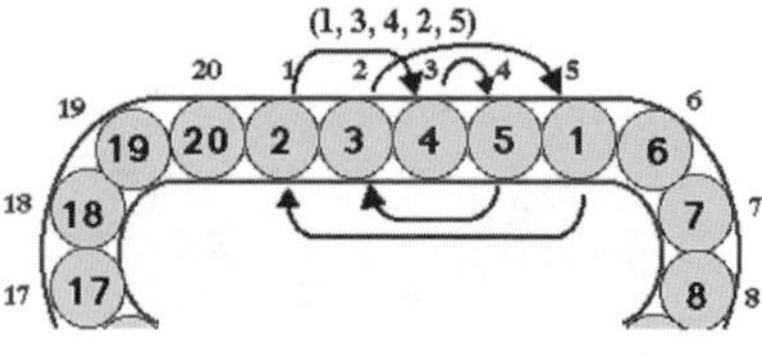

(c) Reverse the setup procedure to get an orderly 5-cycle (5, 4, 3, 2, 1).

Figure 12.2. With n disks, $n \geq 9$, the "Do Arrows," (5, 4, 3, 2, 1), of OT 4 can be performed on OT 15 which has (1, 3, 4, 2, 5) as its "Do Arrows".

Since we have set the location for disk 5, we need to think about how to order the disks with disk 5. Let's see. For our final result, we want disk 5 to go to spot 4. In doing so, it is "chasing disk 4," that is, moving to where disk 4 was. Our twisted 5-cycle has the disk in spot 5 going to spot 1. Thus, in order to have "disk 5 chases disk 4", we will want to set things up so that disk 4 is in spot 1.

Continuing with this line of reasoning, we want to have "disk 4 chasing disk 3." If we have disk 4 in spot 1, since the twisted 5-cycle moves the disk in spot 1 to spot 3, then to achieve "4 chases 3," we need to put disk 3 into spot 3. Next, we want "disk 3 chases disk 2," so since the twisted 5-cycle moves the disk in spot 3 to spot 4, we need to have disk 2 in spot 4. Finally, since we want "disk 2 chases disk 1," and the twisted 5-cycle maps from spot 4 to spot 2, we want disk 1 in spot 2.

This completes the placement of all the disks. Our placement of the final one, disk 1 into spot 2 works out just the way it should to complete the orderly 5-cycle. We want "disk 1 chases disk 5." Disk 5 will be in spot 5 when we run the twisted 5-cycle, and it maps the disk in spot 2 to spot 5, so we will indeed get the outcome we are after.

Let us now organize these placement directives from the top down in terms of the destination spots: disk 5 into spot 5, disk 2 into spot 4, disk 3 into spot 3, disk 1 into spot 2, and disk 4 into spot 1. It is a straightforward process to get such placements using a sequence of applications of R followed by the right number of twisted 5-cycle "Do Arrows" or "Back Arrows." The sequence that does the job is

(12.3) On the Oval Track Puzzle 15 with $T = (1, 3, 4, 2, 5)$ and n active disks with $n \geq 9$, $A = R\,T^{-2}R\,T\,R\,T^{-2}R\,T^{2}R^{-4} = (1, 2, 4)\,(n, n-1, n-3)$.

Check it out for yourself on your copy of the software. The result of running this sequence is shown in Figure 12.2(a). You'll note that we have named this sequence of moves with the letter A. When we run this sequence of steps, we want to program it into button A, so that performing its inverse later will be simpler.

Now follow this set-up permutation with one press of the twisted 5-cycle "Do Arrows" button. This results in moving disks 1 through 5 following the arrows. Note that disk 5 is chasing disk 4, disk 4 is chasing disk 3, etc. The result is shown in Figure 12.2(b). Finally, press the button A^{-1}. This undoes what the set-up procedure did, except for the 5-cycle in the middle. The results are shown in Figure 12.2(c). Thanks to the correctness of our planning of the set-up procedure, we have produced the orderly 5-cycle (5, 4, 3, 2, 1).

To summarize this, we see that

(12.4) On the Oval Track Puzzle 15 with $T = (1, 3, 4, 2, 5)$ and n active disks with $n \geq 9$, and with $A = RT^{-2}RTRT^{-2}RT^{2}R^{-4} = (1, 2, 4)\,(n, n-1, n-3)$, $ATA^{-1} = (5, 4, 3, 2, 1)$.

The consequence of (12.4) is that we can replicate version 4 ($T = (5, 4, 3, 2, 1)$) of the puzzle in version 15 ($T = (1, 3, 4, 2, 5)$), provided that the number of active disks is at least nine. Thus, everything that is possible in version 4 is also possible in version 15. Our knowledge as of now regarding what is possible for version 4 comes from our knowledge of what is possible in version 3 ($T = (3, 2, 1)$). Thus our transmission of knowledge chain lengthens.

Note that the methods we have just used do not give us knowledge transmission from version 4 to version 15 for cases where the number of active disks is between 5 and 8. It might be possible to get such transmission by replicating the "Do Arrows" of version 4 in version 15 in a different way from what we have set forth in (12.4).

There are many other opportunities for transferring knowledge of what is possible between the different versions of the Oval Track Puzzles. If you work with the puzzles enough, likely you will find more of them. Some you will come upon by serendipity as you explore for end game tools in new puzzles. Others you might find by consciously looking for them.

To have it for reference later, let's set forth this general principle regarding knowledge transfer.

(12.5) Suppose that a way is found on version i of the Oval Track Puzzle to replicate the "Do Arrows" permutation of version j of the Puzzle. Then for all numbers of active disks for which this replication method is valid, anything that is possible on version j is also possible on version i. However, there may be things that are possible on version i that are not possible on version j.

Exercises

12.1 A transfer from version 3 to version 5. On the Oval Track Puzzle version 5, which has (1, 2) (3, 4) as its "Do Arrows" permutation, check out what the squared commutator $[R^{-3}, T]^2$ does. Recall that this same procedure is one of the fundamental end game tools for solving version 1, but, with a different "Do Arrows" for version 5, the results will be different. Compare the results with version 3 of the Oval Tracks Puzzle. What conclusions can you draw?

12.2 A transfer from version 3 to version 4. (a) Show that on version 3 of the Oval Track Puzzle, which has $T = (3, 2, 1)$, if the number of active disks is at least 5, then $TR^{-2}TR^2 = (5, 4, 3, 2, 1)$. That is, we can produce the "Do Arrows" permutation of version 4 on version 3. (b) Explain what the claim in (a) means when taken together with the fact set forth in (12.1) which says that we can also produce the "Do Arrows" of version 3 on version 4.

12.3 If you can do my "Do Arrows" and I can do yours? Suppose you have explored versions i and j of the Oval Track Puzzles and have found a way to perform the "Do Arrows" of version i on version j as well as a way to perform the "Do Arrows" of version j on version i. What are the consequences of this?

12.4 Do you do my "Do Arrows"? See if you can find ways to produce the "Do Arrows" of the first version of Oval Track on the second version in each of these cases.

(a) The "Do Arrows" of version 1, namely (1, 4) (2, 3), on version 5.
(b) The "Do Arrows of version 5, namely (1, 2) (3, 4), on version 1.
(c) The "Do Arrows of version 2, namely (4, 3, 2, 1), on version 11.

12.5 Straightening twists and closing gaps. We showed in (12.4) how the orderly 5-cycle (5, 4, 3, 2, 1) could be produced on version 15 which has the twisted 5-cycle (1, 3, 4, 2, 5) as its "Do Arrows." Show that the twisted "Do Arrows" can be straightened out in these other cases, and that in addition, gaps can be removed.

(a) On version 14 with $T = (1, 6, 5, 3, 4, 2)$, produce the 6-cycle (6, 5, 4, 3, 2, 1).

(c) On version 17 with $T = (1, 6)\,(2, 5)\,(3, 4)$, produce the 3-cycle $(3, 2, 1)$.
(d) On version 19, build $T = (1, 4, 2, 5, 3)$ and then produce the 5-cycle $(5, 4, 3, 2, 1)$.

12.6 Some around the track transfers. (a) Show that on version 7 of Oval Track which has $T = (5, 3, 1)$, with 19 active disks, $(R^{-2}T)(R^{-4}T)^3(R^{-2}T^{-1})(R^4T^{-1})^4 = (3, 2, 1)$. Thus, the methods of solving version 2 which has $T = (3, 2, 1)$ can be transferred to solving version 7. Note that this procedure is an "around the track" sort in that it takes disk 3 by all the others one way and then returns disk 2 the other way. This is unavoidable, as we will see in the next chapter. (b) The procedure for getting (3, 2, 1) given in (a) is specifically for Oval Track 7 with 19 active disks, but it can be modified for other odd numbers of disks. Find a generalization of it for n active disks, where n is an odd number.

12.7 Find an around the track yourself. Oval Track 8 with $T = (1, 3)\,(2, 4)$ is another one that will require an "around the track" method to get the 3-cycle (3, 2, 1). (a) When n, the number of active disks, is 19, find a way of getting the 3-cycle (3, 2, 1) on this puzzle. (b) Generalize your method to give a formula for (3, 2, 1) for Oval Track 8 with an n arbitrary odd number.

12.8 Two-generator subgroups of S_n associated with the Oval Track Puzzles. Any of the Oval Track Puzzles, with n active disks, is related to the subgroup of S_n, the group of all permutations on the numbers 1 through n, which is generated by the rotational n-cycle, $R = (1, 2, 3, \ldots, n)$ and the "Do Arrows" permutation T. By the subgroup generated by R and T, we mean the set of all permutations in S_n that one can obtain by repeated compositions of these two generating permutations. Let us denote the "Do Arrows" of Oval Track Puzzle i by T_i, and denote the subgroup of S_n generated by R and T_i by $OT_i(n)$. The observations we have been making about transferring knowledge of one version to another reduces to the fact that if $T_j \in OT_i(n)$, then $OT_j(n) \subseteq OT_i(n)$. Prove this, and explain how it relates to the statement "anything that can be done on version j can also be done on version i."

What a Difference a Disk Makes!

Changing the Number of Disks, and Using *Maple* or *GAP*

13

We have learned by now that it often makes a difference for the solvability of some versions of the Oval Track Puzzle whether the number of active disks is odd or even. Typically, we have seen that the degree of difference is a factor of two. It is a matter of whether or not you can solve all possible permutations or just the half of the total that are even.

In this chapter, we will take a closer look at these issues and will see that sometimes the degree of difference you encounter when adding or subtracting one disk can be much greater. We will explore when this happens and why. In doing so, we will employ a new exploration tool, namely using one of several computer software systems that do permutation group calculations.

13.1 Subgroups Associated with Puzzles

Think about one of the puzzles, say Oval Track Puzzle 1. Suppose that we have it set with 20 active disks. How many different arrangements can we get on the puzzle starting from the identity permutation? Every arrangement we can possibly get is going to result from some sequence of applications of the rotation permutation R and of the "Do Arrows" permutation T. The number of permutations that we can get this way is of interest because it is precisely these permutations that we are able to solve.

The set of permutations that we can get with the puzzle form a subset of S_n called a *subgroup:* a subset of S_n that is nonempty and has the property that it contains $\alpha \circ \beta$ whenever it contains α and β and contains γ^{-1} whenever it contains γ. The study of subgroups is an important part of the theory of groups, so knowing that we are dealing with a certain subgroup of S_n when we are exploring one of the Oval Track Puzzles is useful.

The subgroup of S_n that is associated with one of the puzzles is the subgroup generated by the two basic manipulations, the rotation R and the "Do Arrows" T. For our specific example of the Oval Track Puzzle number 1 with 20 active disks, these generators of the subgroup are the 20-cycle $R = (1, 2, 3, \ldots, 20)$ and the turntable permutation $T = (1, 4)(2, 3)$. The subgroup consists of permutations that can be obtained by repeated compositions involving these two.

The number of permutations in this subgroup is of interest if we want to understand the puzzle. If a permutation α is in this subgroup, then you can solve the puzzle when α comes up as a scramble. If α is not in the subgroup, then you cannot solve the puzzle if α is the scramble. Thus knowing the size of this subgroup tells us how solvable this puzzle is. Comparing the number of permutations in the subgroup with $n!$, which is the size of S_n, tells us the relative solvability of the puzzle. Specifically, the fraction you get by dividing the size of the subgroup by $n!$ is the probability that a randomly generated scramble will be solvable.

We already know these numbers for the first three versions of the puzzle based upon our analysis in Chapter 11. The transfer of knowledge ideas in the last chapter give us lower bounds for the sizes of the subgroups associated with versions 4 and 15. However, there are many other versions yet to be studied. It would be helpful to have some computer assistance with the analysis.

Maple, a commercial, general-purpose computer algebra system, and *GAP*, a freeware program designed for group theory, provide good tools for the job. We will describe the easy steps that you must perform to get either of them to answer puzzle-related questions and will give several examples of *Maple* output. The information presented here will provide adequate instructions to allow you to use either *Maple* or *GAP* for your own explorations. If you do not have access to either, you will be limited to the insight that you get from the examples that we work out here, but you will be aware that computer analysis is available for this sort of problem.

13.2 Using *Maple* to Count Permutations

In this section, we will use *Maple* to count permutations. In the next section, we will discuss more briefly the very similar procedures you use with *GAP*.

Maple is one of several very sophisticated mathematical software packages that are available nowadays. These packages make it possible to bring the awesome speed and accuracy of the computer to mathematical problems that involve symbolic manipulation.

One feature that sets *Maple* apart from other systems, such as *Mathematica* and *Derive,* is that it has a group theory package among its tools. This package, among the other things that it does, can determine the number of elements in subgroups of permutation groups S_n. This is an exact fit with our need for information about our puzzles.

Using *Maple* for this specific purpose is quite easy. When you start the program, it opens a window called a "Worksheet" in *Maple* terminology. You type commands into this worksheet and *Maple* responds to them with its answers.

First, you must tell *Maple* to load its group theory software. (To minimize memory demands, *Maple* does not automatically load the software that performs its advanced processes unless you ask for them.) You request the group routines by typing in the simple command "with(group);" following the prompt symbol "[>." *Maple* uses the word "with" to request the loading of specialized packages of routines. The word "group" is the name for the group theory package of routines. The semicolon following the command is *Maple*'s end-of-command symbol and it must be there for every command you enter. After typing in this command, you press the "enter" key (Windows keyboard) or "return" key (Macintosh keyboard). The screen will show the following *Maple* response:

(13.1)

```
> with(group);
[DerivedS, LCS, NormalClosure, RandElement, Sylow, areconjugate, center,
    centralizer, core, cosets, cosrep, derived, groupmember, grouporder, inter, invperm,
    isabelian, isnormal, issubgroup, mulperms, normalizer, orbit, permrep, pres]
```

Maple produces a list of the package's commands, creates a new prompt for you, and waits for your next command.

Next, enter a description of the subgroup related to the puzzle that you want to analyze. *Maple* refers to subgroups of the group S_n with the word "permgroup." When you define one of these subgroups, you will need to tell *Maple* what the value of n is, and what the generators are for the subgroup. These generators are entered in *Maple*'s version of cycle notation. To illustrate, let us suppose that we want to define $G1$ to be the subgroup for the Oval Track Puzzle 1 with 7 active disks. We tell *Maple* this with the command

$$G1 := \text{permgroup}(7, \{[[1,2,3,4,5,6,7]], [[1,4],[2,3]]\});$$

The ":=" is *Maple*'s symbol for assignment of a value. This is telling *Maple* that the symbol "$G1$" is being assigned the value that comes in the definition that follows. As we said, "permgroup" tells *Maple* that $G1$ will be a subgroup of a permutation group S_n. *Maple*'s syntax calls for a description of the particular subgroup enclosed within a pair of parentheses following this word.

The first parameter, 7 here, tells *Maple* that $G1$ is a subgroup of S_7. The set of generators is enclosed in a pair of braces, "{ ... }." Each generator is specified in cycle notation. *Maple* wants permutations enclosed in a pair of brackets, "[...]," and each cycle of a permutation enclosed in its own pair of brackets with commas separating the cycles if there are more than one. Thus, for our example, the 7-cycle rotation R is denoted by [[1,2,3,4,5,6,7]]. The outer pair of brackets indicates that this is the entire permutation and the inner pair defines the 7-cycle. The turntable permutation T consists of a pair of disjoint cycles. These are denoted in *Maple*'s notation by [1,4] and [2,3]. The permutation consisting of these two transpositions together is thus [[1,4], [2,3]].

When you enter this command into *Maple*, assuming that you've typed it correctly, the result in your worksheet looks like this:

(13.2)
```
> G1 := permgroup(7, {[[1,2,3,4,5,6,7]], [[1,4], [2,3]]});
          G1 := permgroup(7, {[[1, 4], [2, 3]], [[1, 2, 3, 4, 5, 6, 7]]})
```

Maple responds by repeating the definition of the subgroup.

Once we have entered in a definition for the subgroup of interest, we are ready to ask *Maple* to calculate its size. *Maple*'s command for doing this is "grouporder." (Note that this is one of the names in the list that *Maple* produced in response to the "with(group);" command.) The syntax for the "grouporder" command is very simple. Here is the command that we need to enter:

$$n1 := \text{grouporder}(G1);$$

By enclosing the name of the subgroup whose size we want in parentheses, we give *Maple* all that it needs to know. The first part of the command, "$n1$:=", would not be necessary, but by including this, we are telling *Maple* that once it calculates the number of elements in $G1$, we want the name $n1$ assigned to this value. This will make it easier for us to use this number later.

When you enter this command into *Maple*, the result is

(13.3)

```
> n1 := grouporder(G1);
                                        n1 := 2520
```

Maple is telling us that there are 2,520 elements in the subgroup of S_7 associated with Oval Track Puzzle 1 with 7 active disks. This means that there are 2,520 of the 7! permutations of the puzzle that are in fact solvable.

We know what this number ought to be. Since R is a 7-cycle in this case and is even, and since $T = (1, 4)\,(2,3)$ is also even, we can only solve even permutations. Since we can get all 3-cycles on this puzzle, we can get all of the even permutations. Thus, the basic permutations R and T ought to generate a subgroup of size 7!/2.

If 7! is among the factorials that you keep in your memory, you will recognize that 2,520 is indeed 7!/2. In case you want *Maple* to confirm this for you, a handy way to ask it is to simply enter the command "$n1/7!$;". Given an algebraic expression like this, *Maple* responds by calculating its value, assuming that a numerical value has been assigned to the variable $n1$. We assigned a value to $n1$ in our last command (we introduced this notation so we could refer to this number now with the symbol $n1$). When you enter this command, *Maple* produces this:

(13.4)

```
> n1/7!;
```

$$\frac{1}{2}$$

Maple is telling us that 2,520/7! = 1/2, which is of course equivalent to 2,520 = 7!/2. Another interpretation of this fraction is that it is the probability that an arbitrary scrambling of this puzzle is solvable. If you have the computer scramble this puzzle, the probability is 1/2 that you will be able to solve it. It is equally likely that the scramble will be odd as that it will be even. If it is even, you can solve it, and if it is odd you cannot.

Having run *Maple* on a case that we can analyze independently gives us confidence that we are using and interpreting *Maple* correctly. Let's take this a few steps further, doing this same analysis for version 1 of the Oval Track Puzzle with 8, 9, and 10 active disks. Here is *Maple*'s worksheet output after we enter this series of commands.

(13.5)

```
> G2 := permgroup(8, {[[1,2,3,4,5,6,7,8]], [[1,4], [2,3]]});
            G2 := permgroup(8, {[[1, 4], [2, 3]], [[1, 2, 3, 4, 5, 6, 7, 8]]})
> n2 := grouporder(G2);
                              n2 := 40320
> n2/8!;
                                   1
> G3 := permgroup(9, {[[1,2,3,4,5,6,7,8,9]], [[1,4],
  [2,3]]});
            G3 := permgroup(9, {[[1, 4], [2, 3]], [[1, 2, 3, 4, 5, 6, 7, 8, 9]]})
> n3 := grouporder(G3);
                              n3 := 181440
> n3/9!;
                                  1/2
> G4 := permgroup(10, {[[1,2,3,4,5,6,7,8,9,10]], [[1,4],
  [2,3]]});
            G4 := permgroup(10, {[[1, 4], [2, 3]], [[1, 2, 3, 4, 5, 6, 7, 8, 9, 10]]})
> n4 := grouporder(G4);
                              n4 := 3628800
> n4/10!;
                                   1
```

This output shows that we defined $G2$, $G3$, and $G4$ to be the subgroups of S_8, S_9, and S_{10} associated with the puzzle with 8, 9, and 10 active disks respectively. The numbers $n2$, $n3$, and $n4$ are the sizes of these subgroups. The results are as we expect. When we divide the number of elements in the subgroup by the appropriate $n!$, we get 1/2 when n is odd and 1 when n is even. Getting $n2/8!$ and $n3/10!$ both equal to 1 tells us that the puzzle is totally solvable for these cases in which the number of active disks is even. Of course we knew that already.

This is a good place to interject a helpful hint about entering these commands into the *Maple* worksheet. You can save time by using "copy" and "paste" methods. The definitions of these permutation groups are long, so it is tedious to type them in afresh every time. To avoid this, you can copy the command from the previous time it was written and then paste it in at the next prompt. Then you can edit the command to reflect the details of the new puzzle subgroup. For example, the definition of $G4$ above was obtained by copying the definition for $G3$ and pasting it at the next prompt. Then the "3" of "$G3$" was changed to "4" the parameter "9" was changed to "10" and a ",10" was added to the first generator to extend it from a 9-cycle to a 10-cycle.

13.3 Using *GAP*

GAP is an acronym for Groups, Algorithms, and Programming. It is a computer program that is available for Unix, Windows, Macintosh, and DOS operating systems. The School of Mathematical and Computational Sciences of the University of Saint Andrews in Fife, Scotland currently holds the copyright for it. It is distributed freely for noncommercial use.

One can download *GAP* by going to the web page

`http://www-gap.dcs.st-andrews.ac.u./~gap/`

Starting from there, you will find links to pages for downloading versions for the various platforms, and to mirror sights elsewhere from which the software can be downloaded.

GAP, as the terms from which its name derives suggest, is a piece of software devoted to computations related to group theory. As such, it is more limited in its aims than *Maple*, but it does what we need for studying puzzles.

The syntax for using *GAP* to determine the sizes of subgroups of permutation groups S_n is very similar to what we saw with *Maple*. Here is a table of the *Maple* terms and notations that we described in the previous section and their *GAP* equivalents.

	Maple	***GAP***
Initialization	with(group);	none needed
Command prompt	[>	gap>
End of command	;	;
Defining a variable	:=	:=
Cycle notation example	[[1,4],[2,3]]	(1,4)(2,3)
Subgroup of S_n generated by a and b	permgroup(n, {a, b});	Group(a, b);
Number of elements in G	grouporder(G);	Size(G);
Factorial of an integer n	n!;	Factorial(n);

If you have *GAP* available to you, you should be able to use the information in this table to translate everything we have described using *Maple* into the *GAP* equivalents, and get similar results. To illustrate, a *GAP* session equivalent to the *Maple* input and output shown in (13.2), (13.3), and (13.4) looks like this:

(13.6)

```
gap> G1 := Group((1,2,3,4,5,6,7),(1,4)(2,3));
Group([(1,2,3,4,5,6,7),(1,4)(2,3)])
gap> n1 := Size(G1);
2520
gap> n1/Factorial(7);
1/2
```

Note that *GAP* uses the same system for cycle notation that we have been using in contrast to *Maple*'s variation. *GAP* is designed for doing group theory, so it can conform to standard notations of this subdiscipline, but *Maple*, being an all purpose symbolic algebra system, sometimes has to modify notations to avoid ambiguity. *GAP*'s response does not confirm that the value, 2520, calculated for the number of elements in $G1$, has been assigned to the variable $n1$, but the subsequent calculation confirms that this has happened. One way that *Maple* conforms to standard practice and *GAP* does not is in using an exclamation point for factorial notation. With *GAP*, we must enter the function name as a word.

Another way in which *GAP* differs from *Maple* is that its user interface is more basic. *GAP* and *Maple* are both basically command line oriented for input, but *Maple* offers options for more elegant output, and gives options for doing some things with menu commands. Copy and paste functionality is not a part of *GAP*'s basic installation, but the installation instructions refer the user to a web page from which one can get instructions for adding this functionality. Also, you will need to be a somewhat more experienced computer user to understand the instructions for installing *GAP*.

13.4 New Insights Via *Maple* or *GAP*

Now that we have learned the *Maple* and *GAP* basics needed to do the computations we want, and have confirmed that we, *Maple*, and *GAP* agree on some things that we already know about, let's look for information about cases that we do not yet fully understand. We will do the analysis with *Maple* and show *Maple* output, but all of these computations could be done with *GAP* as well.

Recall that for Oval Track Puzzle 1, given that our fundamental 3-cycle needs the wiggle room of at least 7 active disks to work, the cases of 4, 5, or 6 active disks need to be treated by different methods. Let's see what *Maple* has to say about these cases. The output is:

(13.7)
```
> G5 := permgroup(6, {[[1,2,3,4,5,6]], [[1,4], [2,3]]});
              G5 := permgroup(6, {[[1, 4], [2, 3]], [[1, 2, 3, 4, 5, 6]]})
> n5 := grouporder (G5);
                              n5 := 720
```

(13.7) *cont.*

```
> n5/6!;
                                  1
> G6 := permgroup(5, {[[1,2,3,4,5]], [[1,4], [2,3]]});
            G6 := permgroup(5, {[[1, 2, 3, 4, 5]], [[1, 4], [2, 3]]})
> n6 := grouporder(G6);
                               n6 := 10
> n6/5!;
                                  1
                                 ---
                                 12
> G7 := permgroup(4, {[[1,2,3,4]], [[1,4], [2,3]]});
            G7 := permgroup(4, {[[1, 4], [2, 3]], [[1, 2, 3, 4]]})
> n7 := grouporder(G7);
                               n7 := 8
> n7/4!;
                                  1
                                  -
                                  3
>
```


Hmmm …. The results for six active disks are as expected. The puzzle is totally solvable. Our tool transposition $(TR^{-1})^3 = (1, 3)$ works in this case, and there is enough wiggle room to get all other transpositions as conjugates of this one, so we can totally solve the puzzle in this case. However, when there are five active disks, *Maple* is telling us that we can solve only 10 of the possible 5!, or 120, scramblings. This is only one twelfth of the total, a very small proportion. The lack of wiggle room really restricts us here. We also get a new fraction in the case of four active disks. We can solve only 8 of the 4! or 24 scramblings. Can we explain in detail what is going on in these two cases? See Exercise 13.3 for the analysis.

Now let us turn to version 7 of the Oval Track Puzzles, which has as its "Do Arrows" permutation the 3-cycle (5, 3, 1). Let's ask *Maple* about the subgroup of S_{19} associated with this puzzle in the case that there are 19 active disks. Here is *Maple*'s output:

13

(13.8)

```
> R19 :=
  [[1,2,3,4,5,6,7,8,9,10,11,12,13,14,15,16,17,18,19]]:
          R19 := [[1, 2, 3, 4, 5, 6, 7, 8, 9, 10, 11, 12, 13, 14, 15, 16, 17, 18, 19]]
> T7 := [[5,3,1]]:
                         T7 := [[5, 3, 1]]
> G8 := permgroup(19, {R19, T7}):
G8 := permgroup(19,

  {[[5, 3, 1]], [[1, 2, 3, 4, 5, 6, 7, 8, 9, 10, 11, 12, 13, 14, 15, 16, 17, 18, 19]]})
> n8 := grouporder(G8):
                    n8 := 60822550204416000
> n8/19!:
                              1
                              -
                              2
> |
```

Note that we have done something different in the way we have given *Maple* the commands this time. We assigned recognizable variable names to the two generators: $R19$ for the 19-cycle rotation of this puzzle, and $T7$ for its "Do Arrows," including the "7" because this is the "Do Arrows" of version 7. By doing this, we can shorten the command for defining the group.

Maple confirms what we expect. The R and T permutations are both even so we know that we will only have even permutations in the puzzle's subgroup. Since we have a 3-cycle as the "Do Arrows," we would expect that we could get all other 3-cycles as conjugates of this one, and therefore we are not surprised that *Maple* tells us that we in fact get all of the even permutations.

Now let's have *Maple* give us the corresponding information for version 7 of the Oval Track Puzzles for the case of 20 active disks. The results are:

(13.9)

```
> R20 :=
  [[1,2,3,4,5,6,7,8,9,10,11,12,13,14,15,16,17,18,19,20]];
        R20 := [[1, 2, 3, 4, 5, 6, 7, 8, 9, 10, 11, 12, 13, 14, 15, 16, 17, 18, 19, 20]]
> G9 := permgroup(20, {R20, T7});
G9 := permgroup(20,

   {[[1, 2, 3, 4, 5, 6, 7, 8, 9, 10, 11, 12, 13, 14, 15, 16, 17, 18, 19, 20]], [[5, 3, 1]]})
> n9 := grouporder(G9);
                              n9 := 13168189440000
> n9/20!;
                                      1
                                    ------
                                    184756
>
```

Wow! That number for $n9/20!$ is a surprise. This says that when you scramble version 7 of the puzzle with 20 active disks, then there is only 1 chance in 184,756 that the puzzle will be solvable. Something very different is going on here. Recall that at the end of Chapter 11, we alluded to some surprises to come. This is one of them.

Investigating this matter further, we find that whenever n is odd and $n \geq 7$, then we get the same results that we got for $n = 19$, namely the size of the subgroup is half of $n!$. For even values of n, we get the ratios of the size of the subgroup to $n!$ as shown in this table.

(13.10)

Value of n	Subgroup Order	Ratio of subgroup size to $n!$
6	18	1/40
8	576	1/70
10	7,200	1/504
12	518,400	1/924
14	12,700,800	1/6864
16	1,625,702,400	1/12870
18	65,840,947,200	1/97240
20	13,168,189,440,000	1/184756

What can we make of this? The larger an even integer n becomes the less likely it becomes that we will be able to solve an arbitrary scramble of the puzzle. Meanwhile as we move through the odd numbers, the probability that an arbitrary scramble is solvable remains constant at one half.

13.5 Analysis of Version 7 with $T = (5, 3, 1)$

There is a reason for the dramatic differences we see between the odd and even cases for this version of the Oval Tracks Puzzle with "Do Arrows" permutation $T = (5, 3, 1)$ that is rather obvious once you notice it. It has to do with the fact that all of the location numbers involved in the "Do Arrows" permutation are odd. Could this possibly make such a big difference? Indeed it can and does.

Here is the reason why. When you start out with the disks in their home spots, they alternate in parity: odd, even, odd, even, … . A rotation of the disks does not disrupt this pattern of alternation of the parities. Nor does the 3-cycle (5, 3, 1). If the parities of the disks are alternating, then the disks that are in spots 1, 3, and 5 will all have to be of the same parity. Following the performance of the 3-cycle that shifts the disks between these three positions, the parities of the disks in these three spots is the same as it was before.

This parity-alternating pattern breaks down if there is an odd number of active disks. For example, if there are 19 active disks, then we have disk number 19 and disk number 1 adjacent to each other. In contrast, if the number of active disks is even, then the even-odd pattern survives as we cycle back to the beginning following the highest number.

For version 7 of the Oval Track Puzzle, this means that the only rearrangements of disks that you could possibly do if the number of active disks is even are ones that preserve the alternating parity all the way around the track. Thus, if you are presented with a random scramble of the disks, you do not have a prayer of a chance to solve it if the random scramble does not have alternating parity of the disks. If you think about it, having the disks come up at random with alternating parity ought to be extremely rare.

To get a feeling for this phenomenon, you might want to set the number of active disks on version 7 to an even number and then run a number of buttons scrambles. You will see that each time the puzzle will be scrambled in a way that preserves the alternating parities pattern. (Since a buttons scramble is one that can always be solved, you should be able to always solve the puzzle when scrambled this way. It is an interesting problem to try to solve the puzzle in this situation. You will find the opening game rather easy, but there can be a bit of challenge in the end game.)

Now that we have this insight about alternating parity, let's see if we can use it to explain the numbers that we see in (13.10). Let's imagine that we dump all of the disks off of the track and then put them back at random in a way that preserves alternating parity. To simplify matters, let's suppose that we have all 20 disks active.

After dumping them out, separate the disks into two piles, one with the odd-numbered disks and the other with the evens. Let's imagine putting them back onto the track starting with position 1 and counting up. How many ways can we do this? We will be able to tell if we keep track of how many ways we can perform each step along the way. We will get the total number of possible ways of performing the whole job by multiplying these numbers.

The first choice we have to make is whether we pick a disk for spot 1 from the odd pile or the even pile. There are two ways to make this choice. Having done so, we pick a disk from our chosen pile and put it onto spot 1. There are ten ways to make this choice since there are 10 disks numbered between 1 and 20 of whichever parity we chose. Next let us choose a disk to go onto spot 2. There is no choice regarding its parity. It must be the opposite parity from the one we picked for the disk to go onto spot 1. Since we have not removed any disks from this pile yet, there are ten of them from which we may choose, so we again have ten ways to make this choice.

Next let us pick a disk for spot 3. We switch parities again, going back to whatever the parity was for spot 1. Since there are nine disks left in the pile for this parity, we have nine ways to choose this disk. When we go on to spot 4, we switch parities again and pick one of the nine remaining disks in the second pile.

Continuing in this manner we see that the choices of ways of doing the next steps are 8, 8, 7, 7,..., 1, 1. If we multiply all these numbers together, we see that the number of ways of placing the disks on the track with alternating parity is $2 \times 10 \times 10 \times 9 \times 9 \times 8 \times 8 \times 7 \times 7 \times \cdots \times 2 \times 2 \times 1 \times 1 = 2 \times (10!)^2$.

If we think in general about doing this with n active disks, where n is even, then the number of ways of placing the disks is given by the general formula $2 \times ((n/2)!)^2$. Let us see how these numbers compare with those we got from *Maple* for the numbers of solvable permutations. Here is a table that gives these numbers and also their ratios.

(13.11)

Value of n	Subgroup size	Value of $2((n/2)!)^2$	Ratio of subgroup size to $2((n/2)!)^2$
6	18	72	1/4
8	576	1152	1/2
10	7,200	28,800	1/4
12	518,400	1,036,800	1/2
14	12,700,800	50,803,200	1/4
16	1,625,702,400	3,251,404,800	1/2
18	65,840,947,200	263,363,788,800	1/4
20	13,168,189,440,000	26,336,378,880,000	1/2

Now, we are in the ballpark at least. *Maple* says that we cannot solve *all* of the permutations that alternate parities, but you can solve either one fourth or one half of them. It appears that this depends on whether $n/2$ is odd or even, where n is the (even) number of active disks.

Why not all of them, and why this difference that depends on the parity of $n/2$? We will save the details of this part of the analysis for consideration in Exercises 13.5–13.7 and turn our attention now to the dramatically different behavior in the puzzle when n is odd.

Since *Maple* is telling us that when n is odd we can solve $n!/2$ of the $n!$ possible permutations, and since we know that the only permutations we can get are even ones, and since there are exactly $n!/2$ even permutations, it is clear that it is precisely the even permutations that we can solve.

We already have a 3-cycle on this puzzle, the "Do Arrows" permutation (5, 3, 1), so if we could get all other 3-cycles as conjugates of this one, we would be able to get all of the even permutations. However we have reason to be a bit gun-shy because we of course have this same 3-cycle available to us in the cases where the number of active disks is even, and we saw how little good that did us for getting all 3-cycles as conjugates.

In fact, when the number of active disks is odd, we can get any 3-cycle as a conjugate of the "Do Arrows" 3-cycle (5, 3, 1). This is a result of this fact:

(13.12) On the Oval Track Puzzle 7 with $T = (5, 3, 1)$ and with an odd number n of disks with $n \geq 7$, it is possible to place any three disks with numbers in the set $\{a, b, c\}$ into the spots with numbers in the set $\{1, 3, 5\}$.

The way we have stated this claim is purposely general in that we do not want to commit to any particular placement of disks a, b, and c into spots 1, 3, and 5. Any placement will do the job for us, so let's stay flexible.

To see why (13.12) is true, let us take any set $\{a, b, c\}$ of three numbers between 1 and n. We will want to get the disks with these numbers so that they are spaced out with one disk in between each of them. That is, we will want to get them spaced out with a pattern like this: d_1, r, d_2, s, d_3, where the d_i's represent our three target disk numbers a, b, and c in any order and r and s represent any two other numbers. Once we have gotten an ordering like this, clearly the right rotation of the disks can put our three target disks into spots 1, 3, and 5.

Let us first focus attention on disks a and b. Starting from disk a, wherever it may happen to be, count positions in a clockwise direction until you get to the position of disk b. This will be a number i somewhere between 1 and $n - 1$. Now do a count in a counterclockwise direction between the spot where disk a is to the spot where disk b is. This will be a number j that is also between 1 and $n - 1$. It must be the case that $i + j = n$. Since n, the number of active disks, is odd, it must be the case that i and j have opposite parities.

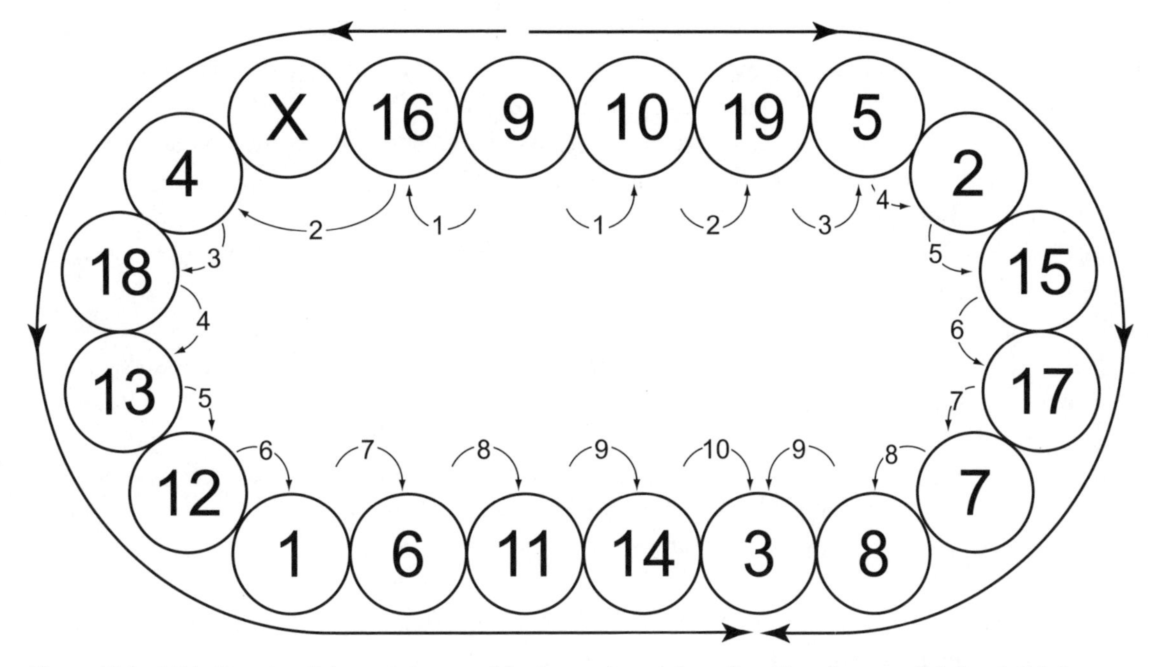

Figure 13.1. With 19 active disks, and the scramble shown, the number of positions between disk 9 and disk 3 counting clockwise is 9 and the number counting counterclockwise is 10. The sum of these two numbers is 19, which is the number of active disks.

Figure 13.1 gives an example that illustrates this idea. Note that with 19 active disks, the distance between disk 9 and disk 3 in a clockwise direction is 9 while the distance in a counterclockwise direction is 10.

If i is even, then it is possible to reduce the distance between disk a and disk b in a clockwise direction to two positions. If the distance is more than six positions, then by rotating disk b to spot 5 and performing "Back Arrows," you can move disk b to spot 1, thereby cutting the distance down by four. If the distance is exactly four, then put disk b into spot 3 and a "Do Arrows" will cut it down to two. Under all circumstances where this clockwise distance is even, you can use these methods to cut the distance down to two. We then have a relationship of the form a, r, b between disks a and b, with some disk r between then.

On the other hand if j is even, you can reverse your perspective and shorten the counterclockwise distance between disk a and disk b to two. The details for this are similar.

Once the right spacing between disks a and b has been established, you can work at moving disk c so that it is two positions away from one or the other ends of this pair involving disk a and b. Count both clockwise and counterclockwise from the ends of this a, r, b or b, r, a group to disk c. The numbers you get will add up to $n - 2$, which is odd, so one of them must be even. Use the "Do Arrows" to bring to a distance of two on the even side.

The method just described will always allow us to get the spacing between the three target disks the way it must be to put them into spots 1, 3, and 5. This shows that claim (13.12) is true. Now that we know this, we can see that if we want to perform any 3-cycle (a, b, c) whatever, we can get the three disks involved into spots 1, 3, 5, in some order. Suppose we record the steps for doing this in button A as we do it. Then, no matter how we place these disks in the spots 1, 3, and 5, one or the other of the buttons "Do Arrows" or "Back Arrows" will give us the 3-cycle on these three that we want. If we then press the button A^{-1}, we will reverse the setup procedure, and the result is the desired 3-cycle (a, b, c).

Since we can perform any 3-cycle on this version 7 of the Oval Track Puzzle, we can perform every even permutation. This is, of course, with an odd number of active disks. We have now fully explained the behavior reported by *Maple* for an odd number of disks and have mostly explained the radically different behavior for an even number of disks.

It needs to be mentioned that this theoretical knowledge about version 7 of the Oval Tracks Puzzles is not all there is to it. We know that it will always be possible to solve the puzzle to within a single transposition if the number of active disks is odd. However, it still would take some work to develop strategies for doing this efficiently. Perhaps you will want to work on this.

Other versions of the puzzles have this radically different behavior depending on how many disks are active. Given this example, you perhaps can discover some of these situations for yourself.

Exercises

13.1 Practice with *Maple* or *GAP*. Use either *Maple* or *GAP* to determine the sizes of the subgroups of S_n that are related to (a) Oval Track Puzzle 4 with $n = 6, 7, 8, 9$, and 10; (b) Oval Track Puzzle 9 with $n = 4, 5, 6, 7$, and 8. Are the results what you would have expected? Can you independently explain each of the numbers you get?

13.2 Another odd-even version. Do *Maple* or *GAP* computations for Oval Track Puzzle 8, which has "Do Arrows" (1, 3) (2, 4) with n from 10 to 20. Compare the results with the numbers for Oval Track 7 with "Do Arrows" (5, 3, 1).

13.3 The small cases on Oval Track 1. Let n denote the number of active disks on the Oval Track Puzzle 1. (a) Show that if $n = 6$, then the puzzle is totally solvable. [Hint: Consider the consequences of the equation $(TR^{-1})^3 =$ (1, 3).] (b) Show that if $n = 5$, then any arrangement of the disks that is possible keeps the disks in the same cyclic order in either a forward or reverse direction. Thus, for example, possible placements of the disks in spots 1 through 5 respectively are 3, 4, 5, 1, 2 (forward) and 3, 2, 1, 5, 4 (reverse). Show that there are only 10 such arrangements possible and that you can in fact get all of them. (c) Show that if $n = 4$, then the possible placements of the disks is subject to the same restriction as when $n = 5$, namely that the disks stay in the same cyclic sequential order in either the forward or reverse direction. Show that there are 8 arrangements of the disks that can be solved in this case. (Note: In the cases of 4 or 5 active disks, the spots on the puzzle can be identified with the four or five vertices of a square or a regular pentagon, respectively. The rotation of the disks corresponds to a rotation of the polygon that maps vertices to vertices, and the "Do Arrows" permutation corresponds to a reflection of the polygon about one of its axes of symmetry. Thus, the permutation groups for these versions of the puzzle are representations of the *dihedral groups* D_4 and D_5 of all symmetries of these two regular polygons. See most any introductory abstract algebra text for more information on the dihedral groups.)

13.4 Further comparison of Oval Track Puzzles 7 and 8. This exercise goes further in comparing Oval Track Puzzles 7 and 8 with "Do Arrows" (5, 3, 1) and (1, 3) (2, 4), respectively. (a) Show that for every n, with $n \geq 5$, you can obtain (5, 3, 1),the "Do Arrows" of Oval Track 7, on Oval Track 8. (b) Show that if n is even, but not a multiple of 4, then you cannot obtain (1, 3) (2, 4), the "Do Arrows" of Oval Track 8, on Oval Track 7. However if n is a multiple of 4, then you can obtain (1, 3) (2, 4) on Oval Track 7. (c) Consider how these results corroborate the *Maple*/*GAP* results of Exercise 13.2.

In the next three exercises, we undertake a detailed analysis of Oval Track Puzzle 7 with T = (5, 3, 1) in order to understand the numbers we get for these puzzles from Maple.

13.5 OT 7: Regarding two disk locations as one "place." As a first step to understanding OT 7 with n even, we rethink how we identify a scrambling of the puzzle with a permutation. For each number i, with $1 \leq i \leq n/2$, define "place i" to consist of the two disk spots $2i$ - 1 and $2i$. Thus, there are $n/2$ "places" on the puzzle when there are n active disks, with n even.

Let's associate a scrambling of the puzzle with three permutations, α, β, and γ with the first two in $S_{n/2}$, and γ in S_2. We will call α and β the *negative permutation* and the *positive permutation*, respectively, and will call γ the *placement permutation*. Define α so that $i\alpha = j$ if the ith odd disk with number $2i - 1$ is in place j, i.e., disk $2i - 1$ is in either spot $2j - 1$ or spot $2j$. Define β so that $i\beta = j$ if the ith even disk with number $2i$ is in place j. Define γ to be (1,2) if the odd-numbered disks are in even-numbered spots and to be the identity permutation ε if not.

(a) For Oval Track Puzzle 7 with n even, why can each place on the puzzle contain only one odd-numbered disk under any possible permutation of the disks? (We need this for the definition above for the permutation triple (α, β, γ) to be unambiguous.)

(b) List the permutation triples (α, β, γ) for each of the scramblings shown in Figure 13.2. (All of these are buttons scrambles, so they are all possible to achieve. Note that the place numbers have been superimposed on the puzzle.)

(c) On Oval Track Puzzle 7, determine the permutations that correspond to these triples (α, β, γ).

(i) $n = 20,\ \alpha = (1, 4, 5, 2, 7)(3, 8),\ \beta = (1, 3, 8, 6, 10)(2, 4, 7)(5, 9),\ \gamma = \varepsilon$

(ii) $n = 14,\ \alpha = (1, 5, 2, 3)(4, 6),$
$\beta = (1, 5)(2, 7)(3, 4, 6),$
$\gamma = (1, 2)$

(iii) $n = 12,\ \alpha = (1, 4, 5)(2, 6),$
$\beta = (1, 3)(2, 6, 4, 5),\ \gamma = \varepsilon$

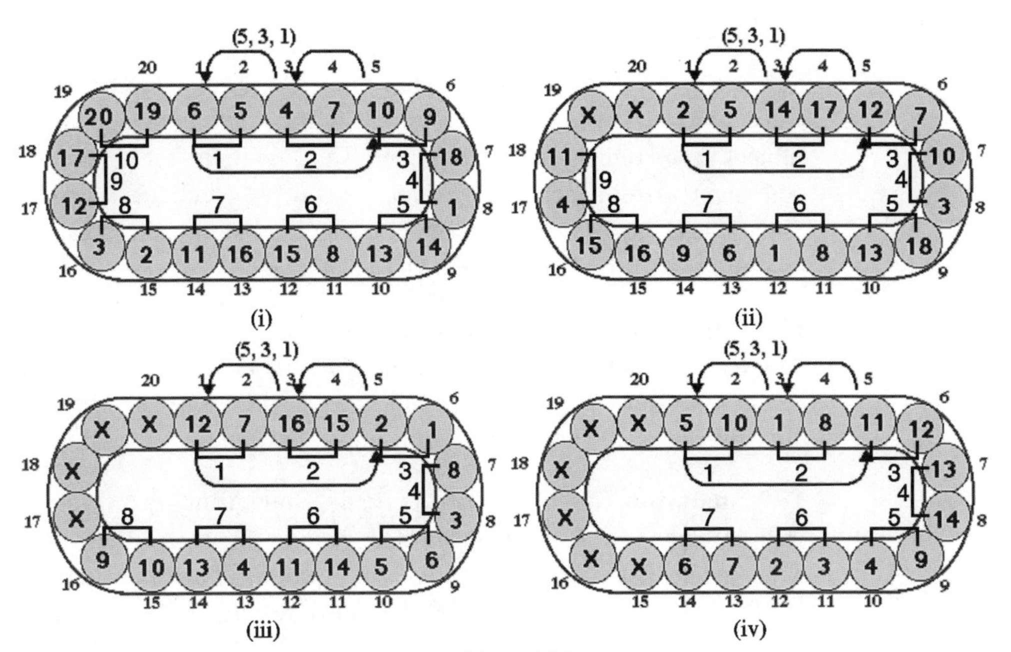

Figure 13.2

13.6 OT 7: More on the triple encoding. (a) Show that the negative permutations resulting from all permutations possible on Oval Track 7 with n disks, n even, is the subgroup of $S_{n/2}$ that is generated by the $n/2$-cycle $(1, 2, \ldots, n/2)$ and the 3-cycle $(3, 2, 1)$. Thus, by our work in Section 11.3, this subgroup is all of $S_{n/2}$ if $n/2$ is even, and is the set of even permutations in $S_{n/2}$ if $n/2$ is odd. (b) Show that the same conditions apply to the positive permutations resulting from permutations possible on Oval Track 7 with n disks and n even. That is, they are all the permutations of $S_{n/2}$ if $n/2$ is even, and just the even permutations in $S_{n/2}$ if $n/2$ is odd. (c) Show that if $n/2$ is even, then the pairs (α, β) of negative and positive permutations that can occur as part of a triple (α, β, γ) resulting from a permutation possible on Oval Track 7 must be such that α and β have the same parity if γ is even and the opposite parities if γ is odd. On the other hand, show that if $n/2$ is odd, α and β must always be even. [Hint: When the puzzle is in the identity state, all three, α, β, and γ, are even. T, being even, does not change the parity of any of the three. Think about how R changes the parities.]

13.7 OT 7: Getting all the arrangements. (a) Suppose n and $n/2$ are both even. Show that for each (α, β, γ) in $S_{n/2} \times S_{n/2} \times S_2$ such that α and β have the same or the opposite parities according as γ is even or odd, the permutation of Oval Track 7 that corresponds to this triple is obtainable on the puzzle. That is, all of the permutations subject to the limits identified in Exercise 13.6 are in fact obtainable. (b) Now suppose that n is even and $n/2$ is odd. Show that for each (α, β, γ) in $S_{n/2} \times S_{n/2} \times S_2$ such that α and β are both even, the permutation of Oval Track 7 that corresponds to this triple is obtainable on the puzzle. That is, all of the permutations subject to the limits identified in Exercise 13.6 are in fact obtainable.

13.8 Analysis of $T = (1, 3)\ (2, 4)$. Oval Track 8 with $T = (1, 3)\ (2, 4)$ is like Oval Track 7 in that it maintains the odd-even alternation of disk numbers when the number of active disks is even. We have seen some relationship between versions 7 and 8 already in Exercise 13.4. Take the new information about Oval Track 7 from the last three exercises into account, and develop a detailed understanding of what is possible on Oval Track 8.

13.9 Analysis of $T = (7, 4, 1)$. (a) Use the Build Your Own Oval Track 19 to make a puzzle with (7, 4, 1) as the "Do Arrows." (b) Run *Maple* or *GAP* data on the order of its subgroup for values of n from 7 through 20. (c) Note that the pattern for the multiples of 3 for this puzzle are similar to what we saw when n is a multiple of 2 and $T = (5, 3, 1)$. Analyze this case, and produce results like we got in Exercises 13.5 through 13.7.

13.10 Study of the low wiggle room cases. Take a few of the Oval Track puzzles of interest to you and generate *Maple* or *GAP* data about their subgroups when the number of active disks is just a bit larger than needed for the "Do Arrows" to work. Determine when a regular pattern sets in, and identify the cases that would need separate attention as was given to Oval Track 1 in Exercise 13.3. With Exercise 13.3 as a model, see if you can describe the subgroups associated with the puzzles in these low wiggle room cases.

Mastering the Slide Puzzle

While looking at conjugates in Chapter 9, we showed that on the Slide puzzle in the familiar "15" configuration, it was possible to get lots of 3-cycles. Let's generalize the main result that we stated in (9.2).

(14.1) On the Slide puzzle in any configuration that has one blank, a two-by-two area of active disks, and no obstacles that prevent moving disks from any box to any other, any 3-cycle which places the blank in its box of origin can be achieved.

We stated the claim in Chapter 9 for the standard "15" configuration of the puzzle, but it can clearly be generalized to this extent. The verification of this is left as an exercise.

Recall that this claim is true because it is very easy to perform a 3-cycle on three disks if you can get them into a two-by-two square with the blank, as illustrated in Figure 14.1. A bit of experience with the puzzle makes it clear that, as long as there are no obstacles that prevent this, something we are assuming here, you can bring any three disks together with the blank in a two-by-two square. Thus, it is possible to perform any 3-cycle as a conjugate of a basic 3-cycle of the two-by-two sort shown in Figure 14.1.

In this chapter we will apply this result and the rest of the theory we have developed to come to a detailed understanding of the Slide Puzzle. Applying what we know, it will be possible for us to look at any scrambling of the puzzle and tell in

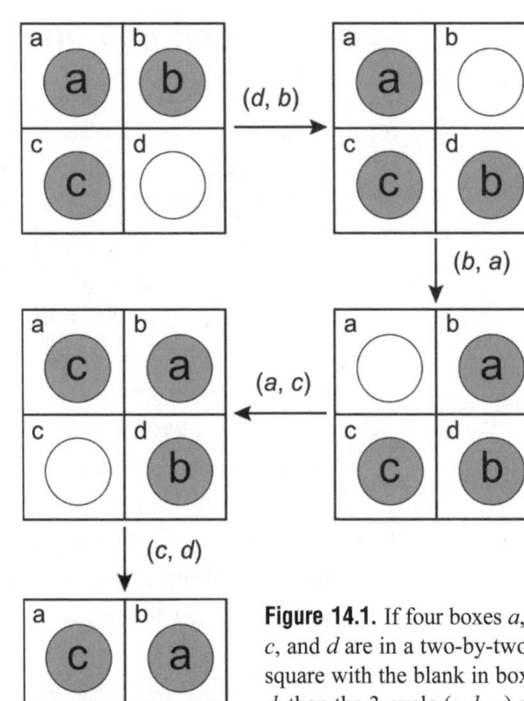

Figure 14.1. If four boxes a, b, c, and d are in a two-by-two square with the blank in box d, then the 3-cycle (a, b, c) can be obtained with four moves of the blank. We have $(a, b, c) = (d, b)(b, a)(a, c)(c, d)$.

advance whether or not the puzzle will be solvable from that scrambling. We will also see what differences result from varying the dimensions of the puzzle, putting in obstacles, using two rather than one blank, and having one or both of the wrapping features in force.

14.1 The Fit with Group Theory

The Slide Puzzle does not fit with the theory of permutation groups as neatly as the Oval Track Puzzles do. It is true that each possible arrangement of the puzzle can be identified with a permutation. The problem is that the way you change the Slide Puzzle always involves one particular object, specifically a "blank disk" that we will introduce as a placeholder in the blank spot. Not only does every change have to involve this specific disk but, in addition, the changes you can make are dependent on the location of this blank disk. You can only perform transpositions involving this blank disk and a disk in an adjacent (above, below, to the left, or to the right) box.

We can represent the moves of the puzzle using the cycle notation for transpositions, and when we use the rules of composition for reducing a sequence of such transpositions to disjoint cycle notation, the resulting expression will represent the permutation produced by the sequence of moves. Thus, the set of all possible configurations that you can create on the puzzle can be identified with a certain subset of S_n where n is the number of active boxes. Let us call this set *SP* (for Slide Puzzle).

The problem is that it is not at all clear that *SP* is a subgroup of S_n. The reason is that it is not obvious that if you compose two of the permutations in *SP* you again get another one in *SP*.
Let's think about it. We'll specialize to the original "15" configuration so that the blank starts off in box 16. We can characterize those permutations that are in *SP* by the way we can get them on the puzzle. We can say that:

(14.2) For the Slide Puzzle in the original "15" configuration, a permutation α is in *SP* if and only if α can be expressed using "Slide Puzzle transpositions." That is $\alpha = (16, a_1)(a_1, a_2)(a_2, a_3)\cdots(a_{m-1}, a_m)$ where m is a positive integer, box 16 is adjacent to box a_1, and for each i with $1 \le i \le m-1$, a_i and a_{i+1} are the numbers of adjacent boxes.

The truth of this characterization should be clear to you after some experience playing with the Slide Puzzle. What we have described here is how a sequence of moves is recorded and presented in the "Show List" window of the Slide Puzzle after you have performed it. If a certain permutation is one that you can produce on the Slide Puzzle, then there will be an expression for it in terms of transpositions of the form indicated. Conversely, if there is such a sequence of transpositions, then you can perform it on the puzzle and produce the permutation that we have denoted by α.

Now that we have described so specifically how we tell if a certain permutation is in *SP*, we can see the problem in telling whether or not *SP* is a subgroup of S_n. One of the criteria is that *SP* ought to be closed under composition. That is, if two permutations α and β are in *SP*, if *SP* is going to be a subgroup, then it must be the case that $\alpha \circ \beta$ is also in *SP*.

However, we have a problem in telling whether or not $\alpha \circ \beta$ is in *SP*. With the other puzzles, if you have representations for two permutations in terms of two sequences of moves to get them, to compose the two permutations, you just string the sequences of moves together. For the Slide Puzzle, this is not always possible. If we have sequences of moves for α and for β as described in (14.2), we cannot just string them together. The reason is that the last transposition in the first sequence must mesh with the first transposition of the second sequence in terms of the adjacency condition, and this is not going to happen most of the time. It is possible that one could find a way of creating a new sequence for $\alpha \circ \beta$ satisfying the conditions in (14.2) whenever there were such sequences for α and β, but this is by no means immediately clear.

We will defer consideration of whether or not *SP* is a subgroup of S_n to Exercise 14.1. However, we will see that if we narrow our focus, then we can easily show that a certain subset of *SP* is a subgroup of S_n. We see that:

(14.3) The subset of *SP* consisting of those permutations that return the (one) blank disk to its original position is a subgroup of S_n. Call this subset *SP**.

As we think about why this is true, we will again think in terms of the original "15" configuration for which the one blank starts out in box 16, but it is clear that any box could take on the role of box 16 for this purpose. The reason (14.3) is true is that we can in this case use the notation of (14.2) as is for expressing compositions of elements in SP^*. If we have a permutation α from SP^* expressed in the notation of (14.2), then the added condition that α takes the blank disk back to its original position translates into a_m being 16. Now let's express β in the notation of (14.2), say $\beta = (16, b_1)(b_1, b_2)(b_2, b_3)\cdots(b_{k-1}, 16)$. If we concatenate the sequences of transpositions for α and β, the junction is between $(a_{m-1}, a_m) = (a_{m-1}, 16)$ and $(16, b_1)$, so we have the necessary match of box numbers. The concatenated sequence of transpositions also represents an element of SP^*. This concatenated sequence also represents $\alpha \circ \beta$, so we have that $\alpha \circ \beta$ is in SP^*. If $\alpha = (16, a_1)(a_1, a_2)(a_2, a_3)\cdots(a_{m-1}, 16)$, we get an expression for α^{-1} in the form, $(16, a_{m-1})(a_{m-1}, a_{m-2})(a_{m-2}, a_{m-3})\cdots(a_1, 16)$. Clearly α^{-1} satisfies the conditions of (14.2) and the added condition of returning the blank disk to its original spot, so we have that α^{-1} is in SP^* whenever α is.

14.2 A Solvability Test

So far, the ideas we have articulated have to do with how a permutation can be shown to be in *SP* by showing a sequence of moves of the blank that produced that permutation. This is theoretically useful, but not very helpful for solving the puzzle. When you do a scramble on the Slide Puzzle, the computer presents you with a permutation. It does not give you any information about the existence or nonexistence of a sequence of moves of the puzzle that would produce this permutation. It would be nice to have a way of telling whether or not an arbitrary permutation is in *SP*, that is, whether or not the scramble is solvable.

What you *can* do easily with a scramble permutation of the puzzle is to express it in disjoint cycle notation, and from this determine its parity. We will see in a moment that simply knowing the parity of the permutation, together with an observation regarding the location of the blank, will give you the information that you need to decide if the permutation is in *SP*.

The outcome is not as simple as having even permutations solvable and odd ones not solvable. What happens is that an even permutation is in *SP* if the blank disk is in an *even box* but not if it is in an *odd box*. In the same way, an odd permutation is in *SP* if the blank is in an odd box but not if it is in an even box.

The system for defining what boxes are even and what are odd is simple. The box where the blank originates is even. Starting from there, we alternate in a checkerboard fashion in labeling the boxes odd and even. Figure 14.2 shows the familiar checkerboard pattern for the standard "15" configuration. The shaded boxes are odd and the plain ones are even.

With this concept of odd and even boxes now defined, we can state our new characterization of the permutations in *SP*. We will state the characterization in the most general terms that we can.

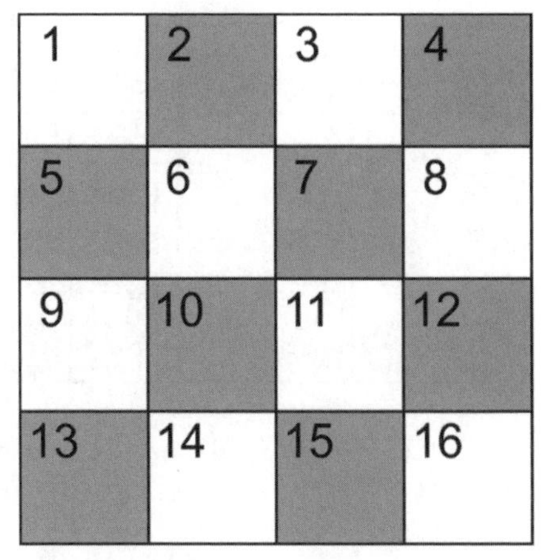

Figure 14.2. Define the boxes to be either odd or even, alternating in a checkerboard pattern. Here the white boxes are even and the shaded ones are odd.

(14.4) Let α denote an arbitrary permutation represented on the Slide Puzzle in any configuration that

(i) Has just one blank disk,

(ii) Has at least one two-by-two array of boxes all of which are in use, and

(iii) Does not have any obstacles that trap disks.

Then the permutation α is solvable if and only if the parity of α is the same as the parity of the location of the blank disk.

The first restriction we place on the way the puzzle is configured is clear, but the second and third need a bit of explanation. We need to have at least one area of the puzzle where there are four boxes, all of which are in use, that are arranged in a two-by-two square. This will be needed so that we can invoke (14.1). This is why we include the second condition.

When we say in the third condition that the configuration has no obstacles that trap some disks, we mean that whatever obstacles are on the puzzle do not make it impossible to move a disk from any given box to any other one. Figure 14.3 shows two configurations that both have obstacles. The obstacles in boxes 7 and 8 cause disks 3 and 4 to be trapped in Figure 14.3(a). There is no way to get disk 3 placed anywhere other than boxes 2 and 3, and there is no way to move disk 4 anywhere other than boxes 3 and 4. (Note, however, that the isolation of these two disks would vanish if we had either horizontal or vertical wrapping turned on.) In contrast, the obstacles in boxes

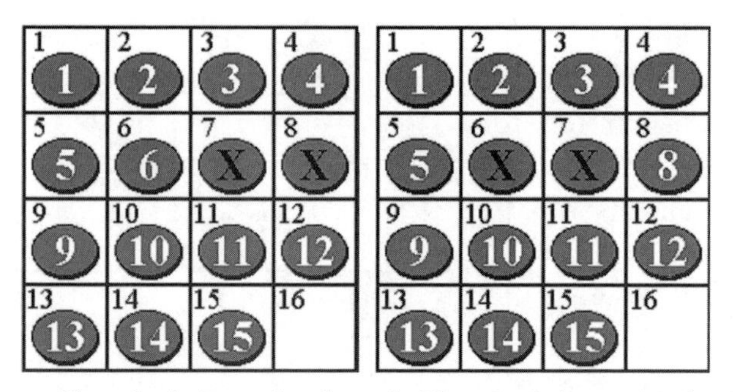

a. Obstacles in boxes 7 and 8 trap disks in boxes 3 and 4

b. Obstacles in boxes 6 and 7 do not trap any disks.

Figure 14.3. Some placements of obstacles trap some of the disks and others do not.

6 and 7 in Figure 14.3(b) do not prevent disks from any of the active boxes from being moved to any other active box. You might want to configure your copy of Slide Puzzle in these two ways and explore them awhile to convince yourself of the truth of these claims.

Note that both of the configurations shown in Figure 14.3 satisfy the second condition in multiple ways. For configuration a, we see that the boxes numbered 1, 2, 5, and 6 form a two-by-two array of active boxes. There are an additional four ways of defining two-by-two arrays of active boxes on this puzzle. For configuration b, we have three ways of defining two-by-two arrays of active boxes using the bottom two rows.

Now that we have clarified what we mean by (14.4), let's see why it is true. First, we will suppose that we have a permutation α on the puzzle that satisfies the condition that the parity of α and the parity of the box containing the blank disk are the same. The first thing that we will do is make it clear why the parities of the boxes were defined as they were. From wherever the blank disk is, do a sequence of moves that bring it to its box of origin. (We know that this is possible because of our assumption that there are no locations in which the disks are trapped.) The number of transpositions that you will have used to get the blank back to its box of origin will be of the same parity as the parity of the box from which it started.

Now consider the permutation that you got after making this change. It must be even. The reason is that if α was odd, then the location of the blank was also odd, so you have extended α by an odd number of transpositions, which results in an even permutation. Similarly if α was even, you have extended it by an even number of transpositions, which again produces an even result.

We now have an even permutation, which has the blank disk in its original box. Let us call it α^*. By (8.9), we can express this even permutation as a composition of 3-cycles. Suppose that $\alpha^* = \tau_1\tau_2 \cdots \tau_m$ is an expression for α^* where all of the τ_i's are 3-cycles. None of these 3-cycles will involve the box number of the blank disk since it is left in place by α^*. Since the inverses of these 3-cycles are also 3-cycles, (Recall that $(a, b, c)^{-1} = (c, b, a)$.) we can cite (14.1), which tells us that we can successively apply $\tau_m^{-1}, \tau_{m-1}^{-1}, \ldots, \tau_1^{-1}$ to the puzzle. This will undo the puzzle and totally solve it.

[The method just described for solving the puzzle is definitely not an efficient method. You would likely not use this method in practice, but it is useful to think in these terms in order to give a rigorous argument as to what is possible.]

Now let us suppose that the puzzle is solvable when it has been scrambled into permutation α. We would like to be able to show from this assumption that the parity of α and the parity of the box holding the blank are the same.

As before, let's first do a sequence of moves that brings the blank to its box of origin. Recall that we argued that, if the parity of α agrees with the parity of the box where the blank was, then the permutation α^* we obtain by moving the blank home is even. We now observe that the converse is true. If α^* is even, then the parities had to agree prior to moving the blank back to its box of origin.

We will see that α^* is indeed even. The reason is that since we are assuming that α is solvable, α^* is also solvable. This means that there is a sequence of allowable moves that completely restores the puzzle to its initial state.

This sequence must involve an even number of transpositions. The reason is that each of the transpositions involves moving the blank either left or right or up or down. Since the blank started in its box of origin when the state of the puzzle was the permutation α^* and since it returns there, we can pair up these moves so that every upward shift gets paired with a countervailing downward shift and every left shift with a right shift. This clearly gives us an even number of shifts, making α^*even.

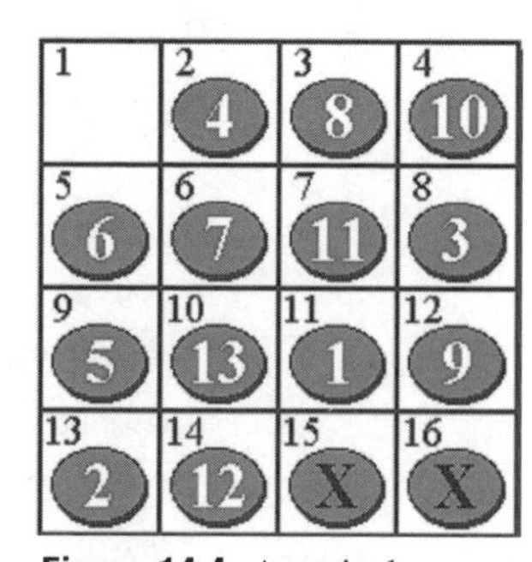

Figure 14.4. A typical scramble of the Slide Puzzle with 14 disks.

Having shown the truth of (14.4), let's put it to use. Figure 14.4 shows a typical scramble of the puzzle configured with fourteen active disks. We clearly have only one blank, have many two-by-two areas of active disks, and have no obstacles that trap disks. Thus, (14.4) applies.

We first observe that box 1, the location of the blank disk, is four transpositions away from box 14, the box of origin for the blank for this puzzle. That is, the blank is in an even box. If we call this permutation α, then we see that, expressed in disjoint cycle notation,

$$\alpha = (1, 11, 7, 6, 5, 9, 12, 14)(2, 13, 10, 4)(3, 8).$$

Thus, α is composed of an 8-cycle, a 4-cycle, and a 2-cycle. This tells us that α can be expressed using $7 + 3 + 1 = 11$ transpositions, so α is odd.

The parities of α and the location of the blank do not agree. This means that we cannot totally solve this puzzle. We can get it down to a single transposition, but that is as close as we can get.

14.3 Practicalities of Solving the Slide Puzzle

Now that we have a way of telling whether or not a particular scrambling of the Slide Puzzle can be solved, let's spend a few moments discussing the practicalities of how to go about solving the puzzle. Again, you do not want to start out drawing upon the theory and chipping away at the puzzle by performing a sequence of 3-cycles. This would be quite inefficient. Instead, as was the case with the Oval Track Puzzles, you start out using heuristic methods.

A typical opening game strategy is to start putting the disks into place with disk number 1 and working your way up in terms of the numbers and down in terms of position on the puzzle, gradually narrowing down to the box of origin of the blank disk. Depending on where the top row disks are in the scramble, you will find yourself planning ahead a bit and perhaps grouping them together and convoying them to the top. You might find that it works better sometimes to move disk 4 to the top and into box 3 before moving disk 3 into the top row. Then you can bring disk 3 into box 7, get the blank into box 8, and then shift disk 4 over into box 4 and disk 3 up into box 3.

Of course, the heuristics change if you have one or both of the wrapping features turned on. If the boxes along the edges are adjacent to those on the opposite side, this gives you more options for moving to and from these spots.

We will not go into detail about the heuristics of this puzzle. Part of the fun of solving puzzles like this is discovering these heuristics on your own.

As with the earlier puzzles, the theory does not have to kick in until you get to the end game. The irony is that for the Slide Puzzle, there is essentially no end game. The heuristic methods that you develop will take you as far as you are going to be able to get in solving the puzzle. The contribution of group theory is that it tells you when you are at a stage where it is pointless to go on.

Specifically, if you get down to the point where all that would remain to solve the puzzle is to exchange two disks, you know that you might as well quit. The theory tells us that it is not possible to perform a single transposition of two non-blank disks.

Think about it. We are assuming that you have the blank disk back in its home spot and two disks in each other's boxes. This is an odd permutation with the blank in an even box. According to (14.4), the parity mismatch tells you that the remaining scramble is not solvable. (There are exceptions to this generalization, which we discuss in Section 14.4 below.)

For developing your strategy for solving the puzzle, you might find it of interest to run a series of scrambles with the same configuration and keep track of the average number of steps it takes you to either completely solve the puzzle or get it to the point where just the highest two numbered disks are out of place and you can go no further. You will find that there is a rather wide variance in how many steps it takes depending on how far the disks have to be moved. We find that for the standard "15" configuration, the average number of moves for a series of 10 or more scrambles is somewhere around 120 to 130, with the minimum and maximum numbers being perhaps 30 or 40 to either side of the average.

14.4 Effects of Two Blank Disks and Obstacles

Up until now, we have been considering configurations of the puzzle that involved just one blank disk. Now let us think about how things change when you have two blank disks. Before reading this, you should experimented with the Slide Puzzle in two-blank mode and formulated some conjectures about the differences.

After doing a few scrambles in two-blank mode, you will discover that you do not end up with a final transposition of numbered disks that you cannot perform. Having two blank disks gives you the freedom to put these last two disks right.

The reason that you can get those last two disks right if there are two blank disks is illustrated in Figure 14.5. We show here a maneuver that is possible when you have the two blanks together with two numbered disks in a two-by-two square. You can perform a pair of transpositions that swaps the two numbered disks and also the two

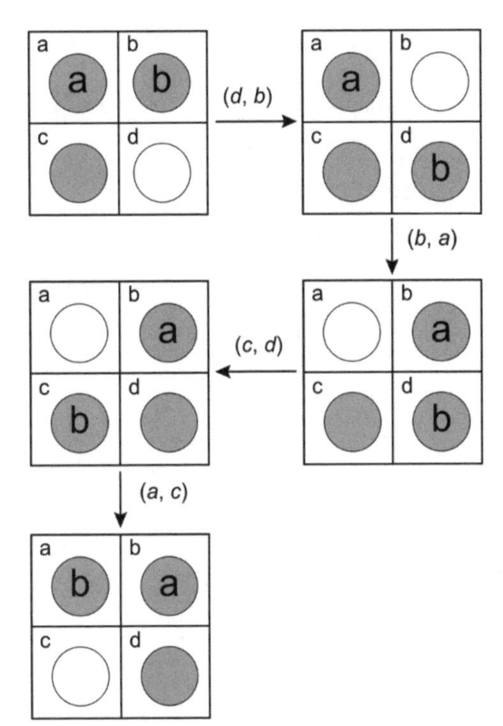

Figure 14.5. Having two blank disks makes it possible to give the appearance of performing a single transposition. With blanks in boxes c and d, one gets $(a, b)\,(c, d) = (d, b)\,(b, a)\,(c, d)\,(a, c)$.

blanks. We have distinguished the two blanks from each other here by showing one of them shaded and the other plain. The shading makes it possible to see that the blanks have traded places as well as the disks labeled a and b. However, if you could not tell one blank from the other, you would think that you had performed a transposition here.

The total solvability of the puzzle when there are two blanks is thus in some respects an illusion. It might be that when the puzzle looks like it is totally solved, you may still be an invisible transposition away. If the two blanks could be distinguished in terms of which was their box of origin, it very well may be that they are in each other's boxes. If, however, the blanks cannot be distinguished, you really do not care if they are swapped and you can regard the puzzle as being solved.

Figure 14.5 indeed shows an illusory transposition, for we achieve this apparent transposition (a, b) with four transpositions. This would violate the parity theorem if we really did have a transposition here. Might there be some other way of getting an actual transposition with two blanks available? The answer is no. The same parity issues apply with two blanks as with one. We leave the details of this analysis to the exercises. (However, things may change if we have wrapping turned on, as we will see in the next section.)

Does having two blanks make much difference for the heuristics of the opening game and allow for significant increases in efficiency in solving the puzzle? This question is left for you to explore.

One additional qualitative difference that having a second blank disk can make is that some disks that are trapped when there is only one blank are not trapped when the second is available. Figure 14.6 gives an example of this. With boxes 1 through 12 in use, except for an obstacle in box 8, the disk in box 4 is trapped if there is only one blank. It can move between boxes 3 and 4 and that is all. However if there are two blanks starting out in boxes 11 and 12 as shown in Figure 14.6(a), then it is possible to move both of the blank disks up near box 4, and, with a second blank available when the first one is spent moving disk 4 to box 3, you can move it beyond the obstacle and into the open field, as shown in Figure 14.6(b).

Some arrangements of obstacles are such that having a second blank is still not enough to free up trapped disks. For example, if the arrangement shown in Figure 14.6 were modified by adding an additional obstacle in box 7, then two blanks would not be enough to free up disk 4. It would be restricted to moving between boxes 4, 3, and 2.

If you experiment with configurations having obstacles, you will discover what sort of placements of them causes entrapments and what sort do not. You will also discover that even when a set of obstacles does not cause entrapments, it likely will cause the average number of steps needed to solve the puzzle to rise dramatically.

The placement of obstacles creates far too many options for us to consider here. Also, dealing with the consequences of them falls more into the realm of heuristics rather than mathematical theory, and we are leaving the heuristics to the reader for the most part.

Summing up this discussion about two blank disks, we observe that:

(14.5) If α denotes an arbitrary permutation represented on the Slide Puzzle in any configuration that
 (i) Has two blank disks,
 (ii) Has at least one two-by-two array of boxes all of which are in use, and
 (iii) Does not have any obstacles which trap disks,
 then α is solvable up to a possible invisible transposition of the blank disks.

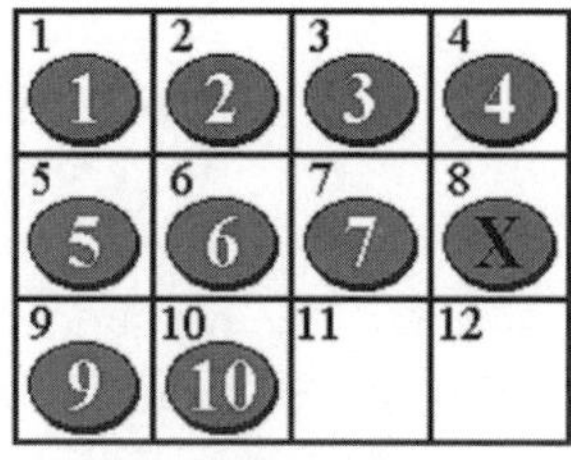

a. In spite of the obstacle in box 8, with two blanks ...

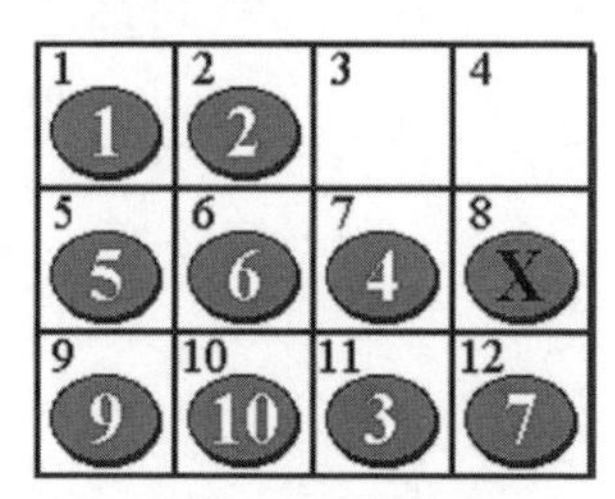

b. ... it is possible to move the disk out of corner box 4.

Figure 14.6. An obstacle that traps a corner with one blank may not if there are two.

14

14.5 Going Cylindrical or Toroidal

Let's give some thought to how the Slide Puzzle changes when you put it into either the cylindrical or toroidal mode by turning on one or both of the wrapping options.

Recall that you can configure the software as a cylindrical or toroidal puzzle by clicking on the appropriate button in the wrapping dialog box during the configuration process. When you choose side-to-side wrapping, the software sets itself so that the left and right edges are considered adjacent. If you choose top-to-bottom wrapping, then the top and bottom edges are adjacent. Making one but not both of these choices has the effect of turning the puzzle into a cylindrical puzzle, although it is still displayed on a flat plane. Turning both wrapping features on gives you a puzzle on a torus. Refer back to Chapter 3 to refresh your memory regarding the details.

What changes if a given version of a planar Slide Puzzle configuration is turned into a cylindrical version? In terms of the basic solvability, it will all depend on parity considerations.

The solvability criteria stated in (14.4) for the one blank case are built on the notion that each box can be unambiguously assigned a certain parity. In order to be able to solve a scrambling, the parity of the permutation it presents must agree with the parity of the box in which the blank is found. If the puzzle is on a cylinder or torus, this notion of defining parity of positions may or may not be possible.

To illustrate, let's consider a four-by-three version of the puzzle (12 active boxes) with vertical wrapping turned on. We show this configuration in Figure 14.7.

Figure 14.7a shows the initial state. We note that we can move the blank from its home in box 12 to box 8 with the single transposition (12, 8). However, since box 4 is adjacent to box 12 because of the vertical wrapping being on, we can get the blank to box 8 by going over the top. That is, the sequence (12, 4)(4, 8) also puts the blank into box 8. Since we can do this using either an odd or an even number of transpositions, we cannot assign a parity to this box. The same ambiguity applies to every location. You will get opposite parities for the number of transpositions used to move the blank to any given location according to whether or not you take the blank over the top in getting it there.

Before we mourn the loss of the parity concept for locations, let's make note of what we gain because of this loss. We gain the ability to perform genuine transpositions. Figure 14.7(b) shows the transposition (4, 8) on this puzzle, which we get by the blank-over-the-top sequence (12, 4)(4, 8)(8, 12).

We can readily see that this puzzle gives us plenty of flexibility to get any two disks together like disks 4 and 8 are in Figure 14.7(a). Thus, by applying conjugacy and the basic transposition we have just noted, we can get all transpositions on this puzzle. As a result the theory of permutation groups tells us that this puzzle is completely solvable.

Without wrapping turned on, this three by four version of the puzzle would only be half solvable in accordance with our earlier criteria. This example illustrates that having wrapping turned on can turn unsolvable puzzles into solvable ones. But is that *always* the case? No, it is not.

For example, had we turned on either horizontal or vertical wrapping with the basic four by four "15" puzzle, we still have a puzzle that is parity bound. This is because both of the dimensions are even, unlike our previous example. Therefore, we do not get paths with different parities depending on whether we take a route over an edge or not. To illustrate, if you move the blank from box 16 to box 12 going directly, you use one transposition (16, 12). If you go over the top, the sequence is (16, 4) (4, 8) (8, 12). Both routes require an odd number.

We can formalize a statement of the principle that extends (14.4) and takes these wrapping-induced differences into account.

(14.6) Suppose we have a configuration of the Slide Puzzle that
- (i) Has just one blank disk,
- (ii) Has at least one two-by-two array of boxes all of which are in use, and
- (iii) Does not have any obstacles that trap disks, and
- (iv) Wrapping makes it possible to move the blank out of its home location and back in an odd number of steps.

Then any permutation α is solvable on this version of the puzzle.

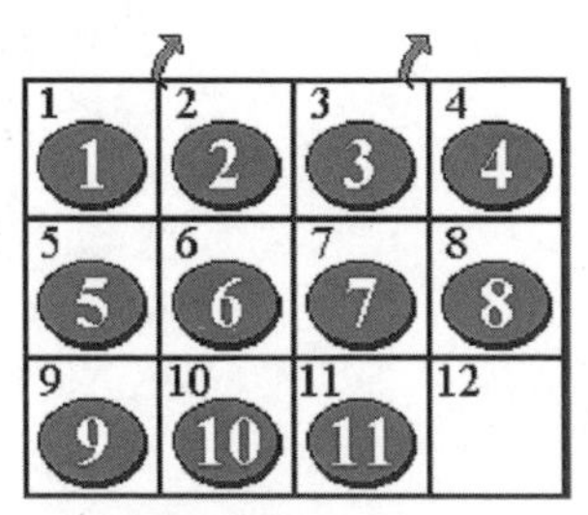

a. Either (12, 8) or (12, 4) (4,8) will move the blank to box 8, so ...

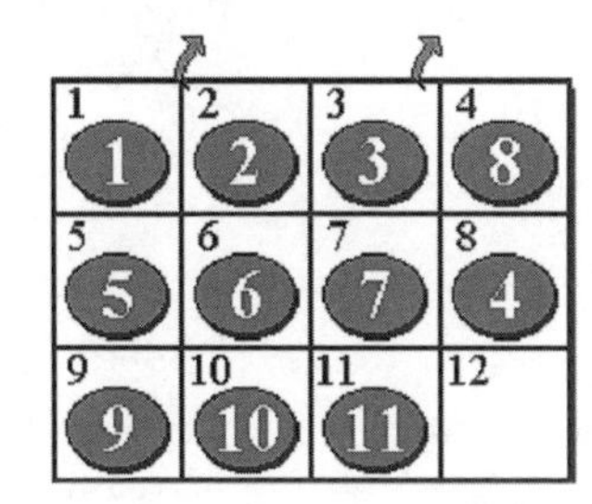

b. ... we see that (4, 8) = (12, 4) (4, 8) (8, 12)! A transposition is possible!

Figure 14.7. Wrapping can thwart location parity and make transpositions possible.

To see why this is true, we observe that we could solve half of the permutations without making use of the wrapping. Now let's consider a scrambling α that we could not solve in the planar version of the puzzle because of a mismatch in the parities of the permutation and the location of the blank. Let us modify α by doing the following:

(a) Run the blank spot to its home location.

(b) Run the blank on an odd-length tour away from and back to its home

(c) Reverse the steps of (a) to get the blank back to its α location.

Condition (iv) of (14.6) tells us that step (b) is possible. Let us denote by α^* the permutation that we now have after modifying α in this way. We have modified α by extending it by an odd number of transpositions. This is so because in step (b) we used an odd number and in steps (a) and (c) we used equal numbers. Thus, the parity of α^* is the opposite of that of α. Since we had a parity mismatch between α and the location of the blank, we now have a parity match for α^*. This means that we can solve α^*, which means that we have solved α.

This result tells us that in theory, we only need to make one pass over an edge to solve any scrambling of the puzzle. In practice, you should allow yourself the freedom to use whatever solves the puzzle most efficiently.

Another way that having wrapping turned on changes things is that it becomes harder for out of action spots to be obstacles. This is so because some locations get more neighbors with wrapping.

This is especially the case for the corners, which in the planar mode have only two adjacent spots. Turn on one of the wrapping features and they each pick up one new neighbor. Turn on the other and they gain yet another neighbor. With both wrapping options turned on, there really is no such thing as a corner. On the toroidal version of the puzzle, all locations have four neighbors, assuming no obstacles.

We have discussed the cylindrical and toroidal versions of the puzzle in general terms. The generalities guide you as you explore specific configurations. There are so many options that it would not be practical to consider them all. Given this background, you ought to be able to make your own discoveries now.

14.6 Forming up Trains: An Heuristic Metaphor

We will end our look at the Slide Puzzle with one additional heuristic insight. As you work at solving the Slide Puzzle, keep in mind the metaphor of forming up trains in the railroad freight yard. The train engineer sometimes has to move his train back and forth along a track in order to position the end near the next car to be added. In a similar way, sometimes you want to think of the disks that you already have placed not as fixed in their locations for the rest of the process, but rather as cars on a forming train that are arranged with respect to each other as they need to be, but which also need to move together around on the track in order to add new cars.

This heuristic metaphor is especially helpful when you are working on a version of the Slide Puzzle that has obstacles that have created narrow corridors. For example, consider the partially solved puzzle shown in Figure 14.8(a). We see that disk numbers 1 through 5 and 9 are already placed. Disk 13 is the next candidate for placement. However, it is a long way from home. Its destination is at one end of the first-row/first-column corridor formed by the obstacles but it is at the other end of this corridor.

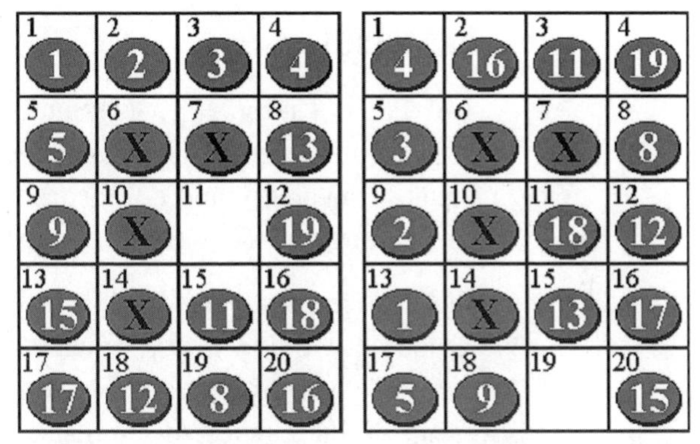

a. Disks 9, 5, 1, 2, 3, and 4 form a correctly placed train. **b. Move the train to add the next car, disk 13.**

Figure 14.8. Moving the train to add a car.

We see that it is easy to bring disk 13 out into the open by moving disk 19 into location 11 and then dropping disk 13 down to location 12. However, when we do this, we have the blank in location 8 at the entrance to the corridor. We can either backtrack, or take the blank all the way around the corridor. Since reversing what we have just done is clearly counterproductive, we run the blank counterclockwise around the corridor. Note that in doing so, we have "backed up the train" consisting of the organized disks 9, 5, 1, 2, 3, and 4.

Once we have brought the blank back into location 19 via the lower left entrance to the corridor, we have it in the open yard again and can use it to move disk 13 into spot 15, where we will park it for the time being. We will next run our train forward in the corridor by running the blank around it in a clockwise direction. We do this several times, until we have gotten disk 9 into location 18, as is shown in Figure 14.8(b). It has taken 80 moves of the blank to get to this stage.

At this point, it is possible to add the next car, disk 13, at the right point in the train. Note that disks 17 and 18 are the ones that need to come next in the train, and they are positioned well to join without any extra work. We run this longer train clockwise around the track, and before long, our task is reduced to solving the lower right corner between locations 11 and 20.

Exercises

14.1 Is *SP* a subgroup? (a) Show that when the Slide Puzzle is in the standard "15" configuration, *SP*, the set of all possible permutations, is *not* a group with composition as the operation. [Consider a 3-cycle which moves the blank horizontally and a transposition that moves the blank vertically.] (b) Do you get the same result for other configurations? (c) Can *SP* ever be a group for the cylindrical or toroidal versions of the puzzle?

14.2 Solving a scrambling from Puzzle Resources. (a) Load the first preset configuration of the Slide Puzzle's Resources window into your copy of Slide Puzzle. (It has the label "1. Exercise 14.2". See Appendix A if you need to review how to access preset configurations from Resources.) (b) Determine the parity of this scramble and the parity of the location of the blank. Use (14.4) to predict whether the puzzle is totally solvable or not. Then try to solve the puzzle and see if the claim of (14.4) holds true for this case. (c) Load entry number 6 of the Slide Puzzle Resources collection of macros into one of the macro but tons. This solution of this puzzle takes 105 steps. Run the solution to see what sorts of strategies are built into it. Set the macro delay to slow the steps down so you can follow the process.

14.3 A second scrambling from Resources. Load the second preset configuration of the Slide Puzzle's Resources window into your puzzle. It is the one labeled "2. Exercise 14.3". Follow the instructions of the previous exercise for predicting the solvability of this scramble. A near solution of this scramble in 107 steps is included as entry number 7 in the collection of macros.

14.4 What effect with wrapping on? (a) The third preset configuration of the Slide Puzzle's Resources has the same scramble as the one for Exercise 14.2, except that this time both of the wrapping options are turned on. Load this preset (labeled "3. Exercise 14.4") into your Slide Puzzle, test its solvability, and solve it to the extent possible. Does having the wrapping make a difference in the solvability? Does it make a difference in how many steps are needed? (b) Load the solution in 63 steps for this puzzle provided as entry number 8 in the collection of macros and run it slowly. Do you see how the wrapping is exploited to reduce the number of steps needed? Can you see ways to solve the puzzle more efficiently?

14.5 More exploration with wrapping on. This exercise relates to Exercise 14.3 in the way Exercise 14.4 relates to Exercise 14.2. The fourth preset configuration of the Slide Puzzle's Resources has the same scramble as the one for Exercise 14.3, except that this time both of the wrapping options are turned on. Load this preset (labeled "4. Exercise 14.5") into your Slide Puzzle, test its solvability, and solve it to the extent possible. Does having the wrapping make a difference in the solvability? Does it make a difference in how many steps are needed?

14.6 Exploration with obstacles. (a) The fifth preset configuration for the Slide Puzzle in Resources presents a scramble of a configuration of the puzzle that has obstacles in locations 6 and 11. Load this configuration, analyze it, and try to solve it. What difference does it make to have these obstacles in terms of the solvability of the puzzle, and how many steps it takes to solve? (b) A solution of this scrambling is also given in the macro collection as entry number 9. Load it and run it slowly enough to follow the approach. Can you see ways to improve on this solution?

14.7 No impossible scramblings? The sixth preset configuration for the Slide Puzzle in Resources presents a three-row and four-column version of the puzzle with top-to-bottom wrapping turned on. This version of the Slide Puzzle is totally solvable. Try several scrambles of it to convince yourself experimentally that this is true. Then justify why it is true. [Hint: Show that you can get a transposition on this version of the puzzle, and think about the consequences of this.]

14.8. The example in Figure 14.4. You will find the scramble shown in Figure 14.4 as the seventh preset configuration in Puzzle Resources for the Slide Puzzle. Load it into your Slide Puzzle. Then verify the claim made at the end of Section 14.2 that this scramble is solvable to within a single transposition by getting all disks into their home spots except for having disks 12 and 13 reversed.

14.9 Getting all possible transpositions on a 3 by 4. Reload the configuration that you used for Exercise 14.7. (You can either load it from Puzzle Resources or set it yourself: 12 active boxes, one blank, no obstacles, and top-to-bottom wrap turned on.) (a) Perform the transposition (1, 2) on this puzzle. (One way to do this is provided as entry number 10 in the collection of macros in Puzzle Resources. Load and run it if you wish.) (b) Now perform several additional transpositions, specifically (4, 9), (2, 11), and (6, 8). (c) Can you describe a procedure for performing *any* transposition whatever? (Being able to do this would allow us to cite (7.3) to justify the claim that every permutation is achievable, and therefore also solvable, on this configuration of the puzzle.)

14.10 How solvable is this one? Using the Configure button (number array icon) from the button bar, configure your Slide Puzzle as follows: Active spots 17, one blank, no obstacles, top-to-bottom wrap turned on. Determine whether or not this puzzle is totally solvable. [Hint: The 4-cycle (1, 5, 9, 13) = (17,13)(13,9)(9,5)(5,1)(1,17) is possible on this version of Slide. Can you make use of this fact?]

Mastering the Hungarian Rings with Numbers

We now turn our attention to the toughest puzzle in the lot— the Hungarian Rings Puzzle. In spite of this, now that we have the theory available to us, we can make a quick determination that the puzzle is completely solvable. We can also use the theory to find ways to solve this puzzle just like the rest. We will begin by exploring the puzzle in its numbers mode in which all of the disks are distinguishable from each other.

15

15.1 The Theory Assures Complete Solvability

Back in Chapter 10, we showed how you can get a certain 3-cycle on the Hungarian Rings puzzle using a compound commutator. Specifically, we showed that

(15.1) On the Hungarian Rings puzzle, if $\gamma = R^{-1}LR$, then

$$[[L^5, R^5], \gamma] = (L^5R^5L^{-5}R^{-5})(R^{-1}LR)(R^5L^5R^{-5}L^{-5})(R^{-1}L^{-1}R) = (6, 11, 12).$$

We can parlay this knowledge of the existence of a single 3-cycle into assurance that the puzzle is completely solvable. Clearly this puzzle gives enough flexibility to get any three disks into spots 6, 11, and 12. (See Exercise 15.4.) Thus, we can get all of the 3-cycles as conjugates of this one. Since we can get all of the 3-cycles, we can get all of the even permutations of the puzzle.

Since there are twenty disks on each of the rings, both of the rotations L and R are 20-cycles, which means that they are odd permutations. Since we can perform all of the even permutations and at least one odd permutation, this means that *the puzzle is completely solvable.*

What we have said so far assures us that, in theory, the puzzle is always solvable, but that does not tell us how to do it. We still have some work to do if we want to know *how* to solve it.

Using the knowledge transfer insights of Chapter 12, we can set forth one approach, albeit not a very efficient one. Since we know we can get all 3-cycles on the Hungarian Rings, we know we can get the 3-cycle (3, 2, 1). Suppose we program it into macro button A. At this point, the left ring of the puzzle, together with button A, becomes equivalent to Oval Track Puzzle version 3, which has (3, 2, 1) as its "Do Arrows" permutation. Since we know how to get a single transposition on this version of Oval Track by an "around the track" method, we could just transfer this strategy to the Hungarian Rings. This would give us a basic transposition. We could do other transpositions as conjugates of this one. Once we can do transpositions, we can solve anything.

This will be a far from efficient way to do the job, but it gives us an initial way to complete a solution. We will look at more efficient ways in later sections of this chapter.

15.2 Some End Game Tools

We commented back in Chapter 4 that the Hungarian Rings puzzle has a very large end game. After you get approximately the first 20 to 23 disks in place, you need to use theory-based end game strategies to place the final 15 to 18 disks. Given that the opening game of this puzzle is rather straightforward, and given the large proportion of the disks that are involved in the end game, we will start our discussion of solving this puzzle with developing some end game tools.

As always, you are encouraged to follow along with this analysis by opening Multi-Puzzle on your computer, going to the Hungarian Rings, and initializing the puzzle with numbers.

Since this puzzle has an end game that has to deal with a rather large number of disks, it will be an advantage to have some end game tools that place as many disks as possible at one time. The methods that we will develop will make heavy use of a particular eight-step 5-cycle that is very easy to remember. Specifically, we have that

(15.2) On the Hungarian Rings puzzle, $(L^5R^5)^4 = (1, 25, 30, 11, 16)$.

The result of this procedure is shown in Figure 15.1. Note that we are giving each of the rings a five-position turn four times. Since $4 \cdot 5 = 20$, this means that all of the disks that never get into the intersection spots and thus are affected by turns on only one of the rings are moved 20 positions clockwise on their home ring. This, of course, brings them back home.

The disks that get moved into the intersection points are the ones that have home positions at distances away from the intersection points that are multiples of five. These are clearly disks with their numbers in the set $\{1, 6, 11, 16, 25, 30\}$. If you run this procedure on the puzzle, you will see that disk 6 just gets moved back and forth four times between the intersection spots and thus ends up back at home. However, the other five of these disks end up getting moved somewhere else, and form the 5-cycle that we have shown.

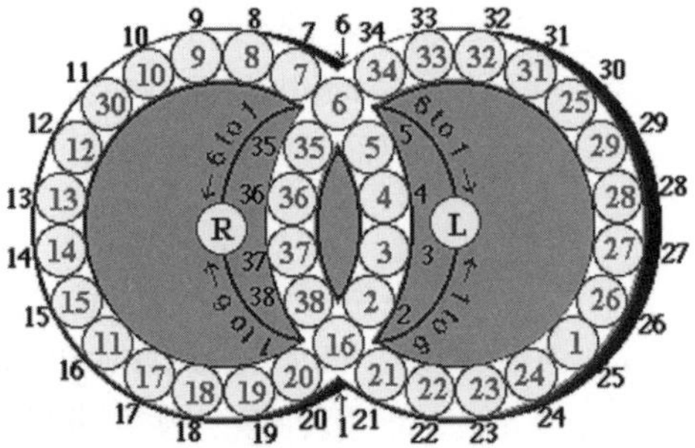

Conjugates of this 5-cycle will be useful to us when there are long cycles that need to be dealt with in the end game. They will allow us to put at least four disks into the right place at one time. We'll get back to the details of this later.

Figure 15.1. A useful 5-cycle end game tool on the Hungarian Rings Puzzle: $(L^5R^5)^4 = (1, 25, 30, 11, 16)$.

Let's consider an end game tool that produces an odd permutation. Theoretical considerations like we used to get claim (11.4) give us some important insight regarding how this must be done. We will need to use an "around the track" method of some sort.

The reason for this is that any procedure that has both of the rings just rocking back and forth and keeping some disks away from the intersection points has to result in an even permutation. We can push this idea a bit further, and, paralleling (11.4), say that

(15.3) On the Hungarian Rings puzzle, any procedure that produces an odd permutation of the puzzle and that returns at least one disk on each of the rings to its home spot must, during the process, move off of the left ring all of the disks from the left ring that ultimately return home or must move off of the right ring all of the disks from the right ring that ultimately return home.

To see why (15.3) is true, suppose that we have a procedure which returns home at least one disk on each ring and in addition, through the entire procedure, keeps at least one of these homeward bound disks from each ring on its home ring. Let us say that disk u is the one from the left ring and disk v is the one from the right ring that stays on its ring through the entire procedure.

Without loss of generality, our procedure can be represented in the form $L^{i_1}R^{j_1}L^{i_2}R^{j_2}\cdots L^{i_m}R^{j_m}$, where m is some positive integer, and all of the i_r's and j_s's are integers. Since disk u stays on the left ring throughout the process, it is affected only by the L's in this procedure. The overall affect on disk u is the same as that of $L^{i_1}L^{i_2}\cdots L^{i_m}$. Clearly this sequence of turns of the left ring moves disk u by $i_1 + i_2 + \cdots + i_m$ positions. (The shift is clockwise if this number is positive and counterclockwise if it is negative.) However since disk u returns to where it started, this total number of shifts must be a multiple of twenty.

This means that the turns of the left ring contribute an even permutation to the entire procedure. The same reasoning applied to the movement of disk v and its ultimate return to its starting place leads to the conclusion that the turns of the right ring also contribute an even permutation. Thus, under these conditions, the permutation resulting from the entire procedure is even.

We must conclude from this that if we are to obtain an odd permutation which fixes at least one disk on each of the rings, then it must be the case that the procedure that does this moves off of one of the rings all of the disks from that ring that ultimately return home.

This insight (15.3) suggests what we have to do in order to get a small odd permutation. We have to devise a procedure that temporarily moves nearly all of the disks off of one of the rings at least once. Let us do this with respect to the left ring. One way to do this is to use repetitions of $RLR^{-1}L$.

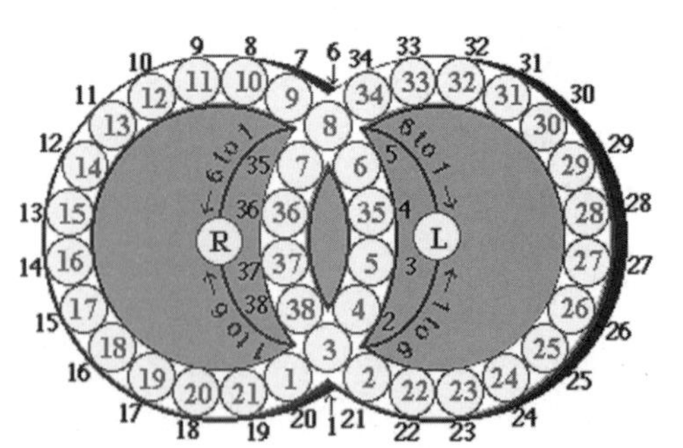

Figure 15.2. The permutation $RLR^{-1}L$ is a first step on the way to a small odd permutation.

Figure 15.2 shows the effect of these four turns. Note that all of the disks on the left ring have advanced two positions clockwise, except for the disks with numbers in the set {1, 2, 6, 7}, and all of the disks from the right ring are in their original places except for those with numbers in the set {21, 35}. Looking at these disks that have been treated differently, we see that disks 1 and 6 have shifted only one position clockwise on the left ring, and disks 2 and 7 are now off of the left ring. Meanwhile, disks 21 and 35 are off of the right ring.

What will a second application of these four turns do? It will, of course, be the same thing as the first time for disks in the corresponding positions. Rather than just thinking this through, let's do them and see. Figure 15.3 shows the result.

We see from Figure 15.3 that most of the disks on the left ring are now shifted four positions clockwise from their home spots. The exceptions are the disks with numbers in the set {1, 2, 3, 6, 7, 8} which have only shifted three positions instead of four. Once again, nearly all of the disks from the right ring are back at home. The exceptions are disks 21 and 35, which were already away from home after the first application of these four moves.

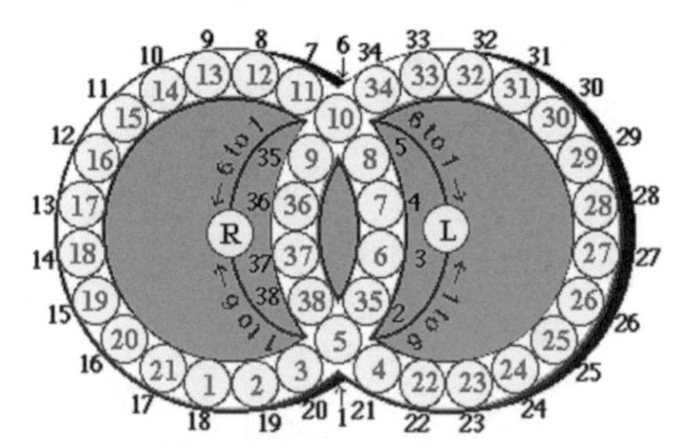

Figure 15.3. Repeat the steps shown in Fig 15.2. Here is $(RLR^{-1}L)^2$.

If you are running these steps on your own puzzle while reading this, which is highly recommended, some additional details may have caught your notice. As of now, the six disks that have shifted only three positions rather than four are exactly the ones that have spent time off of the left ring so far. Each of them was in spot 38 (disks 1 and 3), in spot 21 (disk 2), in spot 34 (disks 6 and 8), or in spot 35 (disk 7) and thus off of the left ring for just one of the applications of L. This explains why these disks have advanced one fewer positions clockwise than the others on the left ring that were there the whole time.

Let's run these four steps a third time and look at the results (Figure 15.4). We have added four more disks to the set that have been off of the left ring for one application of L and thus have shifted one fewer positions clockwise. The new disks to enter this one-fewer category are disks numbered 4, 5, 9, and 10.

We note now that the entire set of disks numbered 1 through 10 on the left ring is in their natural sequence. If we were to continue with another repetition of our basic four-turn procedure, we would disrupt this sequencing. Thus, we'll not do that right away, but instead will apply a five-position clockwise rotation L^5 to the left ring.

Now, let's run $RLR^{-1}L$ another three times. What will the effect be? Well, we have disks 12 through 21 in spots 1 through 10, which is where disks 1 through 10 were before the first three applications of $RLR^{-1}L$. This second time around with $(RLR^{-1}L)^3$ will return disks 12 through 21 to their same order on the left ring, but now shifted five positions clockwise so they are in spots 16, 17,…, 20, 1, 2,…, 5.

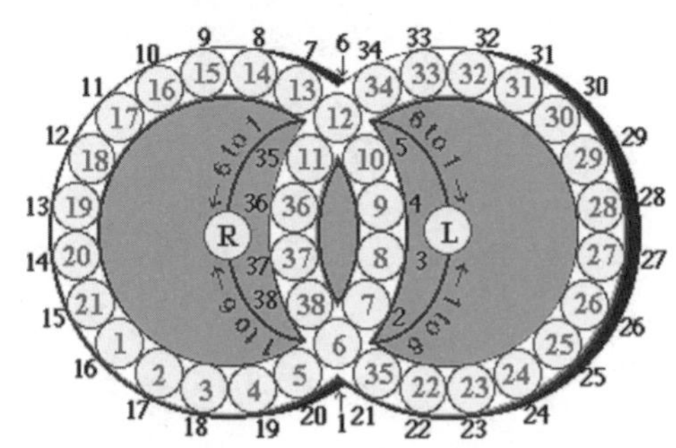

Figure 15.4. Repeat yet again. Here is $(RLR^{-1}L)^3$. Note which disks are in sequence.

Meanwhile, what did this second application of $(RLR^{-1}L)^3$ do to the disks numbered 1 through 10? Since the intermediate L^5 had left them in spots 11 through 20, all of them, except for disk 1, will be too far away from the intersection points to be affected by the applications of R and R^{-1}, so they will be moved six positions clockwise under the six applications of L. This will put disks 2 through 10 into spots 6 through 14.

At this stage, we note that most of the left ring disks, all but disks 1 and 11, are four positions away from their starting spots. From the right ring, we still have only disks 21 and 35 out of position. A final four-position clockwise rotation L^4 of the left ring will get nearly all of the left ring disks back into their starting spots. The final result is that there are only four disks that have moved. Specifically, we have shown that

(15.4) On the Hungarian Rings puzzle, $(RLR^{-1}L)^3L^5(RLR^{-1}L)^3L^4 = (1, 35, 11, 21)$

The final result is pictured in Figure 15.5. We have shown how to get a 4-cycle in 26 steps. This procedure can be collapsed to 24 steps if one joins the intermediate L^5 and the final L^4 with the last L that precedes each of them. We have produced an odd permutation that moves four disks. Given the insight from (15.3), we have done this perhaps as efficiently as we can, because we have done what must be done by moving all of the left ring disks off of the left ring for at least one step

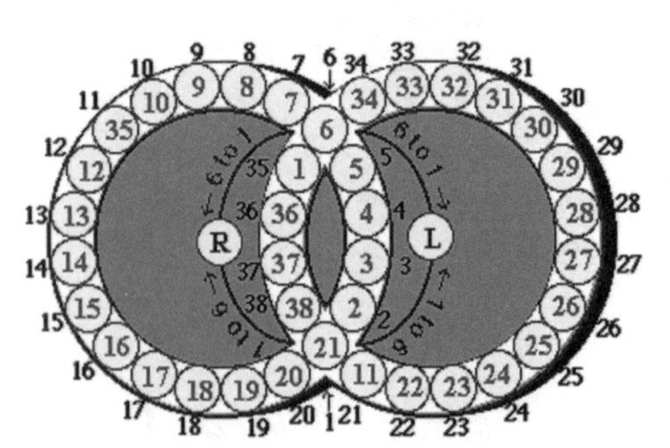

Figure 15.5. The 4-cycle (1, 35, 11, 21) in 26 turns: $(RLR^{-1}L)^3L^5(RLR^{-1}L)^3L^4$.

We now have a 4-cycle to use as a tool for our end game strategy. However there will be times that a single transposition is needed. We can compose this 4-cycle with a conjugate of one of the basic commutators for this puzzle to get a single transposition.

Note what you get in Figure 15.5 if you have disks 1 and 35 trade places, and also have disks 11 and 21 doing the same. You get disks 21 and 35 returning home, disk 1 going to spot 11 and disk 11 to spot 1. We have the transposition (1, 11). We can make this pair of swaps happen as a conjugate of the commutator shown in Figure 10.4, namely $[L^5, R^5] = (1, 25)(6, 11)$.

With the puzzle in the arrangement shown in Figure 15.5, a four-step setup procedure $C = RL^{-1}R^{-6}L$ gets disks 11, 21, 1, and 35 into spots 1, 25, 6, and 11 respectively.

Following this setup with the commutator that produces the pair of transpositions (1, 25) (6, 11), we swap the places for disks 11 and 21 and for disks 1 and 35. If we now undo our setup procedure by applying $C^{-1} = L^{-1}R^6LR^{-1}$, guess what? We've created a fine looking transposition! It's pictured in Figure 15.6.

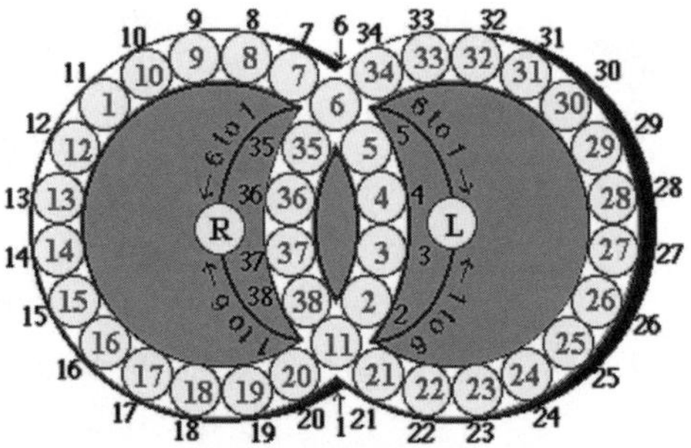

Now that we've calmed down a bit from the thrill of creation, let's take a look at where our new transposition came from. We see that:

(15.5) On the Hungarian Rings puzzle, the following yields the transposition (1, 11):

$$((RLR^{-1}L)^3L^5(RLR^{-1}L)^3L^4)(RL^{-1}R^{-6}L)(L^5R^5L^{-5}R^{-5})(L^{-1}R^6LR^{-1}).$$

Figure 15.6. The transposition (1, 11) is $A(C[L^5, R^5]C^{-1})$, where A is the 4-cycle of (15.4) and $C = RL^{-1}R^{-6}L$.

Note that we have 38 moves when this sequence is listed in a way that reflects its structure. It is possible to reduce this by three given the juxtapositions of rotations of the left ring that occur. Thus, we have produced a single transposition in 35 moves.

We will set forth one final tool that will be of use in solving the Hungarian Rings. Actually, we have made note of it already because it is the simplest type of commutator. Recall that in Chapter 10 we observed that

(15.6) On the Hungarian Rings puzzle $[L, R^{-1}] = LR^{-1}L^{-1}R = (7, 6, 34)\ (2, 1, 38)$.

We record this simple procedure that produces a pair of 3-cycles as yet one more tool in our kit for finishing end games.

15.3 A Numbers Mode Solution Strategy

We now have ways of getting 2-cycles, 3-cycles, 4-cycles, and 5-cycles on the Hungarian Rings. This is a formidable bag of tools for taking on the challenge of solving the puzzle, so let's get on with that task now.

(15.7) **A Hungarian Rings numbers mode solution strategy:**

(i) Starting from a scrambled puzzle, use heuristic means to get 8 consecutively numbered disks, say numbers 11 through 18, into the right order on the left disk. Then get disks 21 through 32 into the right order on the right ring.

(ii) Still using heuristic methods, get a few more disks into place that are nearer to the intersections, taking advantage of opportunities to do so, if any, that do not disrupt the earlier work. Do this whether or not these placements are contiguous to the longer sequences already in order.

(iii) Write down the remaining permutation in disjoint cycle notation.

(iv) Work on cycles that are five disks or longer using conjugates of the 5-cycle of (15.2). If the cycle length is more than five, you will be able to get four disks at a time into their right places. If you have a 5-cycle, you will be able to restore all five of its disks this way.

(v) Use conjugates of the 5-cycle of (15.2) or of the pair of 3-cycles of (15.6) to work on pairs of remaining cycles, such as 4-cycles and 3-cycles. Usually you will be able to get three disks at a time into their right spots.

(vi) If there are 4-cycles that remain, use conjugates of the 4-cycle of (15.4) to put all their disks into place.

(vii) If there are 3-cycles that remain, use conjugates of the 3-cycle of (15.1) to put all their disks into place.

(viii) If there are transpositions that remain, use conjugates of the transposition of (15.5) to put all their disks into place.

You will gain increased understanding of the reasons for suggesting that you work in this order as you solve more scrambles. You will find that limiting yourself to getting eight disks in order on the left ring makes it possible to do sufficient rocking back and forth with the left ring, without disrupting the eight left ring disks already in order, to move right ring disks between the intersection points so that you can place them where they need to go. This will allow you to get at least twelve disks in sequence on the right ring.

Once you get to the point that any further attempts to add disks to your ordered consecutive runs are no longer working out, it still may be possible to put several disks into place because they are close enough to where they need to go so you can get them there without damaging your long runs. Step (ii) tells you to look for such targets of opportunity and take advantage of them. At this point, do not concern yourself with having these new placements contiguous with the long runs. If, for example, you see a way of getting disk 37 into spot 37, even though the disks around it are still out of place, do it. Usually there will be at least a few more that can be put into place this way.

You will find as you work this puzzle that once you have gotten somewhere in the neighborhood of 22 to 24 disks into place, it starts to seem impossible to get more of them into place without messing up the ones you already have. When you get to this stage, you are into the end game. At this point we must bring in our end game tools (developed in advance and based upon group theory). Steps (iii) through (viii) indicate an order for bringing these tools into play that is rather efficient.

The 5-cycle (1, 25, 30, 11, 16) is an efficient one to use since it is so easy to remember and it takes only eight very repetitive moves. Setting up conjugates for it can be a bit tedious, but the result is the ability to get four or five disks into place at a time. Also, programming the setup procedure into a macro button allows us to run its inverse easily.

Enough generalities: It is time to look at a specific example. As usual, we will be leaving you on your own to develop heuristics for the opening game, so we will start our example with a scramble that has been worked through the first two steps. Such an example is given in Figure 15.7. (You can set your puzzle into this configuration using Resources. See Exercise 15.1 for details.)

We see from Figure 15.7 that we have a run from disk 11 through 19 that are all in place on the left ring and our consecutive run on the right ring goes from disk 24 through disk 33. This is the result of step (i) of the strategy (15.7). We also have a few isolated disks in their right places as a result of step (ii) heuristics, namely those numbered 7, 8, and 21.

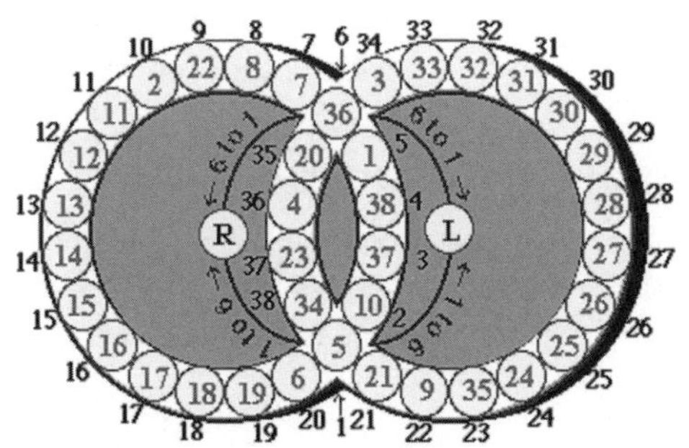

Figure 15.7. A typical scramble, solved through the opening game and ready for end game methods.

There are a total of 22 disks in their right places, which means that we still have 16 of them to put right. (By the way, with only minor attention to efficiency, we ran the counter up to 88 in getting the puzzle to this stage.)

As suggested in step (iii), we write the remaining permutation in disjoint cycle notation. We see that it is

$$(1, 5)\,(2, 10)\,(3, 34, 38, 4, 36, 6, 20, 35, 23, 37)\,(9, 22).$$

We have one long cycle, a 10-cycle, and three transpositions. All four of these cycles are odd permutations, so this permutation turns out to be even. This means that, in spite of the presence of three transpositions, we ought to be able to solve this scramble using our even tools, namely the 5-cycle, the 3-cycle, and the double 3-cycle. These are, of course, preferable because they take fewer steps, not being "around the track" types.

Let us begin by whittling down the long 10-cycle. The 10-cycle shows the state of the puzzle, so what we need to do to solve the puzzle is to apply its inverse (37, 23, 35, 20, 6, 36, 4, 38, 34, 3). For purposes of defining a setup procedure, we find it helpful to write this cycle notation using arrows as follows:

$$37 \to 23 \to 35 \to 20 \to 6 \to 36 \to 4 \to 38 \to 34 \to 3$$

One can interpret this diagram as telling the story in two different ways. On the one hand, we can regard it as telling us that the disk in spot 37 needs to go to spot 23, the disk in spot 23 needs to go to spot 35, the disk in spot 35 goes to spot 20, etc. If we look at Figure 15.7, we see that this describes what needs to happen to get the puzzle solved.

The other interpretation focuses on the numbers *on the disks* as opposed to the locations. Thus, disk 37 needs to go where disk 23 is, which we can say more briefly as "disk 37 chases disk 23." Continuing with this description, disk 23 chases disk 35, which chases 20, which chases 6, etc. Again, Figure 15.7 shows that if we can do this, we get closer to solution.

Focusing on the disk numbers, we will now set things up to use our 5-cycle to get the first four of them into their right places. We write out a diagram that indicates how we want to do our setup procedure:

$$\begin{array}{ccccccccc} 37 & \to & 23 & \to & 35 & \to & 20 & \to & 6 \\ \downarrow & & \downarrow & & \downarrow & & \downarrow & & \downarrow \\ 1 & \to & 25 & \to & 30 & \to & 11 & \to & 16 \end{array}$$

The first row of this array has numbers of disks that are out of place and that need to chase their followers in the list. The second row has numbers that we interpret as location numbers. If we can put the disks with numbers in the first row into the locations with numbers in the second, then our simple 5-cycle will have the first four of them chasing the ones they need to chase to get us closer to a solution.

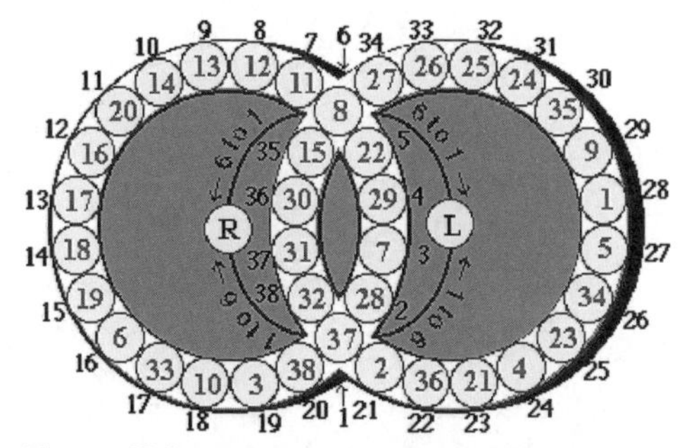

Figure 15.8. The setup for using our tool 5-cycle to get disks 37, 23, 35, 20, and 6 into place.

With this diagram in front of us, we run a setup procedure to place our target disks in this way. We, of course, make sure that we start the process of programming the procedure into button C so that we can automatically reverse it a bit later. A setup procedure that does the job here consists of the 9 steps $R^{-7}L^{9}R^{8}L^{-5}R^{-8}L^{-2}RL^{2}R^{-1}$. The result is shown in Figure 15.8. (See Exercise 15.2 for help in learning how to do this sort of setup.)

When we run the easy 5-cycle $(L^5R^5)^4 = (1, 25, 30, 11, 16)$, we perform a 5-cycle (37, 23, 35, 20, 6) and here we are referring to disk numbers rather than locations. When we reverse the setup procedure, which we can do simply by pressing the inverse macro button C^{-1}, we restore everything except that the disks 37, 23, 35, 20, and 6 have chased after the next in the list. The first four of them, in doing so, have gone to their home spots on the puzzle, as you can see in Figure 15.9.

The state of the puzzle is now four disks closer to solution. Compare Figure 15.9 with Figure 15.7 to see this. The current state of the puzzle is expressed in disjoint cycle notation as $(1, 5)(2, 10)(3, 34, 38, 4, 36, 6)(9, 22)$. This is close to what it was before, except that the last four numbers were removed from the 10-cycle making it a 6-cycle.

Let's take another whack at the longest cycle. Our setup diagram for using our tool 5-cycle this time is

$$\begin{array}{ccccccccc} 6 & \rightarrow & 36 & \rightarrow & 4 & \rightarrow & 38 & \rightarrow & 34 \\ \downarrow & & \downarrow & & \downarrow & & \downarrow & & \downarrow \\ 1 & \rightarrow & 25 & \rightarrow & 30 & \rightarrow & 11 & \rightarrow & 16 \end{array}$$

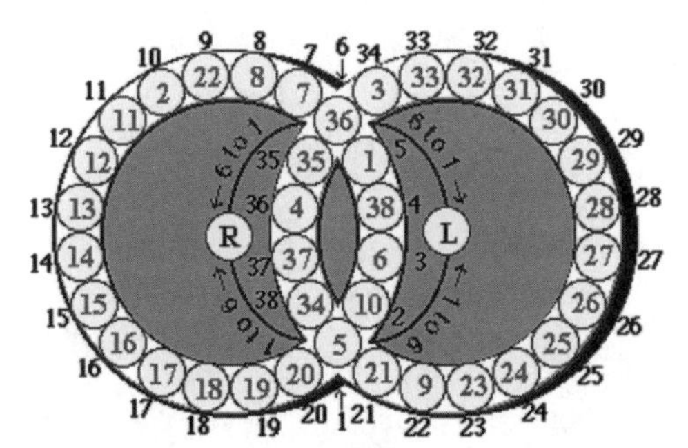

Figure 15.9. Apply our tool 5-cycle and invert the setup and disks 37, 23, 35, and 20 are now in place.

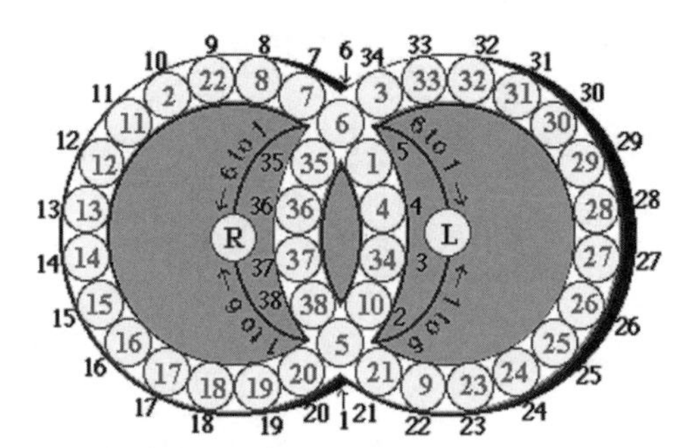

Figure 15.10. A second use of our tool 5-cycle puts disks 6, 36, 4, and 38, into place.

Remember we want to undo it (apply its inverse) later, so we will record it in a macro button. If we do this setup, run our tool 5-cycle, and undo the setup, the result is as shown in Figure 15.10. The remaining permutation is now (1, 5)(2, 10)(3, 34)(9, 22).

At this stage, we could process these transpositions one by one using conjugates of our tool transposition in (15.5). Each one of them would take the 35 steps of (15.5), plus perhaps 3 to 5 steps for the setup and a like number to undo the setup. The counter would go up by about 175 doing it this way. We ought to be able to improve on that.

Given the nature of the remaining permutation, a good tool to use is the very simple commutator $[L, R^{-1}] = (38, 2, 1)(34, 7, 6)$ set forth in (15.6). This tool suggests itself because we can express a pair of disjoint transpositions by using two 3-cycles. Here we are faced with *two pairs* of disjoint transpositions. How about inverting them by using a tool twice that does two 3-cycles at a time?

We could use our pair of disjoint 3-cycles to get one of the disks from each of the first two transpositions into its right place and incorporate the other disk into one of the remaining transpositions. This will leave us with two 3-cycles, which we can solve by using this same tool again.

Had trouble following that? Stay with us while we lay out the details. Let's do a setup as represented by this diagram.

$$\begin{array}{ccccccccccc} 5 & \rightarrow & 1 & \rightarrow & 3 & \quad & 10 & \rightarrow & 2 & \rightarrow & 9 \\ \downarrow & & \downarrow & & \downarrow & & \downarrow & & \downarrow & & \downarrow \\ 38 & \rightarrow & 2 & \rightarrow & 1 & & 34 & \rightarrow & 7 & \rightarrow & 6 \end{array}$$

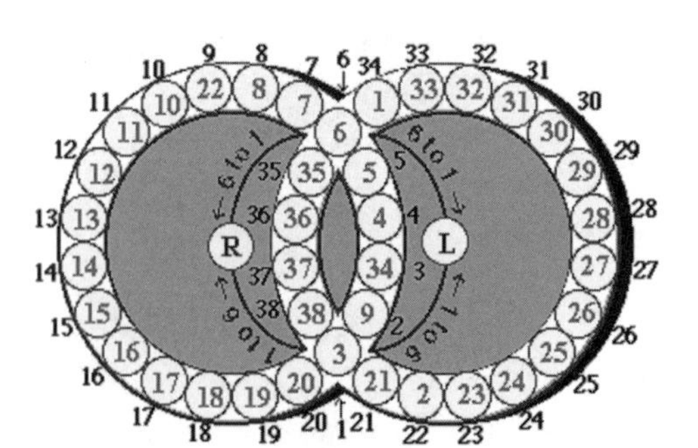

Figure 15.11. A double 3-cycle puts two more disks in place and leaves another pair of 3-cycles.

This shows how we want to place six of our remaining dislocated disks into the spots acted upon by our tool commutator. If we do this setup procedure, then run the commutator $[L, R^{-1}] = (38, 2, 1)(34, 7, 6)$, then undo the setup, the result is shown in Figure 15.11.

We see that disks 5 and 10 are now in place. Also, disk 1 has joined the transposition (34, 3) to form the 3-cycle (1, 34, 3) while disk 2 has joined with the transposition (22, 9) to form the 3-cycle (2, 22, 9).

Was this a worthwhile thing to do? It took 15 moves to set up for doing the pair of 3-cycle, four moves to do this pair of 3-cycles, and 15 to undo the setup. This is a total of 34 moves to get two disks in place. However, we have set up a situation wherein we have a pair of 3-cycles remaining, and we can apply the same methods to get all of these remaining six disks back in place in about the same number of moves. This will be quite a bit shorter than the 175 moves we contemplated if we handled each transposition separately.

We now have the permutation (1, 34, 3) (2, 22, 9) on the puzzle. We will invert this pair as a conjugate of the same tool commutator $[L, R^{-1}]$. Here is the setup diagram that we will use:

$$\begin{array}{cccccccccccc} 1 & \rightarrow & 3 & \rightarrow & 34 & \quad & 2 & \rightarrow & 9 & \rightarrow & 22 \\ \downarrow & & \downarrow & & \downarrow & & \downarrow & & \downarrow & & \downarrow \\ 38 & \rightarrow & 2 & \rightarrow & 1 & & 34 & \rightarrow & 7 & \rightarrow & 6 \end{array}$$

From the state that is shown in Figure 15.11, getting these six disks into the six desired positions is a bit trickier than you might think at first. It takes a bit of planning ahead to avoid inadvertently messing up parts that you had already gotten right.

One plan that can get the job done without too much trouble builds on the fact that four of the positions are on the left ring (1, 2, 6, and 7) and two are on the right, namely 34 and 38. Here is the plan.

1. Get the disks for spots 7 and 6 together in the right sequence and on the left ring.
2. Make sure that these spot 7 and 6 disks stay away from the intersection points while you get the disk for spot 2 onto the left ring so that it is five positions clockwise away from the disk for spot 7. Then put the disk for spot 1 next to it.
3. Rotate the left ring so that these four disks are well out of the way. Then get the disk for spot 34 into spot 30. When you have done this, then get the disk for spot 38 into spot 6.
4. Rotate the right ring so that the disks for spots 38 and 34 are in their right spots.
5. Rotate the left ring so that the four disks on it are in their right spots.

A sequence of moves that can be used to carry out this plan is:

$$C = L^{-4}RL^{8}R^{-1}L^{-7}RL^{-2}RL^{4}R^{-1}L^{-1}R^{-1}L^{-1}R^{9}L^{-2}R^{6}L^{8}$$

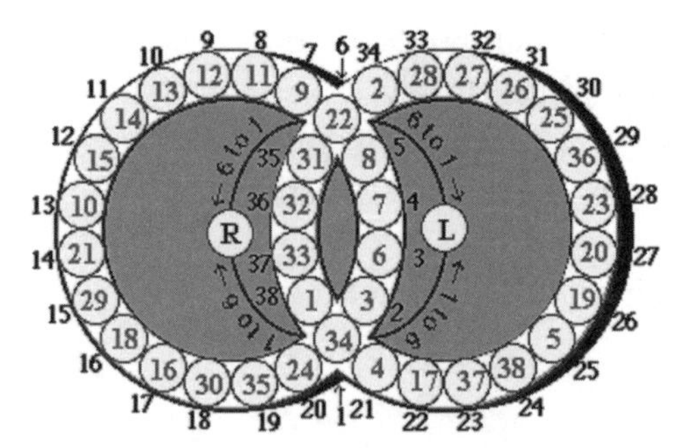

Figure 15.12. A setup for the last double 3-cycle puts disks 1, 3, 34, 2, 9, 22 into spots 38, 2, 1, 34, 7, and 6.

We carry out this sequence, programming it into macro button C, and the result is shown in Figure 15.12. It is disquieting to see the disks appearing so mixed up again, but we are really close to being done now. We apply the commutator $[L, R^{-1}]$ and then let the computer invert the setup procedure C by pressing the macro button C^{-1}, and *voila*! It is now fully solved!

This worked out example with numbers will be all that we do in such detail. The exercises take you through several more, and give you guidance for loading certain scrambles from the Resources collection so that you can work on specific ones with assistance that is specific to them.

In addition, the exercises provide practice with the setup steps for using the basic tool procedures. With practice and tenacity, you can learn how to solve this very challenging puzzle.

Exercises

15.1 The Figure 15.7 scramble. Load the Exercise 15.1 preset configuration from Puzzle Resources into the Hungarian Rings puzzle. (See Appendix A if you need to learn how to do this.) This will put it into the state shown in Figure 15.7. Now work through for yourself the solution procedure for this example described in this chapter.

Exercises 15.2 through 15.5 give practice with doing the setup procedures for performing conjugates of the tool moves used for solving the Hungarian Rings Puzzle in numbers mode. In each case, you load a configuration from Puzzle Resources, which has the disks for the key spots labeled with numbers and, all of the rest with a single color. With this initialization, you can scramble the disks and practice putting the target disks where they need to go in order to use the tool being studied. Placing disks into the spots acted upon by our several tool procedures is a necessary skill if you want to get small changes as conjugates of these tool procedures.

15.2 Practice with the 5-cycle tool. Load the Exercise 15.2 preset configuration and practice setting up for conjugates of the tool 5-cycle $(L^5 R^5)^4 = (1, 25, 30, 11, 16)$. To do this, have the computer scramble the disks. Then return the numbered disks from where they have landed to their home spots. [Hint: After scrambling, first get disks 25 and 30 placed properly on the right ring, then do the same with disks 11 and 16 on the left ring, and finally work on disk 1.]

15.3 Practice with the 4-cycle tool. Load the Exercise 15.3 preset configuration and practice setting up for conjugates of the tool 4-cycle, (1, 35, 11, 21), of (15.4). [Hint: First get disks 21 and 35 spaced five spaces apart on the right ring, with disk 35 being clockwise from disk 21 going the shorter way, and then get disks 1 and 11 ten spaces apart on the left ring.]

15.4 Practice with the 3-cycle tool. Load the Exercise 15.4 preset configuration and practice setting up for conjugates of the tool 3-cycle, (6, 11, 12), of (15.1).

15.5 Practice with the double 3-cycle tool. Load the Exercise 15.5 preset configuration and practice setting up for conjugates of the tool pair of 3-cycles, (38, 2, 1) (34, 7, 6). [Hint: First get the disks into the counterclockwise order 38, 1, 2, B, B, 34, 6, 7 on the left ring. Then rotate the left ring to get disk 38 into spot 1 and perform $R\ L$.]

Exercises 15.6 through 15.9 present end games which require, in each case, just one of the end game tools. Load the end game settings from Puzzle Resources and use conjugates of these tools to solve the puzzles. The skills developed in Exercises 15.2 through 15.5 should help you do this.

15.6 A 5-cycle end game. Load the Exercise 15.6 preset configuration that presents an end game requiring the 5-cycle tool. Solve this end game.

15.7 A 4-cycle end game. Load the Exercise 15.7 preset configuration that presents an end game requiring the 4-cycle tool. Solve this end game.

15.8 A 3-cycle end game. Load the Exercise 15.8 preset configuration that presents an end game requiring the 3-cycle tool. Solve this end game.

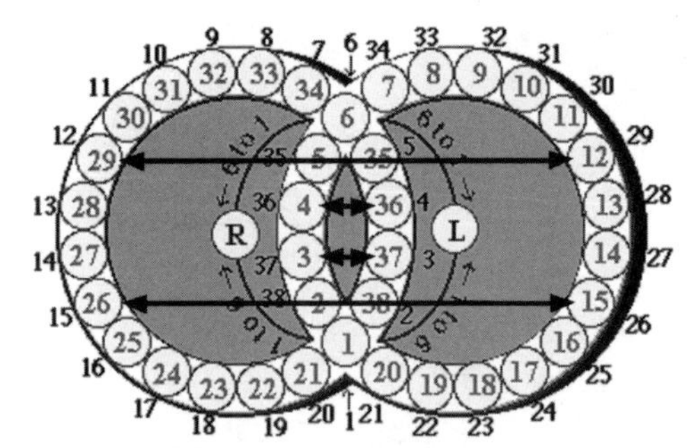

Figure 15.13. An arrangement that reflects disks through the vertical centerline.

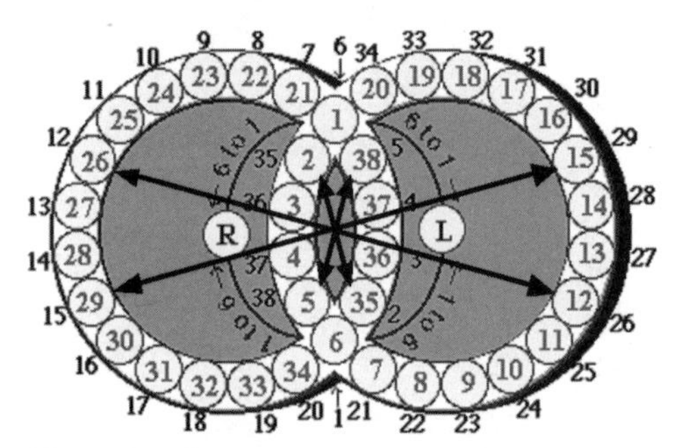

Figure 15.14. An arrangement that reflects disks through the center.

15.9 A double 3-cycle end game. Load the Exercise 15.9 preset configuration that presents an end game requiring the two 3-cycles tool. Solve this end game.

15.10 Reflection through the vertical axis of symmetry. Figure 15.13 shows an arrangement of the disks that has each of them moved from its home spot to the spot that is opposite it when reflected through the vertical axis of symmetry of the puzzle. Find a way of putting the puzzle into this pattern. (If you need help, you will find a way to do this as entry 8 in the Locally Resident Macros collection of Puzzle Resources.)

15.11 Reflection through the center. Figure 15.14 shows an arrangement of the disks that has each of them in the spot that is opposite its home spot when reflected through the center of the puzzle. It is very easy to do this with a very long shuffle. (See the next chapter for the definition of a shuffle.) Entry 9 in the Locally Resident Macros is a procedure for getting to this arrangement in 261 steps.

(a) Can you explain why this particular number of shuffle steps is what works?

(b) Can you find a shorter way to get to this arrangement?

15.12 Reflection through a horizontal axis of symmetry. Find an efficient way to produce an arrangement of the disks that reflects each of them through the horizontal axis of symmetry. (We know that every permutation is possible, so this can be done. The question is how efficiently can it be done.)

15.13 Improving your average. Scramble and solve the Hungarian Rings Puzzle in numbers mode about 10 times. Keep a record of how many steps it takes to solve the puzzle, and then determine your average. See if, with practice, you can lower your average, and formulate some ideas on what is the ultimate average for the most efficient way to solve the scramblings. (On a series of 33 scrambles, the author solved them with the number of steps ranging from 154 to 233, with an average of 185.5.)

Mastering the Hungarian Rings with Colors

Having set forth the details for solving the Hungarian Rings Puzzle in numbers mode in Chapter 15, it is time now for us to consider it in its four-color mode. We will see that with the puzzle configured in this different way, we need different strategies for solving it.

16.1 Comparison of Solvability and the Number of Distinguishable Arrangements

We showed in the previous chapter that the Hungarian Rings Puzzle is always solvable when it is in numbers mode. Its solvability in numbers mode assures its solvability in four-color mode as well.

To see this, imagine that you start with the puzzle initialized in four-color mode and with each of the colors grouped together in the initial configuration. Suppose you add numbers to each of the colored disks, making the numbers agree with those of the spots in which you initially find the disks.

Now scramble the disks. By our work in Chapter 15, we know that we can get every disk back to where it came from according to their numbers. Of course, this would return them to positions that restore the initial pattern of colors as well. Clearly, in solving the puzzle number-wise, we have also solved it color-wise.

In returning each colored disk to the spot from which it came, we have in fact been far more precise than we need to be to solve the color-mode puzzle. Since we cannot distinguish between disks of the same color, it does not really matter that we return them to their individual starting spots. We just need to place every one of the ten red disks into one of the ten spots where red disks should go, and the same for each of the other three colors.

This flexibility makes a huge difference in terms of the numbers. When the puzzle is in numbers mode, there is only one correct solution. After a scramble, you must return disk i to spot i for $i = 1, 2, 3, \ldots, 38$.

How many ways are there to put the ten red disks into the ten spots that need red disks? The number is $10!$. You have 10 choices for the red disk that you place into the first red spot, then 9 choices for the second one you place,

16

etc. Since there are ten spots for green disks, you have a similar 10! ways to place the green disks. There are nine each of blue and yellow disks, so the number of ways of placing these colors is 9! each.

We get the total number of ways to solve the puzzle in color mode by multiplying these four factorial numbers. This gives us the fact that

(16.1) In color mode, there are $(10!)^2 (9!)^2 = 1{,}734{,}012{,}131{,}277{,}275{,}136{,}000{,}000$ different, but indistinguishable ways to solve the Hungarian Rings puzzle.

That is a whopping 1.7 septillion (Is that really a word?) ways to do the job! It seems especially large when compared with one way in number mode. It ought to make some difference in the ease of solving the puzzle. We will see that it does, but not as much as the size of the number might lead us to expect.

Perhaps some feeling for why having 1.7 septillion ways to solve the puzzle in colors mode in comparison with one way in numbers mode does not make such a difference will come from looking at some other numbers. The number of scramblings that are possible on the puzzle in numbers mode is of course 38!, which is on the order of 5.23×10^{44}. If we divide this number by $(10!)^2(9!)^2$, we will get the number of distinguishable scramblings in the four-color mode. This works out to 3.02×10^{20}. We still have plenty of different scramblings that we need to master.

16

16.2 Some Colors Mode End Game Tools

The fact that there are sets of disks that are indistinguishable from each other makes things harder in some ways while making it easier in others. The intricate setup procedures that we described for the numbers mode become impractical. Because you cannot distinguish between the disks, you tend to get lost after a few steps. At least I do. The response is to adopt strategies that do not require lengthy *ad hoc* setups but instead use *repetitive* setups.

Double 3-Cycles: One strategy that works well makes use of the simple double 3-cycle commutator $[L, R^{-1}]$ of (15.6). A typical way to use it is to get three disks that we need to rearrange with a 3-cycle into the spots 1, 2, and 38 while getting three disks of the same color into spots 6, 7, and 34. We then run this commutator or its inverse to perform the desired 3-cycle on the disks in spots 1, 2, and 38. We also get a 3-cycle on the disks in spots 6, 7, and 34, but since they are all of the same color, this 3-cycle goes unnoticed and does not cause us any problems.

Typically, when we are using this strategy, we will have two colors of disks in the spots 1, 2, and 38, let us say, for example, two blue and one yellow. When we do the 3-cycle on these spots, invariably one of the blue disks will move to where the yellow one was located, and the yellow disk will have to move to where a blue one had been. The visual effect will be a transposition between a blue and the yellow disk, even though we really have done a 3-cycle. Since we know that transpositions are fundamental, we can infer that if we have this way of doing such pseudo-transpositions, we ought to have a tool that could by itself lead to a solution.

Since this four-step commutator and its companion $[R, L^{-1}]$ are of considerable importance in mastering the four-color version of Hungarian Rings, we will discuss them in a bit of detail now.

Figure 16.1 shows the patterns of these two basic commutators. We see that $[L, R^{-1}]$ does its 3-cycles on what we will call the *upper outer triangle* and the *lower inner triangle*, while $[R, L^{-1}]$ does its 3-cycles on the *lower outer triangle* and the *upper inner triangle*. The definitions of these are:

The upper outer triangle: Spots 7, 6, 34
The upper inner triangle: Spots 5, 6, 35
The lower inner triangle: Spots 2, 1, 38
The lower outer triangle: Spots 20, 1, 21

Notice also that the rotation of the 3-cycles on the inner triangles is clockwise in both cases while the rotation is counterclockwise on the outer triangles in both cases. If we use the inverse of these commutators, then we reverse all of the rotations, making them go counterclockwise on the inners and clockwise on the outers.

Shuffling: As we said earlier, we are typically going to exploit these commutators by getting three disks of the same color into one of the outer triangles while getting three disks for which we want a visible change onto the inner triangle. We could do the opposite, namely get the identical disks onto an inner triangle, but there is a reason why the same-on-outer strategy is natural. The reason is that "shuffling" tends to favor this approach.

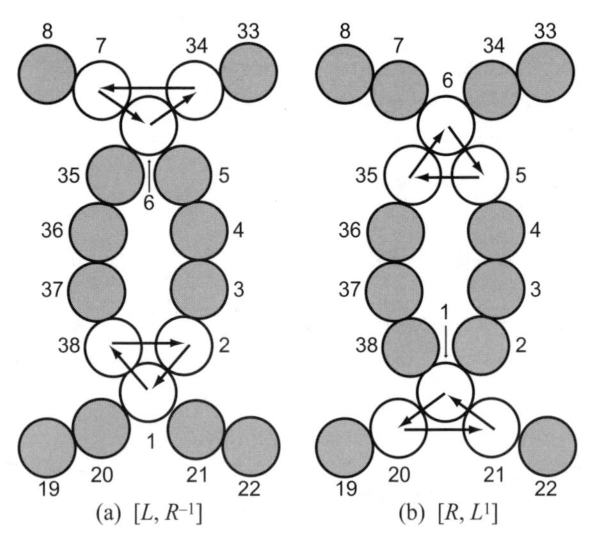

Figure 16.1. Two basic commutators used as tools for solving in four-color mode.

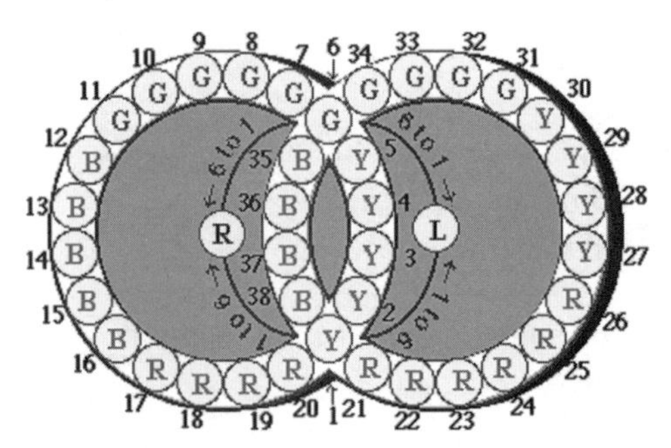

Figure 16.2. Four steps of a left-first, clockwise shuffle, $(LR)^4L$.

So what do we mean by shuffling? Go to your puzzle and initialize it in four-color mode. Now, perform a sequence of four pairs of clockwise rotations, first on the left ring and then on the right one, followed by one more left rotation. That is, do the move $(LR)^4L$. The result is shown in Figure 16.2.

We see that four of the green disks from the left ring have been shuffled over to the right ring on the top and, on the bottom, four red disks have been shuffled over to the left ring. While this was happening, there has been a clockwise rotation of disks on what we will call the *inner loop*.

By the inner loop, we mean the portions of the rings that are between the intersection points and on the inside. One of the intersection points will be on the inner loop and the other not, depending on the situation. We will also use the term *outer loop* to refer to the loop of locations that is disjoint from the inner loop.

Using this terminology, our sample shuffle shown in Figure 16.2 has produced a clockwise rotation of the disks on the outer loop and a clockwise rotation of disks on the inner loop. Spot 1 is on the inner loop in this case, and spot 6 is on the outer one. To get a better feeling for this, you might want to return to your puzzle, convert to numbers mode, and run these steps again. Having each disk numbered will make the rotation pattern stand out in minute detail.

If this is an example of a shuffle move, exactly what is it that we mean by this term? We mean that:

(16.2) A *shuffle* is any sequence of steps on the Hungarian Rings Puzzle of the form $(LR)^iL$ or $(RL)^iR$ or $(L^{-1}R^{-1})^iL^{-1}$ or $(R^{-1}L^{-1})^iR^{-1}$ where i is a positive integer. We will call these, in order, *left first clockwise*, *right first clockwise*, *left first counterclockwise*, and *right first counterclockwise* shuffles.

Shuffles will be very standard procedures for setting up to use the double 3-cycle commutators shown in Figure 16.1. Typically we will have groups of the same color on the outer loop. We will shuffle them around to one of the outer triangles while shuffling three distinguishable disks from the inner loop onto the opposite inner triangle. When we run the right double 3-cycle, we can make a desired change on the three disks on the inner triangle.

Figure 16.3 summarizes the way the disks move under the various types of shuffles.

Jump Rotations: One more useful end game procedure we will need is called a *jump rotation*. Basically, a jump rotation is a procedure that causes the 18 disks on one of the rings, not counting those in the intersection points, to rotate while jumping over the intersection points. (There is just a bit more clutter that we are ignoring for the moment.)

Here is the formal definition.

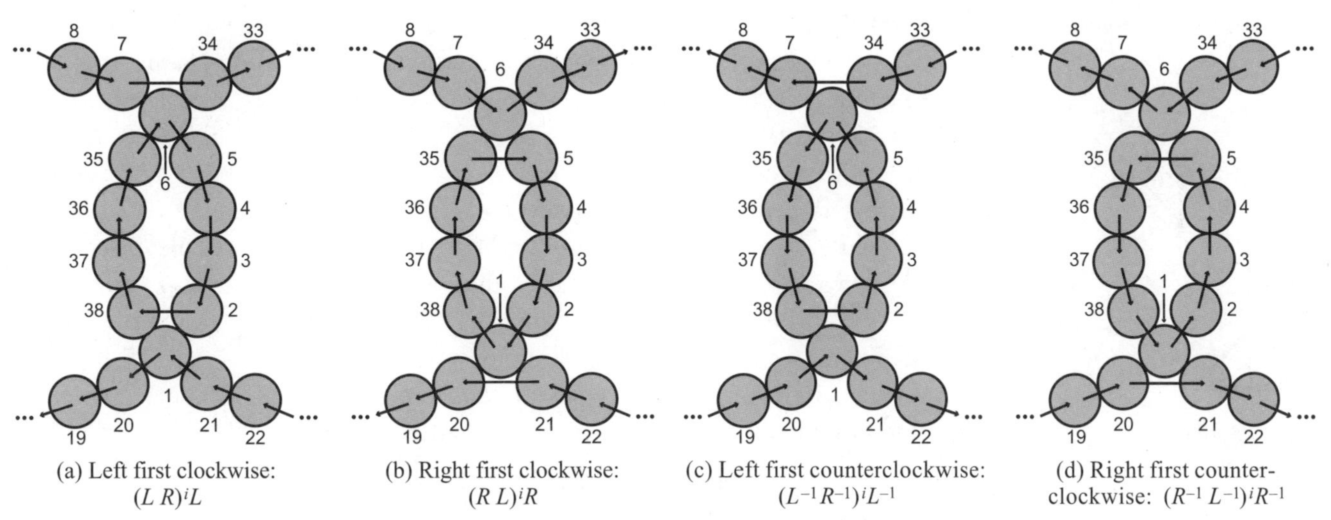

(a) Left first clockwise: $(L\,R)^i L$

(b) Right first clockwise: $(R\,L)^i R$

(c) Left first counterclockwise: $(L^{-1}R^{-1})^i L^{-1}$

(d) Right first counterclockwise: $(R^{-1}\,L^{-1})^i R^{-1}$

Figure 16.3. Rotation patterns for the four shuffles

(16.3) A *jump rotation* is a sequence of steps on the Hungarian Rings Puzzle of the form
$[L, R^{-1}]L = LR^{-1}L^{-1}RL = (1, 38)(6, 34)(20, 19, \ldots, 7, 5, 4, 3, 2)$, or
$[R, L^{-1}]R = RL^{-1}R^{-1}LR = (1, 20)(6, 5)(38, 37, 36, 35, \ldots, 22, 21)$, or
$[L^{-1}, R]L = L^{-1}RLR^{-1}L^{-1} = (1, 21)(6, 35)(2, 3, 4, 5, 7, 8, \ldots, 19, 20)$, or
$[R^{-1}, L]R = R^{-1}LRL^{-1}R^{-1} = (1, 2)(6, 7)(21, 22, 23, \ldots, 37, 38)$.

Each of these procedures results in a rotational 18-cycle being done on the disks on one of the rings not including the two at the intersection points. In addition, each of the disks at the intersection points is involved in a transposition with one of its neighbors on the other ring. (This is the bit of clutter that we alluded to earlier.)

Jump rotations are useful when it is necessary to move disks on one of the rings from the outer to the inner loop or vice versa without making other visible changes. We will see some of their uses when we start working out an example.

16

Looking at the notation for jump rotations that we just presented, you might think it would be difficult to commit jump rotations to memory and use them automatically. If you rely on notation, this is the case, but we can point out some features which make it easy to use "muscle memory" to remember what you need to do in a given situation. We recommend that you do the rotations by pressing rotation buttons rather than dragging when you are doing jump rotations because there is a nice pattern that can guide your actions.

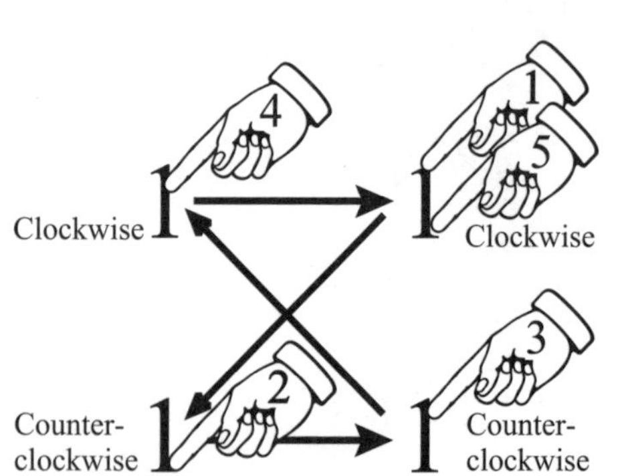

Figure 16.4. The crisscross button-pushing pattern for the "jump-rotate" procedure $[R, L^{-1}]R$ that does a clockwise jump-rotation on the right ring.

Suppose you decide that you need to do a jump rotation that moves the disks on the right ring clockwise. Remember this: Start with a rotation that goes the way you want on the ring you want. For this example, start with *R*, which you do by pressing the Right Clockwise button number 1. The next thing to remember is that you press all the other buttons numbered 1 one time, and return to your first button for a fifth and final press. You want to move between buttons with this pattern:

Diagonal, Horizontal, Diagonal, Horizontal

Figure 16.4 shows the button-pressing sequence for the right clockwise jump rotation.

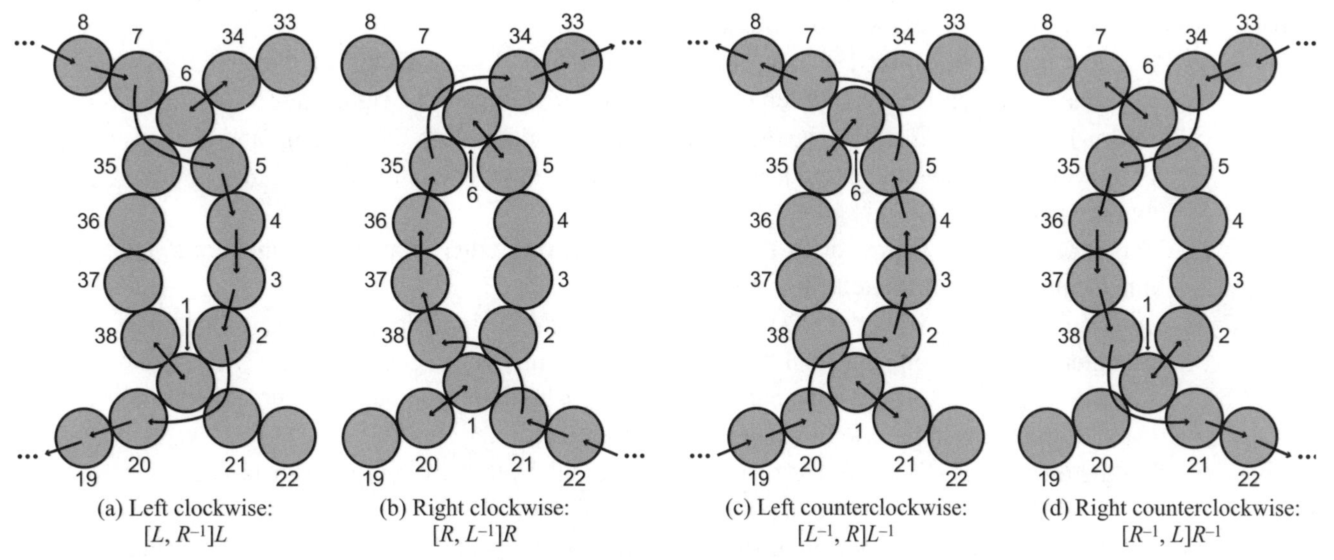

(a) Left clockwise: $[L, R^{-1}]L$

(b) Right clockwise: $[R, L^{-1}]R$

(c) Left counterclockwise: $[L^{-1}, R]L^{-1}$

(d) Right counterclockwise: $[R^{-1}, L]R^{-1}$

Figure 16.5. The four jump rotation patterns.

You get the rest of them by starting at a different one of the one-position buttons and following the arrows, starting with the diagonal one, until you return to where you started. If the diagonal arrow is pointing toward you, as it is if you start with a left ring rotation, just go against the grain.

Figure 16.5 shows with arrows the way the disks move under the four jump rotations.

16.3 A Colors Mode Strategy

Now that we have set forth the end game tools that we will use, we are ready to describe a strategy.

(16.4) A Hungarian Rings four-color mode solution strategy:

(i) Starting from a scrambled puzzle, use heuristic means to get all ten of the green disks together on the left ring and all ten of the red disks together on the right ring.

(ii) Still using heuristic methods, get as many of the blue disks onto the left ring and as many of the yellow disks onto the right ring as you can without breaking up the groupings of green and red disks.

(iii) Target one or more yellow disks that are on the left ring outer loop and get them onto the inner loop portion of the left ring using jump rotations.

(iv) Do a shuffle to get two yellow disks on the left ring and the inner loop onto one of the inner triangles along with one blue disk that they need to jump over to get to the right ring. At the same time, get disks of the same color into the opposite outer triangle. Then do a double 3-cycle to get the yellow disk closer to the right ring. Undo the shuffle.

(v) Do the analogs of steps (iii) and (iv) to move blue disks that are on the inner loop portion of the right ring to the inner loop portion of the left ring.

(vi) Once all of the blue disks are on the left ring and all of the yellow disks are on the right ring, use jump rotations to bring them together.

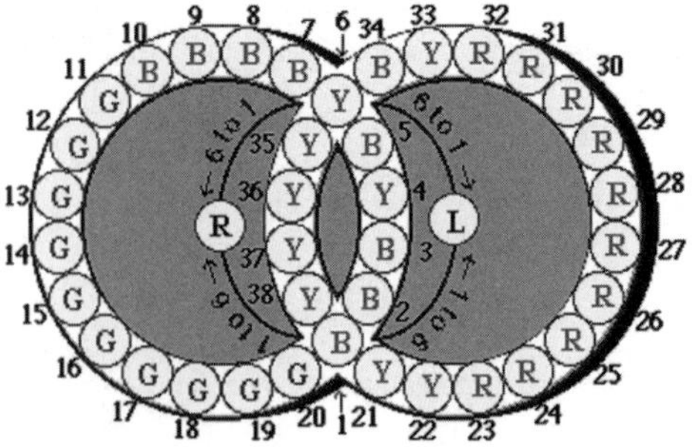

Figure 16.6. This is what remains of our sample scramble after the first two opening game steps.

16.4 A Detailed Worked Out Example

Let's work through an example in detail to illustrate how this strategy plays out in practice. You can load this example into your puzzle from Puzzle Resources and follow along step by step. See Exercise 16.1 for the details.

Since the opening game of this puzzle is really rather straightforward, we will leave it to you to work through the first two steps of the strategy, and we will look at an example that picks up after these steps have been performed. Figure 16.6 shows an arrangement of the puzzle after a scramble and the application of the first two steps. (For this example, it took 29 steps to process the green and red disks, getting them together on the left and right rings, respectively. This is fairly typical. We also used 5 more steps to rearrange a few of the blue and yellow disks that moved easily.)

We now must plan our end game strategy. We see that we have just one blue disk that is on the right ring, in spot 34, and we need to move it to the left ring. Using the step (iii) suggestion, we will do two right counterclockwise jump rotations to get this blue disk onto the inner loop. Since we are going to do two of these jump rotations, we do not have to worry about the transpositions that will happen on the left ring because when we do the jump rotation twice, these transpositions will be performed twice, and that will return the disks that are affected to their original spots.

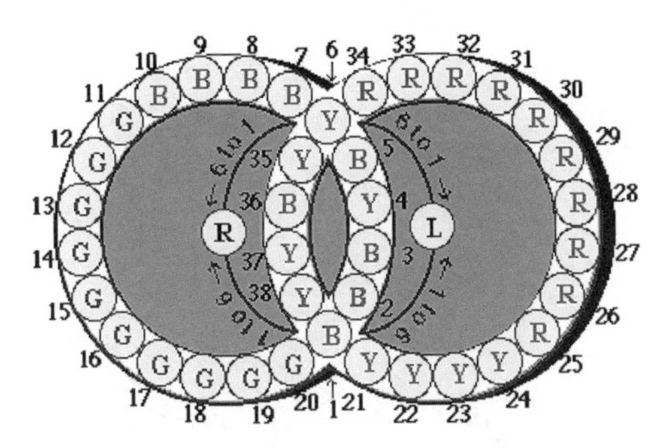

Figure 16.7. After a two-position right counterclockwise jump rotation, the puzzle looks like this.

After doing these two jump rotations, the puzzle is as shown in Figure 16.7. We reevaluate again and make the decision that we are going to move the blue disk now in spot 36 in a clockwise direction around the inner loop until we are able to get it into spot 4. The way we will be able to do this is to do a counterclockwise shuffle that gets this blue disk into spot 2. The shuffle that does this for us is $(R^{-1}L^{-1})^3R^{-1}$. The result of doing this shuffle is shown in Figure 16.8.

The blue disk that was in spot 36 before this shuffle is now in spot 2. Also, the yellow disk in spot 36 is the one that was in spot 4 before the shuffle. We want to move our blue disk, now in spot 2, into the spot that this yellow disk occupies.

Recall that we need to move the blue disk now in spot 2 in a clockwise direction around the inner loop. The commutator $[R^{-1}, L] = [L, R^{-1}]^{-1} = (1, 2, 38)(6, 7, 34)$ will move it in the right direction. Notice also that all of the disks in the upper outer triangle spots, numbered 6, 7, and 34, are red. Thus, the change that will happen there when we make our desired change on the lower inner triangle will not be noticeable.

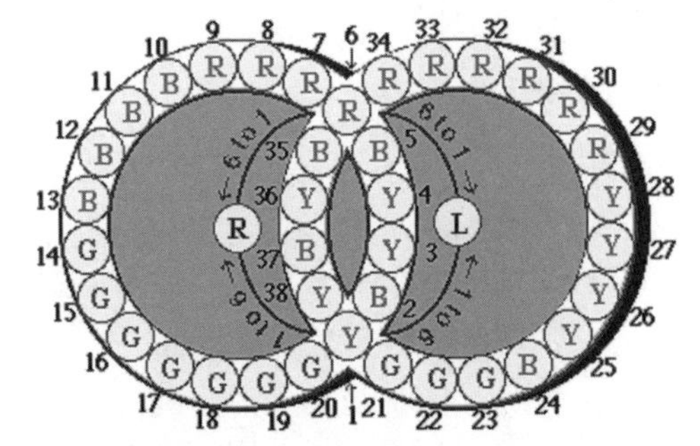

Figure 16.8. A counterclockwise shuffle moves the blue disk that was in spot 36 to spot 2 while putting disks on the upper outer triangle that are all the same color.

Figure 16.9 shows the results of applying this double 3-cycle. We have the blue disk that needs relocation closer to its destination, but it is not there yet. We need to do

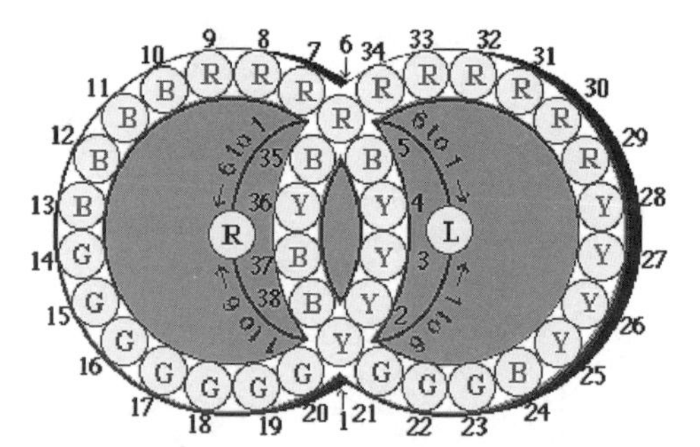

Figure 16.9. Commutator $[R^{-1}, L]$ produces a desired 3-cycle at the lower inner triangle and an unnoticeable 3-cycle at the upper outer triangle.

another shuffle to get the disks now in spots 36, 37, and 38 into the lower inner triangle. The shuffle $(L^{-1}R^{-1})^2$ will do the job. When this is done, the puzzle looks like Figure 16.10.

At this point, the blue disk that has been the focus of our attention is in spot 2 and we want it in spot 38. The tools available to us are the 3-cycle (38, 2, 1) and its inverse (38, 1, 2). (38, 1, 2) puts the blue disk from spot 2 into spot 38 but this 3-cycle also moves the blue disk out of spot 1 and into spot 2, which is *not* something we want.

But do we really need *the* blue disk in spot 2 to go into spot 38? No, we just need *a* blue disk to go there. The inverse 3-cycle (38, 2, 1) moves the blue disk from spot 1 to spot 38 and the blue disk from spot 2 to spot 1 while shifting the yellow disk from spot 38 to spot 2. This will serve our purposes just fine, so we perform $[L, R^{-1}] = (38, 2, 1)$ (34, 7, 6). This also produces a 3-cycle in the upper outer triangle but who cares because we cannot see it.

After performing this pair of 3-cycles by means of $[L, R^{-1}]$, we undo the shuffle that has shifted half of the red disks onto the left ring. This is done with the clockwise shuffle $(RL)^5 R$. After doing this, we have the puzzle in the state shown in Figure 16.11.

Note that the shuffle that rotated the five red disks on the outer loop back onto the right ring also rotated the four blue disks that were on the right ring portion of the inner loop onto the left ring portion of this loop. We now have all of the disks on the rings that they should be on. The only remaining problem is that we should have

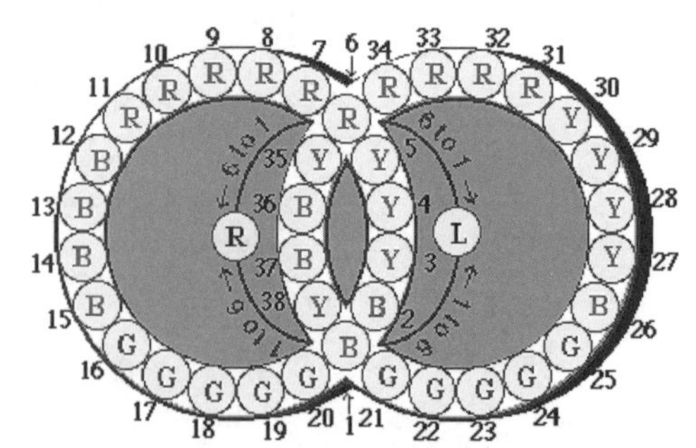

Figure 16.10. Another counterclockwise shuffle gets our target blue disk back into spot 2, and again, puts three red disks into the upper outer triangle.

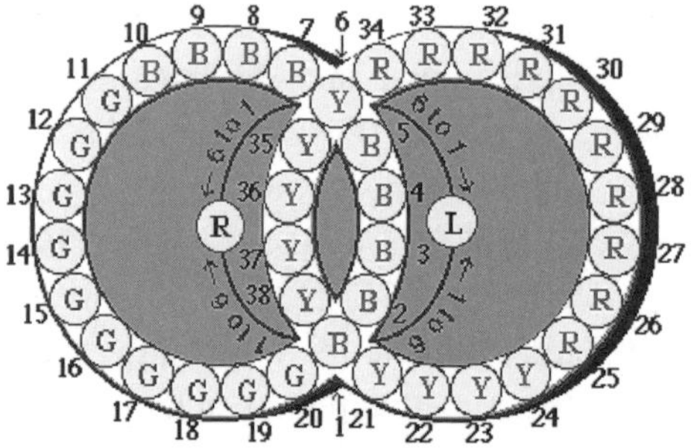

Figure 16.11. After a final commutator $[L, R^{-1}]$ and a clockwise shuffle, the puzzle is in the state shown.

the green disks in spots 7 through 16, but they are rotated four positions counterclockwise from this. This placement of the green disks also causes a separation of the blue disks by the yellow disk in spot 6. Similarly the red disks need to be rotated four positions clockwise.

These are problems that are easily remedied by applying four clockwise jump rotations on each of the rings. Since we are applying an even number of jump rotations, the transpositions that result also will be done four times, which means that they will produce the identity permutation on the disks they move. Once we have applied these four jump rotations on each of the rings as just described, the puzzle will be solved.

Figure 16.12 shows the final solved state of the puzzle. Including the 34 steps to get to the state in Figure 16.6, it took us 107 steps to solve it.

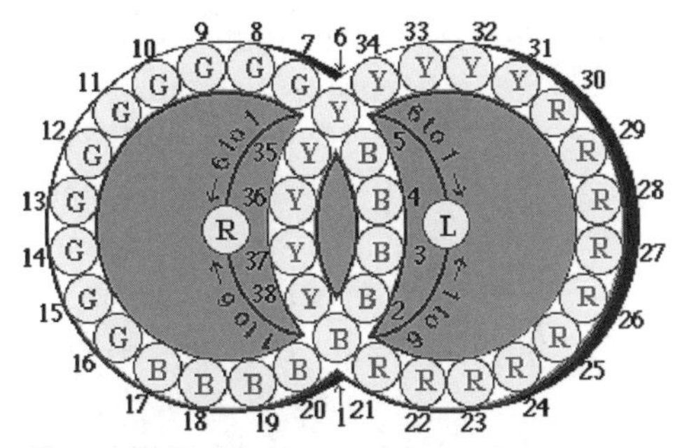

Figure 16.12. Sequences of jump rotations on each of the rings completely solve the puzzle.

16.5 A Few More Pointers

The example we just considered in detail illustrated most of the principles that apply to solving the Hungarian Rings in four-color mode. However, there are a few situations that often arise (but did not in this case) which are worth mentioning.

Odd numbers of jump rotations. Situations that call for an odd number of successive jump rotations need a bit more care because of the pair of transpositions that accompany the target change. For example, suppose in the end game for a scrambling you encounter a situation as in Figure 16.13. What we see is that one of the blue disks, the one in spot 7, is on the wrong side of the yellow disk in spot 6. This is a situation that can be corrected by a clockwise jump rotation on the left ring. This would have all of the disks on the left ring, except for those in the intersection locations 1 and 6, moving one position clockwise, with the disk in spot 7 jumping across spot 6 to spot 5.

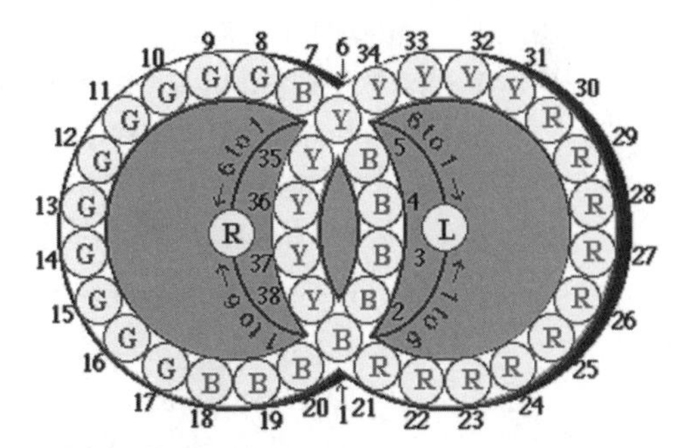

Figure 16.13. A left clockwise jump rotation is needed, but not the unwanted pair of transpositions that come with it.

16

This jump rotation would get all of the blue and green disks the way we want them, but for the pesky pair of transpositions that comes along. In the case of a left clockwise jump rotation, we will also get the transpositions (1, 38) and (6, 34). The transposition (6, 34) would swap two yellow disks, which we would not notice. However (1, 38) would swap the blue disk in spot 1 with the yellow one in spot 38. This is not good.

The simple solution is to neutralize the effect of this transposition by rotating the right disk a few positions clockwise so that we have red disks in both spots 1 and 38, but not so far that we lose the feature of having two yellows in spots 34 and 6.

If we do this, then the transpositions that come along with the jump rotation will go unnoticed. Thus, a way to finish off the solution is with a conjugate of the jump rotation by, say R^2. Apply $R^2\,([L, R^{-1}]L)R^{-2}$ and the puzzle is solved in seven moves.

Shortening repeated jump rotations. After playing with Hungarian Rings for a while, you would likely notice this for yourself, but we will call your attention to the phenomenon now so that you can train your muscle memory to use it right from the start.

Repeated sequences of jump rotations can be shortened by one move per jump because of the successive turns of the same ring that they require. To illustrate, let's consider the left clockwise jump rotation $[L, R^{-1}]L = LR^{-1}L^{-1}RL$. If you perform two of them successively, this yields $(LR^{-1}L^{-1}RL)\,(LR^{-1}L^{-1}RL)$. The two L's in the middle can be brought together and expressed as L^2, yielding the expression $(LR^{-1}L^{-1}RL^2R^{-1}L^{-1}RL)$.

Extending this observation, we note that if one writes out the basic notation for n successive left clockwise jump rotations, namely $([L, R^{-1}]L)^n$, then this can be rewritten in the form

(16.5) An expression for n successive left clockwise jump rotations is $(LR^{-1}L^{-1}R)(L^2R^{-1}L^{-1}R)^{n-1}L$.

From a muscle memory point of view, this has the effect of turning a sequence of jump rotations from a sequence with five steps repeated to a sequence with four steps repeated. What you do is modify the crisscross pattern illustrated in Figure 16.4 so that every time we previously were returning to the one-place rotation button for the direction we are going, we press the two-place button instead. This two-place-press rather than one-place-press alternative applies for all but the first and last rotations in the fundamental direction.

I find it handy to keep my count of how many jump rotations I am doing by speaking my count on the first press of a one-place button, and then on each successive press of the two-place button. Between the counts, muscle memory can have you doing the right thing with rest of the sequence.

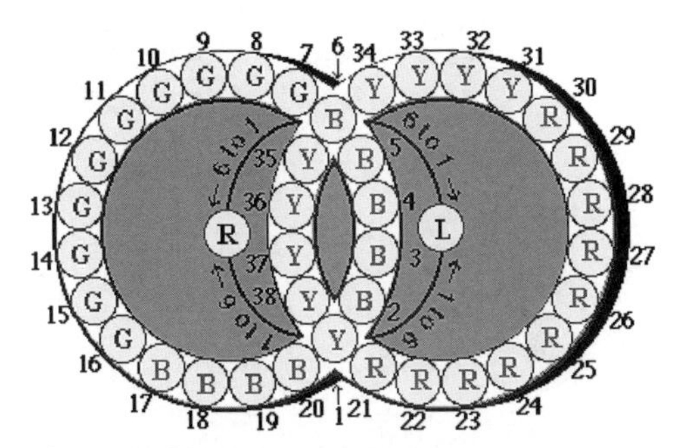

Figure 16.14. Swapping blue and yellow between the intersection points is all that is needed.

A harder than expected end game detail. Let us look at one final end game situation where just a limited amount of change remains to be done. Consider the end game shown in Figure 16.14. At first glance, you might think that the puzzle is solved, but if you look carefully, you see that, even though we have the blues and greens on the left ring and the reds and yellows on the right, we have the wrong colors at the two intersection points. We really ought to have a blue disk at spot 1 and a yellow disk at spot 6. This way the blues and yellows would all be contiguous instead of separated into two groups each.

What is needed is a swap of blue and yellow between spots 1 and 6. Of course some other yellow disk could move into spot 6, not necessarily the one from spot 1, and some other blue disk could move into spot 1. This seems like an easy enough change, but it proves to be more challenging than it looks. One reason for this is that the distance between the two intersection spots seems to thwart strategies that try to get the job done just with jump rotations. The first strategy that we will demonstrate uses a sequence of four jump rotations after using a shuffle procedure to move the blue disk from spot 6 to spot 35.

We will start out with a four-step left first clockwise shuffle, $(LR)^4L$. This will move the blue disk that was in spot 6 to spot 1, while bringing the yellow disk from spot 35 to spot 2. This shuffle has also put all green disks into the upper outer triangle. If we now do the double 3-cycle $[L, R^{-1}]$, we see from Figure 16.1(a) that the 3-cycle in the lower inner triangle moves the blue disk from spot 38 into spot 2 while moving the yellow disk from spot 2 to spot 1 and the blue disk from spot 1 to spot 38. From the perspective of colors, we have traded blue for yellow between spots 1 and 2.

If we now undo the shuffle by applying $(L^{-1}R^{-1})^4L^{-1}$, the result is as shown in Figure 16.15. Visually, we have traded between spots 35 and 6, so that the blue disk that was in spot 6 is now in spot 35 and the yellow disk that

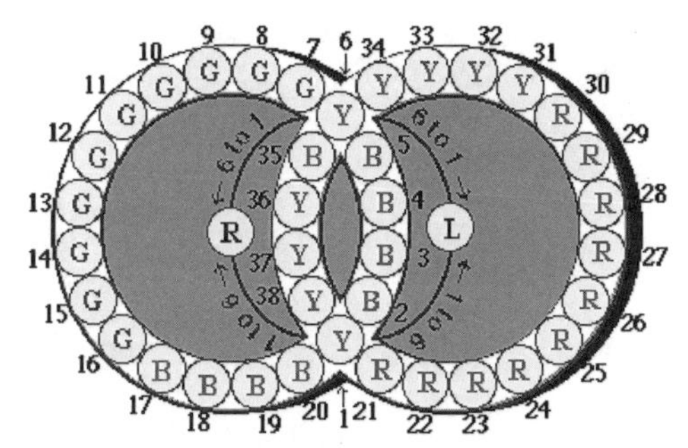

Figure 16.15. The blue disk that was in spot 6 is now in spot 35.

was in spot 35 is now in spot 6. In fact, what has really happened is that, since we have done a conjugate of a double 3-cycle, we really have performed another double 3-cycle. To be precise, we have performed (6, 5, 35) (12, 11, 10). If you look back at Figure 16.14, you will see that, given the benefits of indistinguishability of the disks of the same color, this gives us the visible result that we wanted—a transposition.

Now that we have, in effect, moved a blue disk one position from spot 6 to spot 35, we are now able to finish the job by means of jump rotations. If we now do a rotation R^{-4} to move the blue disk from spot 35 to spot 1, we are set up for doing four right clockwise jump rotations to jump the four yellow disks now in spots 21 through 24 over to the other side of the blue disk now in spot 1. This procedure at the same time is jumping the yellow disks in spots 38 through 35 over the yellow disk in spot 6, but that is not noticeable. The result is that the yellow disks are now all together. In the meantime, since we are doing an even number of jump rotations, the pesky pair of transpositions are being done an even number of times, putting the disks involved back where they started. We are done with the job.

To summarize,

(16.6) A procedure for solving from the configuration shown in Figure 16.14 is
$$\{(LR)^4L\}\ \{[L, R^{-1}]\}\ \{(L^{-1}R^{-1})^4L^{-1}\}\ \{R^{-4}\}\ \{(RL^{-1}R^{-1}L)(R^2L^{-1}R^{-1}L)^3R\}.$$

We have written this out reflecting the steps of the strategy we used. It took 40 moves to do the job, but we see two points where the steps had us doing successive rotations on the same disk. We could merge these and reduce this down to 38 steps.

Never be satisfied! While this procedure developed for solving the end game of Figure 16.14 has some virtues, in that it builds on some of the basic manipulative tools that we hope to become skilled in using, it is not satisfying in that our intuition tells us that it really ought not take that many steps to visually swap the contents of two spots. We will show that in fact one can cut this job down from 38 steps to 9.

We will start from a commutator that fits the structure of the puzzle well:

(16.7) $[L^5, R^{-5}] = L^5R^{-5}L^{-5}R^5 = (1, 6)\,(11, 30)$.

This commutator performs a pair of transpositions, one of which is (1, 6), the one we want. This would make the swap that we need between the blue and yellow disks in spots 6 and 1 respectively, but it would also swap a red disk from spot 30 with a green one from spot 11. This is not desirable.

However, we can do a simple setup procedure that puts disks of the same color into spots 11 and 30 while keeping the same disks in spots 1 and 6. The setup is $R^{-6}L^{-5}R^6$. The conjugate by this setup permutation of the commutator (16.7) will give us what we need. Specifically,

(16.8) A much shorter procedure for solving from the configuration shown in Figure 16.14 is
$\{R^{-6}L^{-5}R^6\}\,[L^5, R^{-5}]\,\{R^{-6}L^{-5}R^6\}^{-1} = (1, 6)(29, 30)$.

We have written this out so as to show its structure as a conjugate of a commutator. In this form, it is ten steps, but this can be compressed to nine by combining two successive right rotations.

That is a satisfying reduction—from 38 steps down to nine. There is a moral here. Even after you have solved a given problem on these puzzles, there is almost always going to be further challenges and satisfaction waiting if you continue to look for the most efficient way to do it.

Exercises

16.1 The Figure 16.6 end game scramble. The end game scramble shown in Figure 16.6 is available in the Locally Resident Presets collection of Puzzle Resources for the Hungarian Rings. Load this configuration, and carry out the solution procedure as described in Section16.4.

16.2 Another end game scramble. Go to the Puzzle Resources window for the Hungarian Rings and load the Exercise 16.2 end game that is shown in Figure 16.16. You will find it listed in the Locally Resident Presets field.

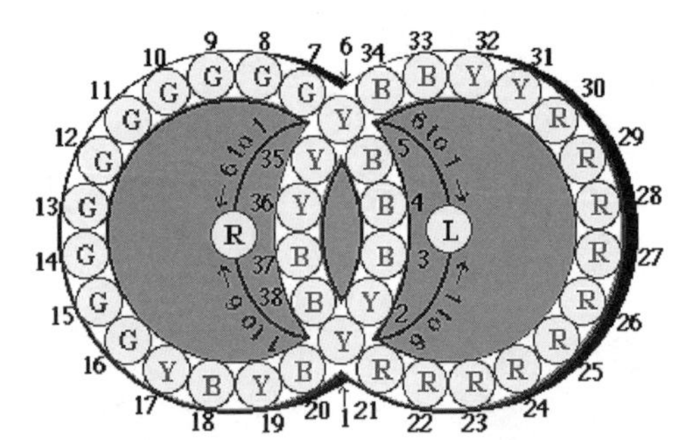

Figure 16.16. Another end game for the Hungarian Rings in four-color mode.

Solve this end game. [Hint: A solution is available in the Locally Resident Macros collection. Run it slowly to learn more about strategies for solving Hungarian Rings in color mode.]

16.3 A complete scramble. Load the Exercise 16.3 total scrambling of the puzzle from the Locally Resident Presets collection of Puzzle Resources. Solve this scrambling. [To see a worked out solution, load and run the Exercise 16.3 program from the Locally Resident Macros collection of Puzzle Resources.]

16.4 Alternating the colors. It is possible to start from the initial arrangement in four-color mode and then get an arrangement in which two of the colors alternate on one of the rings and the other two colors alternate on the other ring. See if you can find a way to do this. [You will find a way to do this in the macro collection if you want to look there.]

16.5 What is your average? Do ten scrambles on the Hungarian Rings Puzzle in four-color mode and solve it, each time making note of how many steps it takes you. Based on this experience, make an estimate of the ultimate average number of steps to solve the puzzle if one could be as efficient as absolutely possible. [A series of ten scrambles that the author solved required anywhere between 70 and 146 steps, with an average of 98.9.]

16.6 A three-color version. Load the Exercise 16.6 setting in the Locally Resident Presets collection of Puzzle Resources. It presents a three-color version of the Hungarian Rings puzzle. Explore solving several scramblings of this puzzle, paying attention to ways in which the strategies of the original four-color version apply, and ways in which modified strategies are needed.

16.7 A five-color version. Puzzle Resources contains a five-color configuration of the puzzle. It is labeled as being for Exercise 16.7. Load it, and work with it. Compare the strategies you develop for this version with the ones you have used for the three- and four-color versions.

16.8 A six-color version. We increase the number of colors by one yet again. Load the configuration for Exercise 16.8 from Puzzle Resources and work with it. Again, compare strategies.

16.9 A two-color version. Let's go in the other direction and reduce the number of colors. (a) Load the Puzzle Resources preset for Exercise 16.9, which configures the puzzle with just two colors. Play with this version for a while and form an opinion as to whether having two colors instead of four makes solving the puzzle significantly simpler. (b) Arrange the disks so that the two colors alternate and are symmetrically place around the vertical axis of symmetry. [Hint: You might find it easier to solve the inverse problem. Load in one of the presets provided that has two colors alternating, and find a way to get the colors together, recording the steps as you go in a macro button, say A. Then run the inverse of that macro, recording it into another macro, say B. See the author's best effort using this approach as entry 12 in the Locally Resident Macros collection, a 36-step solution. Can you improve on this?]

Advanced Challenges

We have developed an arsenal of mathematical tools sufficient for mastering our puzzles and have spent what we hope is enough time showing how they can be applied. You are now ready to follow your imagination into new directions. To orient you to the possibilities, we will discuss some of the advanced challenges that we have enjoyed.

17.1 Types of Challenges

The permutation puzzles advanced challenges divide into several categories:

(a) Open questions about the level of solvability of some of the puzzles.

(b) Efficiency of solution procedures.

(c) Thinking up interesting patterns, and:

  (i) Determining if they are possible, and

  (ii) Finding ways to do them efficiently.

(d) Exploring the numbers of permutations that are possible on the puzzles, and following up on number-theoretic issues that sometimes arise.

(e) Looking for optimality results, that is, looking for arguments as to why a given way for doing something is the best that is possible.

We have, either explicitly or implicitly, determined what is and what is not possible for some of the puzzles, but by no means have we done so for all of them. There are many questions of type (a) that are still to be considered, and for all the puzzles, you can always look for better ways to do things, exploring questions of type (b). Once you have gotten what you think is an efficient way of solving a puzzle, you can convert your exploration to category (e) and think about whether it is possible to prove that your methods are indeed the best possible.

Looking for interesting patterns can be a fascinating recreation with the puzzles. The theory will often settle the question of whether an idea we have for an arrangement can be realized, but even if you know a pattern is possible, finding an efficient way to create it might still be a challenge. Thus, challenges of type (c) in all their aspects can be a source of hours of exploration.

Back in Chapter 13, we saw how a computer algebra system like *Maple* or *GAP* can be used to answer questions of type (d). It does so in the sense of giving an exact number of permutations that are possible on a given puzzle, but in a sense that only whets our appetite to know why this number is the one. This can be the case especially when we see dramatic differences as the number of active disks changes, as was the case with Oval Track 7 which has a "Do Arrows" of (5, 3, 1). Exploring questions like this can lead to interesting discoveries.

17.2 Reversing the Disks: A Type (c) Example

Let's begin by exploring a type (c) challenge. Is it possible to arrange the disks on the various puzzles in reverse order?

Even though we have not looked at this question before, in a sense it is already implicitly answered in the theory we have presented. We just need to apply the theory to see what the answer is. Let's start with Oval Track Puzzle 1.

Consider placing the disks in decreasing order on OT 1 with n active disks. If n is an even number, then we know that all permutations are possible. Thus, arranging the disks into decreasing order has to be possible.

The outcome is different if n is odd. In this case, we can only arrange the disks in decreasing order if the permutation represented by this arrangement is even. We need to determine its parity.

Let's do the analysis. Placing the disks into decreasing order means that we are placing disk n into spot 1, disk $n-1$ into spot 2, disk $n-2$ into spot 3, etc. In general, we are placing disk k into spot $n+1-k$. If we use α to represent the permutation corresponding to this arrangement of the disks, then a rule defining α is $k\alpha = n+1-k$.

Since we are assuming that n is odd, we can say that $n = 2m+1$, for some integer m. The number that is in the middle of the list $1, 2, 3, \ldots, 2m+1$ is $m+1$. This number maps to itself under the reversal placement since we see that $(m + 1)\,\alpha = n + 1 - (m + 1) = (2m + 1) + 1 - (m + 1) = m + 1$. Thus, we can see that $\alpha = (1, n)\,(2, n - 1)$, $(3, n - 2) \cdots (m, m + 2)$. It turns out that α can be expressed in terms of m disjoint transpositions.

Since α is expressed using m transpositions, we see that the parity of α agrees with the parity of m. If m is even, say $m = 2k$, then we can display α on Oval Track 1. (Recall that we have to assume also that the odd number n of active disks is at least 7.)

Now if $m = 2k$, then $n = 2m + 1 = 2(2k) + 1 = 4k + 1$. Thus, n is one more than a multiple of 4. On the other hand, if m is odd, say $m = 2k + 1$, then $n = 2m + 1 = 2(2k + 1) + 1 = 4k + 3$. In this case, α is an odd permutation, so we cannot display it on the puzzle. This case has the odd number n being 3 more than a multiple of 4.

Our conclusion is:

(17.1) On Oval Track 1 with n active disks and $n \geq 7$, it is possible to arrange the disks in decreasing order if n is an even number or is of the form $4k + 1$, but it is not possible if n is of the form $4k + 3$.

The impossible cases are those in which the number of active disks is 7, 11, 15, 19, …. With any other number of active disks, the reversed order pattern is possible. Try it yourself and see.

Having settled the reversal question for one of the Oval Track puzzles, let's consider it for the Slide Puzzle in its "15" configuration. The question of placing the disks into reverse order on this puzzle needs a bit of refinement because of the blank. We have to decide where we want it to be in the end. Let's first say that we want the blank to be in its original location 16 while the numbered disks are in reverse order in locations 1 through 15.

This means that the permutation α that is represented this way is $(1, 15)(2, 14)(3, 13)(4, 12)(5, 11)(6, 10)(7, 9)$. Disk 8 must be in location 8. We see that α, being expressed with seven transpositions, is odd. Since the blank is in spot 16, which is an even spot, we have the parities of α and the location of the blank being opposite. Thus, by (14.4), α is not solvable.

This is a fact that has been known for well over 100 years, and it is central to the lore of the "15" puzzle. Back in the 1870s, Samuel Loyd, who was a showman and inventor of the time, offered a prize of $1,000 to anyone who could solve one of the copies of the puzzle that he was giving out that had the numbers in reverse order and the

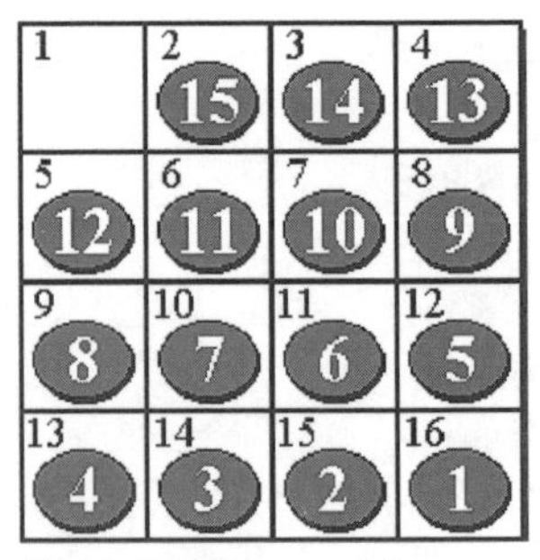

Figure 17.1. It *is* possible to reverse the disks of the "15" puzzle if you put the blank into spot 1 rather than spot 16.

blank in spot 16. He knew that this was impossible, but the general public did not, and his stunt created quite a stir as people tried to do the impossible.

If we cannot arrange the disks of "15" in reverse order with the blank in its original spot, are there other placements of the blank that allow for the reversal? Yes. For example, the arrangement shown in Figure 17.1 is very possible. Not only can we do it, but we can show why the theory says that this configuration is possible.

The permutation that this arrangement represents is $\beta = (1, 16)(2, 15)(3, 14)(4, 13)(5, 12)(6, 11)(7, 10)(8, 9)$. This is a product of eight transpositions, so β is even. Spot 1, which is where we want the blank to be, is an even location. Since these parities agree, by (14.4), the arrangement is possible.

We have made good progress here on part (i) of this reversal question, but what about part (ii)? What is an efficient way to reverse the order of the disks? We will illustrate the sort of results you can get in this direction by showing the best result we have so far for reversing the disks on Oval Track 1 with twenty active disks.

(17.2) On Oval Track 1 with $n = 20$, a procedure that reverses the order of the disks producing $(1, 20)(2, 19)(3, 18) \cdots (10, 11)$ is

$$(R^2T)(R^{-3}T)(RT)^3(R^3T)(R^{-1}T)^6(R^{10}T)(RT)^6(R^{-6}T)$$
$$(RT)^3(R^{-1}T)(RT)(R^{-1}T)(R^{-2}T)^2(R^{-1}T)(RT)R^{-7}.$$

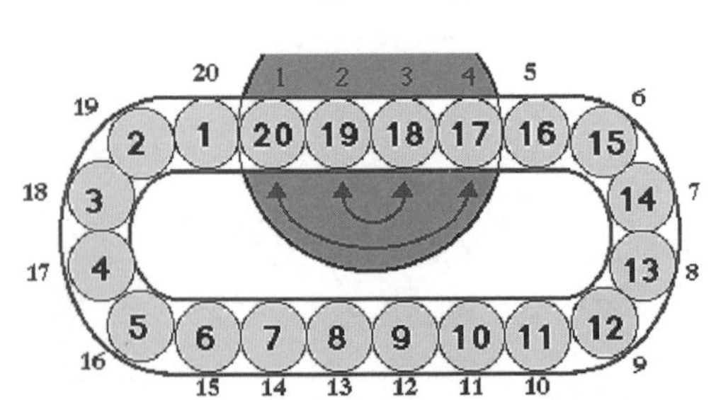
Figure 17.2. Our best way to reverse all disks on OT 1 with $n = 20$ takes 61 moves. Can you find a shorter way to do this?

Figure 17.2 shows the result of applying these moves starting from an initialized puzzle.

The strategy that lies behind this sequence makes use of the fact that the turntable of version 1 allows for moving three disks at a time to the other side of one of the disks on the turntable. Each time it does this, it reverses the direction of these three disks. Making repeated use of this characteristic, we have the following strategy:

1. Reverse the direction of disks 19, 20, 1, and 2. This puts these disks into the reverse order needed and will serve as our reference point. (Steps 1 and 2.)

2. Get disks 3, 4, and 5 to the other side of this group, but not in the order that they will eventually need to be. (Steps 3 to 10.)
3. Move the group 16, 17, and 18 counterclockwise through the group {2, 1, 20, 19} and into the right places in the right order. They end up in the right order naturally. (Steps 11 to 24.)
4. Work with the group 13, 14, and 15, moving it clockwise and into position. Without any extra work, they end up in the right order. (Steps 25 to 38.)
5. Work with disks 10, 11, and 12, taking them clockwise into position. They need some tweaking when they get into the neighborhood to put them into the right order. (Steps 39 to 52.)
6. Get disks 8 and 9 into position. They get there effortlessly in two steps. (Steps 53 and 54.)
7. Work on getting disk 7 into position. As you do this, you also reorient disks 3, 4, and 5 that were left earlier in reverse order, and get them, as well as disk 6 aligned correctly. (Steps 55 to 60.)
8. Rotate everything so that disk 20 is in spot 1. (Step 61.)

This is the shortest procedure we have found so far to get from the identity to this total reversal arrangement. It seems to be rather efficient in that it moves the disks three at a time for the most part and takes heavy advantage of the turntable's natural reversal behavior. But is it the most efficient method? We don't know. You could show that it is not by finding something shorter. Let us know if you do.

As a final example of this reversal theme, let us look at a method for getting the reversing permutation on Oval Track 2 which has $T = (4, 3, 2, 1)$ as its "Do Arrows" permutation, and with 20 active disks. The most efficient way we have found for getting the disks into reverse order takes 76 steps. It is:

(17.3) On Oval Track 2 with $n = 20$, a procedure that reverses the order of the disks producing $(1, 20)(2, 19)(3, 18) \cdots (10, 11)$ is

$$(RT)(RT^2)^2(RT)(R^{-1}T^{-1})^3(R^{-3}T)(RT)^6(R^4T)(RT)^5(RT^{-1})$$
$$(R^{-1}T^2)^2(R^{-4}T)(RT)^2(RT^{-1})(R^{-1}T^2)^2(R^{-1}T^{-1})(R^{-1}T^2)^3R^{-7}.$$

17

There is a strategy guiding this method that you can discover by working through the steps and carefully observing what is happening. Do you think that this is the best that can be done?

17.3 The Shortest Transposition?: A Type (e) Example

As part of a solution strategy for any of the puzzles, you want to know how to perform a transposition, assuming that one is possible. Let's look at an example of one version of the Oval Track Puzzle that offers a bit of a challenge for getting a transposition. The version is number 5, which has $T = (1, 2)\,(3, 4)$ as its "Do Arrows."

A first important fact about this puzzle that we want to mention is that

(17.4) On Oval Track 5 with $T = (1, 2)\,(3, 4)$ and $n \geq 7$,

$$[R^{-3}, T]^2 = (5, 4, 3).$$

We see that the same commutator squared that produces the 3-cycle (7, 4, 1) on Oval Track 1 produces a 3-cycle on this version. However, it is a different 3-cycle because of the different way that this "Do Arrows" permutation moves the disks in locations 1 through 4.

This 3-cycle is essentially the same as the (3, 2, 1) "Do Arrows" of version 3. Thus, an obvious way to get a transposition on this puzzle, when the number of active disks is even, is to borrow and modify what works on version 3. We can use the procedure (17.4) repeatedly to take disk 3 around the track until it is on the other side of disk 2, thus producing (2, 3). The formula is:

(17.5) On Oval Track 5 with $T = (1, 2)\,(3, 4)$, $n \geq 7$ and n even,

$$(2, 3) = ([T, R^{-3}]^2 R^{-2})^{n/2-2}([T, R^{-3}]^2 R^{-3}).$$

This procedure is easy to understand and certainly easy to remember and carry out once you have programmed the 3-cycle (17.4) into one of the macro buttons. It is also rather long. If $n = 20$, this procedure requires 81 steps to do the transposition. The general formula for the number of steps works out to $(9/2)\,n - 9$, which is on the order of 4.5 times n.

In spite of the author spending lots of time looking for something shorter, for many years this was the best way he knew to do a transposition on this puzzle. Then a student, Michael Kowalczyk, came up with a procedure that does a transposition in less than half the number of steps. Here is Kowalczyk's transposition:

(17.6) On Oval Track 5 with $T = (1, 2)\,(3, 4)$, $n \geq 8$ and n even,

$$(3, 5) = (TR^{-2}TR^{-1}T)(A^{n/2-3})(R^{-6}TR^{3}TR), \text{ where } A = R^{-3}TRT.$$

Since there are five setup steps, then four for each application of the repetitive step A, then five wrapup steps, the total number of steps for this procedure is $4\,(n/2 - 3) + 10 = 2n - 2$. If $n = 20$, this yields a total of 38 steps, which compares with the 81 steps in the (17.5) procedure. That is a significant improvement.

How does this procedure work? Figure 17.3 shows six stages along the way for the $n = 20$ case. We see that it takes awhile for a pattern to become clear. One thing that is going on is that disk 5 is staying in the "Do Arrows" area, and specifically is returning to spot 4 after the fifth move and then every fourth thereafter. The setup steps move disks 6 and 7 to the other side of disk 5 and do some rearrangement of disks 1 through 4. Then each application of the repeated group $A = R^{-3}TRT$ moves two more disks to the other side of disk 5. Disks 7 and 8 make the move after the first application of A, disks 9 and 10 after the second, etc.

Starting with disks 9 and 10, we see that the application of A also gets them into the proper order. Each successive application of A after the third gets the next two disks into the right sequence with respect to their predecessors. Meanwhile, the disks 2, 1, 3, 6, 4, 8, 7 march around the track in that order in a counterclockwise direc-

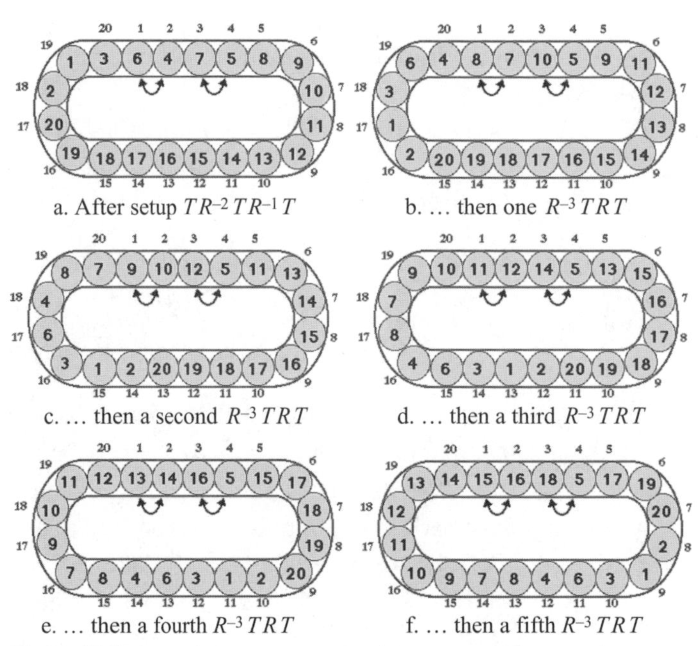

a. After setup $TR^{-2}TR^{-1}T$ b. … then one $R^{-3}TRT$
c. … then a second $R^{-3}TRT$ d. … then a third $R^{-3}TRT$
e. … then a fourth $R^{-3}TRT$ f. … then a fifth $R^{-3}TRT$

Figure 17.3. Several steps of Kowalczyk's transposition procedure. A pattern emerges after 3 applications of the repetitive 4 steps.

tion, progressing two positions with every application of A. All of these disks, except for disk 3, will in the end get put back into the right sequence by the five-step wrapup procedure.

The setup and wrapup are a bit complex, but they seem to get the job done efficiently. Each repetition of the four-step group A in the middle moves two more disks across disk 5 and back into sequence.

Is this procedure the best possible? If it is not, then there are reasons to believe that it is at least rather close. Since questions like this are among the more interesting but difficult challenges that the puzzles present, we are going to explore the matter in some detail to give you an idea of the techniques available for investigating such questions. The end result of the investigation is an improved process which uses two fewer moves and for which there is even more evidence of optimality.

A standardized format. We are looking for an optimal transposition on Oval Track 5, and, because of Kowalczyk's work, we know that its length is at most $2n - 2$ if n is an even number and $n \geq 8$. We note that we can standardize the format for the optimal transposition. Indeed.

(17.7) There must be an optimal transposition for OT 5 expressible in the form

$$(TR^{i_1})(TR^{i_2})(TR^{i_3})\cdots(TR^{i_k})$$

for some positive integer k where the i_j's are all integers.

First of all, it is clear that any permutation on any of the Oval Track Puzzles can be expressed as a sequence of alternating powers of R and T. Since the order of the "Do Arrows" move of Oval Track 5 is 2, the only power of T that we need consider is one. Any even power of T equals the identity and could be dropped from the expression so clearly no such powers are present in an optimal expression. Also, T raised to any odd power is equal to T.

Next, we observe that any sequence of alternating T's and powers of R that has either a power of R or a T at both ends cannot be an optimal for any given cycle structure. If we suppose that something of the form $R^{i_1}TR^{i_2}TR^{i_3}T\cdots R^{i_k}TR^{i_{k+1}}$ is optimal, let us take the conjugate of this expression by R^{-i_1}. We get an expression of the form $TR^{i_2}TR^{i_3}T\cdots R^{i_k}TR^{i_{k+1}+i_1}$, and given our way of counting so that each term of the form R to a power adds just one to the counter, this expression is shorter than the previous one. Also, by (9.1), this shorter conjugate expression has the same cycle structure as the longer one, which refutes the optimality of the former expression.

A similar taking of a conjugate can convert an expression which has a T at both ends to one that eliminates both of them, thus reducing the count by two but retaining the same cycle structure. This again refutes the optimality of the sequence under consideration.

To finish the justification of (17.7), we need to make the observation that any optimal sequence of steps that starts with a power of R and ends with a T can be converted into one of the form given in (17.7). Indeed, if one has such a sequence, just take its conjugate by T^{-1} ($= T$). This has the effect of moving a T from the right end of the sequence to the left end, and thus keeping the count the same. Since we are taking a conjugate, we have the same cycle structure.

To illustrate (17.7), let us standardize Kowalczyk's transposition into the T-first format. One can show that

(17.8) On Oval Track 5 with $T = (1, 2)(3, 4)$, $n \geq 8$ and n even,

$$(3, 6) = (TR^{-6})(TR^{3})(TR)(TR^{-2})(TR^{-1})(B^{n/2-3}), \text{ where } B = TR^{-3}TR.$$

We get to the expression in (17.8) from the one in (17.6) by taking the conjugate by $(R^{-6}TR^{3}TR)$. This rearranges things so that there is no longer a wrapup step but only a somewhat longer setup step followed by the required number of repetitions of the sequence $A = R^{-3}TRT$. Since we are taking the conjugate of a transposition, we know that we will get back a transposition. After having done this, we simply reverse the applications of the T's and the powers of R. Intuition tells us that this ought to still yield a transposition, and we can check that this is indeed the case. Specifically, we see that we get (3, 6).

A consequence of the standardized structure set out in (17.7) is that we can look for an optimal transposition that has an even number of steps. This means that if we are to get any improvement on our transposition of (17.6) or (17.8) that uses $2n - 2$ steps, then we will have to reduce the number of steps by multiples of two. To make this specific, in the case of $n = 20$, we will have to drop the number from 38 steps down to 36.

A Parity Consideration. Let us look at the structure of our optimal transposition from a parity perspective.

Since T is even and R is odd when the number of active disks is even, one can see that the parity of an expression of the form given in (17.7) is determined by the exponents i_j on the various R's. Specifically,

(17.9) The parity of an expression of the form given in (17.7) is the same as the parity of $\sum_{j=1}^{k} i_j$. For such an expression to yield a transposition, this sum must be odd.

To illustrate, note that the sum of the exponents on our transposition in (17.8) is

$$(-6) + 3 + 1 + (-2) + (-1) + ((-3) + 1)\,(n/2 - 3) = -5 + (-n + 6) = -n + 1.$$

Since n is even, $-n + 1$ is odd.

We note that the sum of the exponents works out to be negative in our case. The sign of this sum is something that we can control. If we had an optimal transposition for which the sum of the powers on the R's is positive, then the inverse of the expression also yields the same transposition, and its sum of the exponents is the opposite, or negative. This inverse of the original expression is of the R-first form when the formula $(x_1 x_2 \cdots x_k)^{-1} = x_k^{-1} x_{k-1}^{-1} \cdots x_1^{-1}$ is applied. However, we can put this into the T-first form by the conjugation methods described earlier while keeping the same length.

17

In the same way, if there were a minimal expression for a transposition with the sum of the exponents negative, we could modify it to get a transposition of the same length with the sum positive. Thus, we can say that

(17.10) There is a minimal transposition of the form given in (17.7) which has $\sum_{j=1}^{k} i_j$ odd and negative and another which has this sum odd and positive.

We can carry farther this specification of the form of a minimal transposition. Not only can we determine the parity and sign of the sum of the powers of R, we can also determine its value modulo n. Specifically, we have that

(17.11) For a transposition of the form in (17.7), that beats the one in (17.8), it must be that

$$\sum_{j=1}^{k} i_j \equiv \pm 1 \pmod{n}.$$

The "mod n" notation used here may be unfamiliar so we will explain it. When we say that this summation is equivalent to $\pm 1 \bmod n$, we mean that it is of the form $rn \pm 1$ for some integer r. That is, it is one more or one less than a multiple of n.

The details of verifying this claim are too messy to lay out here, so we will just give an indication of the things one considers along the way. To simplify the notation, we will denote this summation by m. Since we are adding the exponents on the R^i's, this sum m represents the number of places that each disk is displaced from its home by the action of the rotations alone, ignoring for the moment the actions of the T's.

If one is producing a transposition as a result of the series of steps, then $n-2$ disks return to their home spots when the procedure is done. The displacement by m places that the R^i's give these disks needs to be offset somehow, and the only way that this can be done is by the T's.

Since $T = (1, 2)(3, 4)$, each time that T is applied, it gives two disks a one-position clockwise displacement, namely the disks that are in spots 1 and 3, and it gives two disks a one-position counterclockwise displacement, namely the ones in spots 2 and 4.

If we are to have a transposition that improves on the one in (17.8), then the number of moves needs to be less than $2n - 2 = 2(n - 1)$ by an even amount. This means that the number of moves is at most $2n - 4 = 2(n - 2)$, so there will be at most $n - 2$ applications of T to work with.

To complete the argument, one needs to show that, working with this few T's, one can compensate for the displacement done by the R^i's on $n-2$ disks only if the displacement is small, namely by one place to either side.

The (17.8) transposition is not minimal after all. Armed with insights such as in (17.11), and a few more that we will not set forth here, we tried for several weeks to show that it would not be possible to improve upon Kowalczyk's transposition procedure. When it began to seem that perhaps our lack of success was due to the existence of a shorter procedure, we changed our tack.

We began to look for transpositions using the computer algebra program *Maple* to check all possible configurations of the puzzle for the $n = 8$ case that can be reached in fewer than the 14 moves that Kowalczyk's procedure requires in this case. (See Exercise 17.12 for additional discussion of this approach.) The computer found some transpositions that used only 12 moves. Thus, at least in this case, Kowalczyk's procedure is not the best. *Maple* did not find any transpositions with fewer than 12 moves, however.

We pressed on to the $n = 10$ case, but the size of the search space grows so fast that a general purpose symbolic manipulator like *Maple* becomes hopelessly slow already for this case.

A colleague, Barry Peterson, who is more adept at programming computers than the author, became intrigued with the problem at this point, and wrote a Java program specifically designed for the search. His program corroborated our $n = 8$ result in a matter of a few seconds. He wrote his program to list the optimal transpositions of the form in (17.7) which satisfy the condition $\sum_{j=1}^{k} i_j \equiv 1 \pmod 8$. (There are 128 of them.) It also confirmed that there are no transpositions that are shorter.

His program went on to handle the $n = 10$ case in an overnight run, finding 320 transpositions that use 16 moves, which again improves on Kowalczyk's procedure by two moves. Again, the exhaustive search found no transpositions that take fewer than 16 moves.

The $n = 12$ case is on the edge of being too large for exhaustive search even for an optimized program. A first run at this case was not exhaustive, but it did again come up with some results that beat the Kowalczyk procedure by two. Also, once the $n = 12$ case had been done, there were now formulas for efficient transpositions for three values of n. This proved to be enough for Peterson to "triangulate" on a pattern that always produces transpositions. Put into our standardized format, the pattern he discovered is:

17

(17.12) On Oval Track 5 with $T = (1, 2)\,(3, 4)$, $n \geq 8$ and n even,

$$(1, 6) = (TR)\,(TR^{-6})\,(TR)\,(TR^{2})\,(TR)\,(TR^{-6})\,(C^{n/2-4}),$$

where $C = TRTR^{-3}$.

We can verify this formula on the puzzle for the early cases of n, and we could prove it in general if desired. We note that the number of moves is the twelve of the setup procedure, followed by the specified number of repetitions of the four-step group C. Specifically, if there are n active disks, the number of moves given by this formula is $12 + 4(n/2 - 4) = 2n - 4$. This is two fewer than the number $2n - 2$ that we saw in the Kowalczyk procedure of (17.8).

We note further that if one adds up the powers of R here, one gets

$$1 + (-6) + 1 + 2 + 1 + (-6) + (n/2 - 4)(1 + (-3)) = -7 + (n/2 - 4)(-2) = -n + 1.$$

The sum is equivalent to 1 modulo n as predicted in (17.11).

There is now good reason to believe that we have the optimal transposition for Oval Track 5 in the Peterson procedure of (17.12). Exhaustive search verifies that you cannot do better for the three initial cases of $n = 8$, 10, and 12. See Exercise 17.13 for additional analysis that points in this direction.

A serendipitous solution. Although we have not definitively settled the question of the minimal transposition for Oval Track 5, our methods lead to a definitive answer regarding versions of the puzzle that are similar to this one. In particular, we can state with certainty the minimal number of moves needed to get a transposition on a version of the puzzle which has as its "Do Arrows" a pair of transpositions of adjacent disks that has these pairs diametrically opposed.

We present this solution as an example of the serendipity that can make solving challenging problems fun. Sometimes while seeking the solution of one problem, you stumble across the solution of another almost by accident. That was the case here.

Suppose one has a version of the Oval Track Puzzle that has as its "Do Arrows" two pairs of adjacent transpositions. It is no loss of generality to assume that one of the transposition is (1, 2). Thus, let us suppose that for this Oval Track Puzzle, $T = (1, 2)(k, k + 1)$, where $3 \leq k \leq n - 1$. Here n is the number of active disks, and in order that a single transposition be possible, we will assume that n is even.

Here is a definitive result concerning the shortest possible transposition for a puzzle of this form.

(17.13) On an Oval Track Puzzle with $T = (1, 2)(n/2 + 1, n/2 + 2)$, $n \geq 6$ and n even,

$$(1, n/2 + 1) = (TR^{-1})^{n/2-2}(TR^{n/2-1}).$$

This is the shortest possible transposition on this puzzle, and also the shortest possible for any Oval Track Puzzle that has T being a pair of transpositions, each moving a pair of adjacent disks.

17

Note that we have the adjacent-pair transpositions $(1, 2)$ and $(n/2+1, n/2+2)$ diametrically opposed in this case. That is, there are equal numbers of disks between these pairs in either direction.

None of the built-in versions of Oval Track are of the type identified here, but you can use the Build Your Own version 19 to make this "Do Arrows" and then confirm that this procedure does indeed produce a transposition. We see that the procedure has the count $2(n/2-2)+2 = n-2$.

It is easy to see why this procedure works. The first application of T moves disks 2 and $n/2+2$ counterclockwise to the other side of disks 1 and $n/2+1$ respectively. The R^{-1} puts disks 3 and $n/2+3$ into position so that the second application of T moves these disks to the other side of disks 1 and $n/2+1$, respectively.

After the $n/2-1$ applications of T of the procedure, all of the disks numbered 2 through $n/2$ have jumped over disk 1, and all of those numbered $n/2+2$ through n have jumped over disk $n/2+1$. Before the final rotation, the shift that each disk has undergone is the result of $n/2-2$ one-position counterclockwise shifts due to the R^{-1}'s and one due to the single time the disk was acted upon by T, for a total of $n/2-1$ positions counterclockwise. Thus, the final $R^{n/2-1}$ has the effect of moving each of the disks, other than disk 1 and disk $n/2+1$ back to its home spot. Meanwhile, disk 1, which has been moving between spots 1 and 2, is in spot 2 prior to the final $R^{n/2-1}$, and thus gets moved to spot number $n/2+1$. Similarly, disk $n/2+1$ moves from spot $n/2+2$ to spot 1.

17

This shows that the procedure of (17.13) results in the transposition $(1, n/2+1)$ in $n-2$ steps. Now let us consider why this is the shortest possible way to get a transposition for any puzzle with T being a pairs of transpositions, each moving a pair of adjacent disks. Let's think about the $n-2$ disks that get returned to their home locations as a result of a procedure that produces a transposition. The argument that led to claim (17.11) for Oval Track 5 applies equally well to this generalization, which has T being any pair of transpositions each acting on a pair of adjacent disks. Thus, based upon the actions of the rotations alone, each of the $n-2$ homebound disks would be displaced by one position from its home position, and all of them in the same direction. Since they all must be brought home by the procedure, each must be acted upon once by T to get this extra one-position shift. Since there are $n-2$ of these home-bound disks, and since each application of T can act on only two of them to produce a shift in the desired direction, it will take $(n-2)/2 = n/2-1$ applications of T to do all of the needed shifts. When one adds in the necessary rotations following each application of T, we get to a minimum number of moves of $2(n/2 - 1) = n - 2$. Thus, the number of moves used in the procedure in (17.13) is minimal.

17.4 Cycling All Four Colors on the Hungarian Rings?: A Type (c) Color Example

We conclude now with an example of a color mode challenge on the Hungarian Rings puzzle.

Since we know that every permutation of the Hungarian Rings puzzle is obtainable, a general principle is that

(17.14) On the Hungarian Rings Puzzle, any pattern for arranging the colors that you can dream up will work. That is, if the number of disks of each color that you need for the pattern you want agrees with the number that are available, then you can, somehow or other, achieve the pattern.

The problem still remains to get to the pattern efficiently.

One question that occurs is

(17.15) Is it possible to arrange the four-color disks so that we repeatedly cycle through all four colors as we go around each ring, and to do so symmetrically?

A little thought about this question in the light of (17.14) shows that the answer is yes.

Let us begin by thinking about what color must go into spot 3 on the left ring. If we are going to cycle through the four colors as we go around this ring, then this same color must appear in spots 7, 11, 15, and 19 on the left ring. We need this color in five of the 20 left ring positions, or one fourth of the total, which is of course natural if one is cycling four colors through 20 locations. By symmetry, this same color must appear in five positions on the right ring, namely locations 37, 34, 30, 26, and 22. This means that we need ten disks of the same color for these locations, and the color used in them cannot occur anywhere else. We choose either red or green for these locations since we have ten disks of each of these colors.

A similar analysis shows that we need ten disks of the same color also in locations 4, 8, 12, 16, and 20 on the left ring and in 36, 33, 29, 25, and 21 on the right ring. We use the other color from the red-or-green choice that we did not use for the previous locations.

We note that the locations for which we have already made color choices do not include the intersection locations numbered 1 and 6. Whatever colors go into these locations must fit into the cycling pattern for each ring. Furthermore, since the space between the intersections is five positions, they will need to have different colors in order to have a four-color rotation pattern.

These factors all work together well for using our two remaining colors, blue and yellow, for which there are nine disks each. We put one of them into locations 1, 5, 9, 13, and 17 on the left ring and in the symmetrically placed locations 35, 32, 28, and 24 on the right ring. We put the other color for which there are nine disks into locations 2, 6, 10, 14, and 18 on the left ring, and into locations 38, 31, 27, and 23 on the right ring.

The availability of colors needed for a four-color cycling works out right, so by the principle set forth in (17.14), it is possible to get such an arrangement of the colors. The question remains, is it possible to do so in a reasonable number of steps? We will show that it is by giving a way to get this arrangement in 70 steps. In particular, we have

(17.16) A pattern that cycles through all four colors around both rings of the Hungarian Rings puzzle is obtainable by the 70-step procedure

$$[(L^2R^2)^9L^2(RL)^5(R^{-1}L^{-1})^5][(L^4R^4)^4L^4R^2LR^{-1}][R^6L^{-1}R^{-2}L^{-5}R^{-3}]$$
$$[L^5R^5L^{-5}R^{-6}LR^6L^5R^{-5}L^{-5}R^{-1}L^{-1}R][R^3L^5R^2LR^{-6}].$$

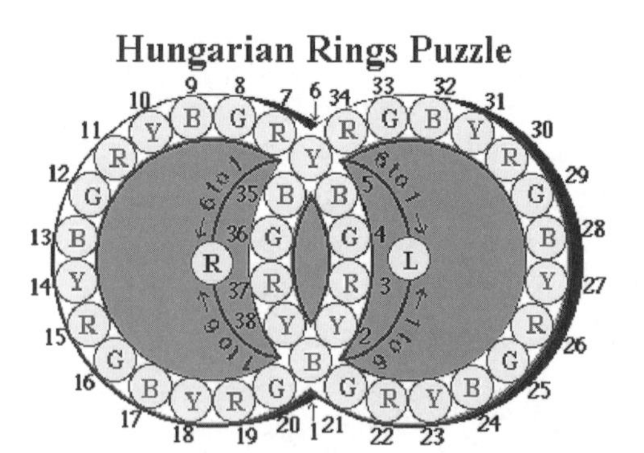

Figure 17.4. An arrangement that cycles all four colors on each ring can be obtained in 70 moves.

The result of performing this procedure is shown in Figure 17.4. Note that the procedure given in (17.16) uses 73 steps. It is given in this way to reflect the structure. However, this number can be reduced by three by combining the steps where the first and second, the second and third, and the fourth and fifth bracketed groups meet.

Let us examine the structure of this procedure. The first bracketed group is of interest in its own right. It is a 39-step procedure that results in a pattern having two of the colors, red and yellow, alternating on the left ring and the other two colors, green and blue, alternating on the right ring. This alternating two pattern is shown in Figure 17.5.

We get close to the four-colors-cycled arrangement by doing a series of four-position shuffles on this pattern, followed by a little tweaking. These are the steps listed in the second bracketed group. Following these steps, the puzzle is in the state shown in Figure 17.6. We see from this figure that we have gotten the desired pattern on the left ring. Starting at location 1, we repeat the cycle "blue, yellow, red, green" five times as we go counterclockwise around the circle.

Also, we see this pattern in a reflected manner at least for part of the right ring. Note that the color in location 34 is red as it is in the opposite position 7. Starting at location 34 and going clockwise, we see the sequence "red, green, blue, yellow" repeated three times. This reflects the counterclockwise pattern on the left ring.

A careful analysis of the colors on the right ring in Figure 17.6 reveals that a single 3-cycle would given us the desired sequence all the way around the right ring. If we could move the blue disk in location 21 to location 35, the yellow disk from there to location 38, and move the green ring from there into location 21, we would have the pattern we want. That is, we need to do (21, 35, 38).

The final three bracketed groups of the procedure in (17.16) gives us this 3-cycle as a conjugate of our tool 3-cycle (6, 11, 12) given in (15.1). The first of these groups is the setup procedure, which moves disks from spots 35, 38, and 21 into spots 6, 11, and 12, respectively. The second is the 3-cycle (6, 11, 12) expressed as in (15.1), but compressed. The final bracketed group is the inverse of the setup procedure.

Once this final 3-cycle is performed, the result is as shown in Figure 17.4, and we have the cycling of all four colors. This way to get the pattern, upon examination, reveals some strategy, but is this the shortest way to get to this pattern? We have no idea, but probably not.

The challenge is there for you, dear reader, to find a better way to do this pattern. Other interesting challenges await discovery if you look for them.

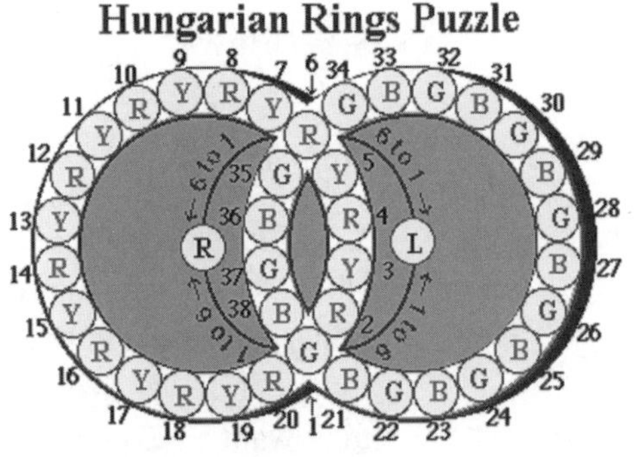

Figure 17.5. The first 39 steps of (17.16) yield this pattern, which alternates two colors on each ring.

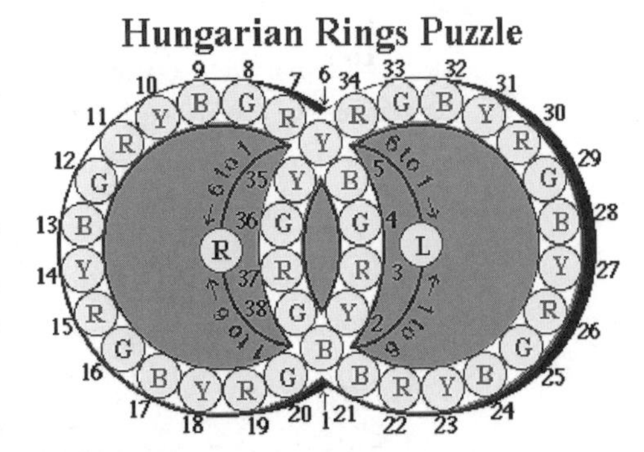

Figure 17.6. A series of four-position shuffles gets us close to the desired four-color-cycle pattern on each ring—a 3-cycle away.

Exercises

The exercises in this section suggest some additional advanced challenges. Many of these problems will be very difficult to solve. Write to the author if you get some results you find interesting.

17.1 The shortest transposition. On any of the puzzles, it is of interest to get transpositions as efficiently as possible. (a) Try to find transpositions on any of the puzzles that you can get in fewer steps. In particular, are there transpositions on Oval Track 1 that take fewer than 34 steps? Is there a transposition on the Hungarian Rings Puzzle in fewer than 35 steps? (b) If your search for a shorter transposition on a puzzle is proving fruitless, you may want to explore whether the best one you know is provably the best. Look for arguments that allow you to say that "a minimal transposition on this puzzle will need to have (*some specific number of*) steps."

17.2 The shortest 3-cycle. Given the fundamental role of 3-cycles as generators for the even permutations, it is of interest to know how to get 3-cycles as efficiently as possible. Explore 3-cycles in ways suggested for transpositions in the previous exercise.

17.3 Reversal of all the disks. On Oval Track 7 with $T = (5, 3, 1)$, the permutation which puts all of the disks into reverse order satisfies the necessary condition that it has the disk numbers alternating in parity as one moves around the track. Prove that in spite of this, with $n = 20$, this particular arrangement is not possible to achieve. [See Exercises 13.5 through 13.7.]

The next four exercises guide you through an exploration of Oval Track 6 with $T = (1, 11)(4, 14)$, *which is one of the most challenging of the Oval Track Puzzles. This puzzle presents type (d) number-theoretic challenges, as you will discover if you study it.*

17.4 Oval Track 6 with all 20 disks. If all 20 disks are active, this puzzle is highly constrained because the possible moves preserve the sequence of the units digits of the numbers. (a) *Maple*/*GAP* tells that the number of possible permutations on this puzzle is 5,120, which is 5×2^{10}. Explain why this number is correct. (b) Practice solving buttons scrambles on this puzzle, which will always be solvable. In spite of the relatively small number of possible arrangements, it still is an interesting and moderately challenging puzzle to solve.

17.5 Oval Track 6 with 19 or 17 disks. When the number of active disks is changed to either of the odd numbers 19 or 17, the number of possible arrangements changes dramatically. (a) Show that with $n = 19$, $(TR^{-3})^2(TR^6) = (1, 11, 10)$. Use the first configuration of the Locally Resident Presets of the Puzzle Resources collection for Oval Track 6 to explore what other 3-cycles can be obtained as conjugates of this particular 3-cycle. The configuration has the letter B in all locations except for these three, which are numbered. Scramble the disks, and see if you can get each of the numbered ones back to its home location from wherever it is. (b) Based upon your experience in (a), prove that on Oval Track 7 with 19 disks active, the permutations that are solvable are exactly all of the even ones. (c) Show that with $n = 17$, $[T, R^{-4}]^2 = (1, 8, 11)$. Explore the consequences of this, ultimately showing that exactly the even permutations are again the ones that are possible.

17.6 Oval Track 6 with 15 active disks. With 15 active disks, this puzzle again becomes highly constrained. (a) Prove that with this version of the puzzle, the only permutations that are possible are ones which keep the disk numbers, when reduced modulo 5, in the cyclic sequence 1, 2, 3, 4, 5, 1, 2, 3, 4, 5, 1, 2, 3, 4, 5. Explain why this means that $5 \times 6^5 = 38{,}880$ is an upper bound for the number of possible permutations. (b) Use *Maple* or *GAP* to determine exactly how many arrangements are possible on this puzzle, and explain why the number that is given is true. (c) Practice with this puzzle to develop ways of solving it after doing buttons scrambles.

17.7. Oval Track 6 with 16 or 18 disks. With an even number of disks, Oval Track 6, like version 7, is subject to the constraint that adjacent disks must always alternate in parity. Analyze Oval Track 6 with 16 and 18 disks and determine exactly how many permutations are possible and why. Practice solving the puzzle with these numbers of disks, noting the similarity and differences in strategy in comparison with the other number of active disks already studied.

17.8 Oval Track 12 gives big end game practice. Oval Track 12 has a wide span for its "Do Arrows," which is $T = (11, 7, 4, 2, 1)$. As a result, it has a very large end game. This means that the sort of strategy developed for the

Hungarian Rings, which also has a large end game, can be modified for use in solving this puzzle. (a) Load the Exercise 17.8a configuration, which is the first in the Locally Resident Presets collection of the Puzzle Resources for this version. It has numbers in the "Do Arrows" locations and the letter B everywhere else. Scramble this configuration and then return all of the numbered disks home. (This will give you practice at setting up to use conjugates of the "Do Arrows" to get up to four or five disks back home at one time.) (b) To test your mastery of the Oval Track 12 end game, load the Exercise 17.8b configuration, which is the second in the Puzzle Resources collection. This is an end game scramble. See if you can solve it. (There is a solution given in the macro collection.) (c) Do other end game scrambles on this puzzle to see if you can improve your efficiency.

17.9 Oval Track 13: Find a more efficient transposition. The "Do Arrows" of Oval Track 13 is a 4-cycle (1, 5, 7, 3). Given that this is an odd permutation, there ought to be a way of getting a transposition by rocking methods rather than around the track. One way to do this by using prior knowledge is to untwist the 4-cycle into a conjugate, $A = (7, 5, 3, 1)$, then modify the transposition (2, 3) of Oval Track 2. This gives us that on this puzzle, $(3, 5) = R^{-2}AR^2A^2R^{-2}AR^2A$, where A is the 4-cycle given above. This produces a transposition in 59 steps. Surely there is a more efficient way to get a transposition on this puzzle. See if you can find it. Explore what might be the shortest transposition possible.

17

17.10 A number-theoretic study of a class of puzzles. This exercise guides you into a study of versions of the Oval Track Puzzle with a single transposition as the "Do Arrows," and explores the number-theoretic patterns. (a) Let k be a positive integer. Consider a "Build Your Own" version of the Oval Track Puzzle that has $T = (1, k+1)$ as its "Do Arrows." Build several versions of this form using the "Build your own" Oval Track Puzzle 19, and work with them. Take values of k in the range of 1 to 6, and values of n, the number of active disks, from $k+1$ to 20. Pay attention to how the degree of solvability of the puzzle changes as n changes with a fixed k. [See the Exercise 17.10 preset configurations in Oval Track 19 Puzzle Resources for some ideas on exploration.] (b) On the basis of your experience in (a), formulate a conjecture regarding the order of the subgroup of S_n that is generated by the permutations $\rho = (1, 2, \ldots, n)$ and $\tau = (1, k+1)$ where k is a fixed positive integer less than n. Prove the conjecture if possible. [Hint: Generate data about the orders of these subgroups using *Maple* or *GAP* in order to formulate or corroborate your conjecture.]

17.11 More color arrangements on the Hungarian Rings. There are an unlimited number of interesting ways to arrange the colors on the Hungarian Rings when it is in a color mode. You can look for different arrangements in the standard four-color mode, or in any other color mode that can be created in Puzzle Resources. Here are just a few more color arrangements to try for. (a) In the standard four-color mode, switch the places of the green with the blue and the red with the yellow. That is, have the colors for which there are ten each on the inside and the ones for which there are nine each on the outside. (b) On the three-color version of Exercise 16.6, switch the green disks from the left to the right ring and the red ones from the right to the left. (c) On the three-color version of Exercise 16.6, alternate the red and the green disks on the outer loop while keeping the blue disks on the inner loop. Have the arrangement of the reds and greens so that it is symmetric about the vertical axis of symmetry. (d) On the five-color version of Exercise 16.7, switch the places of the red (R) disks and the firebrick (F) disks. (e) Also on the five-color version, switch the blue (B) and the green (G) disks.

17.12 Exhaustive search with *Maple*. This exercise introduces the approach of using exhaustive search to find the shortest possible transposition for Oval Track 5 using *Maple*. The program given for this is certainly not the most efficient way to do an exhaustive search on a computer, but it is easy to set up if you have *Maple*. (If you are familiar with *GAP*, you may want to take the same approach with it.) (a) Enter the four *Maple* commands listed below into your computer and let it run overnight. If run fully, it will check all 14-step sequences of the form given in (17.7) for 10 active disks and print out any that are transpositions. Are there any? (b) Below the display of *Maple* commands, there is some explanation of them. Drawing upon this, modify the commands to do other similar exhaustive searches, such as for 12-step transpositions.

```
> with(group);
> a := [[1,2],[3,4]];
> b := [[1,2,3,4,5,6,7,8,9,10]];
> repeet := proc(c, n) if n = 0 then [] else accum := []; for i from 1 to n do accum
```

17

```
:= mulperms(accum,c) od fi end;
> for i1 from 1 to 9 do for i2 from 1 to 9 do for i3 from 1 to 9 do print("i1, i2,
i3 are now ",i1, i2, i3); for i4 from 1 to 9 do for i5 from 1 to 9 do for i6 from
1 to 9 do k7 := 1 - (i1+i2+i3+i4+i5+i6) mod 10; d := mulperms( mulperms( mulperms(
mulperms( mulperms( mulperms( mulperms( mulperms( mulperms( mulperms( mulperms(
mulperms( mulperms( a, repeet(b,i1)),a), repeet(b,i2)),a), repeet(b,i3)),a),
repeet(b,i4)),a), repeet(b,i5)),a), repeet(b,i6)),a), repeet(b,k7)); if repeet(d,2)
= [] then print(i1, i2, i3, i4, i5, i6, k7,d); fi; od od od od od od;
```

Explanation. (i) `mulperms(x,y)` is *Maple*'s way of expressing the composition of permutations x and y given in cycle notation. (ii) The "`repeet`" command defines a *Maple* procedure for raising a permutation c to a positive integral power n. (iii) The final *Maple* command sets up a nest of six "do" loops for calculating seven-fold compositions of terms of the form ab^i, where a and b represent the generators of OT 5. The last of the powers on the seventh b is calculated so that the sum of all of the powers is equivalent to 1 modulo 10. If *Maple* finds one of these compositions that yields the identity when squared, it prints it out, together with the exponents on the b's. This produces some extraneous output, namely compositions of three or five disjoint transpositions, but it will also find any single transpositions of the targeted form if they exist. (iv) The first "print" in the looping program is there to mark progress after every $9^3 = 729$ cases.

17.13 Two reasons why display (17.12) might show a shortest OT 5 transposition. (a) Suppose that P is an m-step procedure yielding a transposition (a, b) on OT 5 with n active disks. Suppose that all of the $n - 2$ disks that ultimately return home are moved by the "Do Arrows" operation at least twice. Prove that this means that $m \geq 2(n - 2)$. (b) With the same assumptions about P, suppose that the disks a and b are never acted upon by "Do Arrows" at the same time. Prove that this also means that $m \geq 2(n - 2)$.

Using Puzzle Resources

Puzzle Resources makes it possible for you to store, retrieve, and create an unlimited number of configurations and macro programs for your puzzles. Having this feature available opens the puzzles to possibilities limited only by your imagination. In this appendix, we will give you the background you need to use this part of the package.

To open Puzzle Resources, go to the small Puzzles Home window and press the "To Resources" button.

A.1 Getting Acquainted with the Screen

Figure A.1 shows you the Puzzle Resources screen for Oval Track Puzzle Number 1. Within this same window, you can navigate between similar screens for each of the versions of Oval Track, and for all three puzzles of Multi-Puzzle.

You can scroll through the top two fields of the Resources screen to see the currently stored configurations for the current puzzle and the currently stored macro button programs. The configurations, also called presets, are displayed in the field on the left and the macro programs are displayed in the right field.

These fields are touch sensitive. You initiate the process of doing something with one of the resources listed by clicking the mouse button while the browser hand is on its entry in the list. When you do this, you get a dialog box offering you several choices.

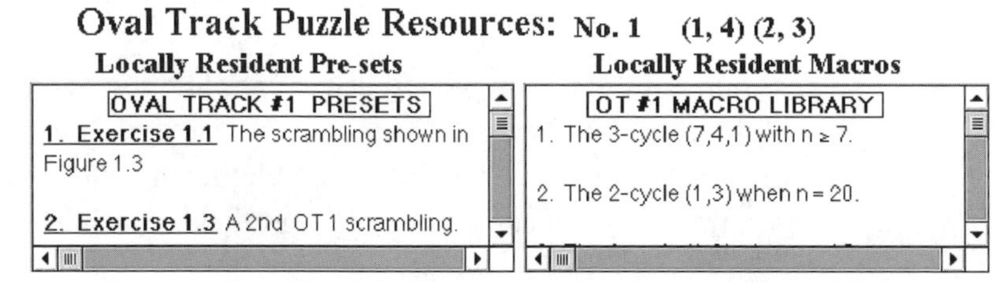

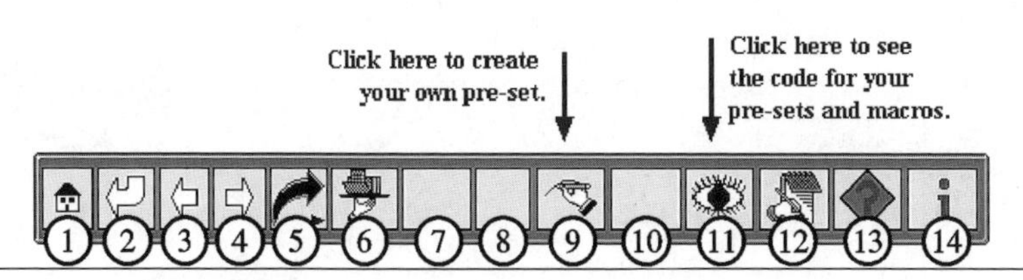

Figure A.1 A typical Puzzle Resources screen. Numbers on buttons keyed to explanations below.

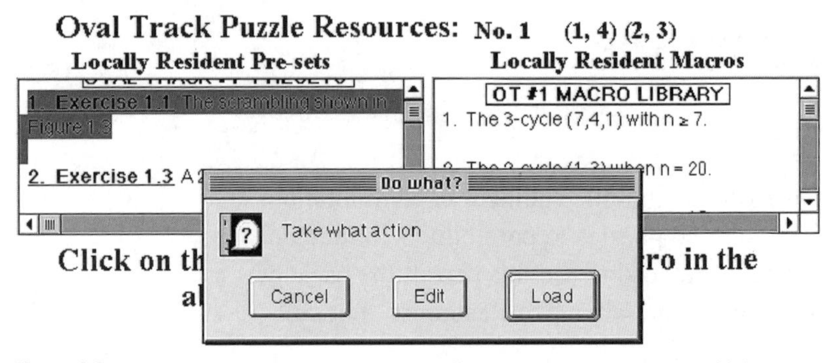

Figure A.2 Clicking on a preset entry brings up a dialog box asking what to do with it.

In the case of presets, these choices are to either load the configuration onto the puzzle or to edit it. For example, having clicked on the Exercise 1.1 entry in the list, as shown in Figure A.2, we could load this into our puzzle and thus have it set in the way that is discussed in this exercise. If we choose to edit the entry, then it will be loaded into an editing display within the Resources window where we can make modifications to the configuration. We will discuss this editing capability in detail later.

There are some entries at the top of the list for each puzzle that are part of a permanent collection. The program does not allow you to delete these. Following these permanent entries, there is the start of a list that you can add to or delete from as you wish. When you click on an entry beyond the permanent ones in the list, the dialog box also offers you the option of deleting the entry.

If you click on an entry in the right field, which lists descriptions of macros, then you see the dialog box shown in Figure A.3. With the macro programs, we have the options of loading them into one of the macro buttons, or deleting them if they are not part of the permanent collection.

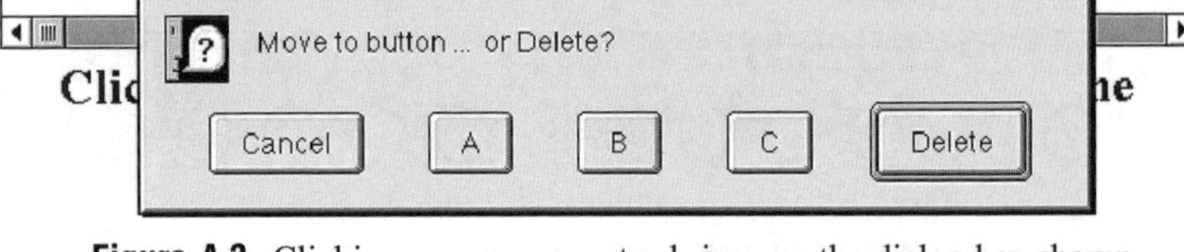

Figure A.3 Clicking on a macro entry brings up the dialog box shown.

A.2 The Resources Button Bar

The button bar for Puzzle Resources looks similar to the button bar for the puzzle windows. The buttons also operate like those of the puzzles, with a few differences that relate to the different nature of this window. We will explain what each button does, referring to them by the numbers that are shown by them in Figure A.1.

1. **The Home Button** (house icon) brings the Puzzles Home window to the front.
2. **The Return Button** (turning arrow icon) takes you to the puzzle that corresponds to the current Resources page.
3. **The Previous Button** (left arrow icon) takes you to the Resources screen for the version of the puzzle that is previous in the ordering, which starts with the three puzzles of Multi-Puzzle and then runs through the versions of Oval Tracks.
4. **The Next Button** (right arrow icon) takes you to the Resources screen for the next puzzle in the sequence.
5. **The Go Anywhere Button** (vaulting arrow icon) opens a pop-up menu that lists all of the puzzles' Resources screens on submenus. By selecting one of them, you can go directly to it. You can also use it to go to Puzzles Home.
6. **The Server Button** (hand-holding-a-tray icon) will start the process of downloading new information about the puzzles via the Internet from the author's Permutation Puzzles server at Northern Michigan University. This will be in the form of an electronic newsletter that displays in its own window. The newsletter, which is updated periodically, contains information on the latest results that users have gotten with the puzzles, such as the most efficient ways of getting to interesting patterns, and efficient or clever solution tools.

   If your computer is connected to the Internet, then a press of this button should be all that it takes to start the download process. If your computer is not hooked up, then what happens will depend on your operating system. With some, the system will establish a connection by, for example, automatically dialing up to your service provider. On some systems, you will have to manually establish the connection in the way that you usually do for checking your e-mail or for using your web browser.
7. **Blank Buttons.** Since Resources does not have as many functions that are accessed using buttons as the puzzles do, there are some blank buttons on the Resources button bar. By having a few blanks like this, we keep the locations of the rest of the buttons the same as they are on the puzzles. As we have noted, with only two exceptions, the buttons of Resources perform the same or analogous functions to the corresponding ones on the puzzles.

8. **Blank Button.** See 7 above.

9. **The Write/Edit Button** (writing hand icon) opens a display between the list fields and the button bar, which presents you with tools for building or editing configurations for the current puzzle. More detailed discussion about what you can do from this display is in section A.3 below.

10. **Blank Button.** See 7 above.

11. **The View Lists Button** (eye icon) has a function for Resources that is analogous to its function in the puzzle windows. In Resources, it makes visible displays of the codes for the preset configurations and macro programs that are currently stored. These displays appear between the fields that show the descriptions of the resources and the button bar. Figure A.4 shows a typical display, this one for the Hungarian Rings.

The coding system for configurations is simple: The configuration's disk numbers (or letters) are listed in order, starting with location number 1. Each entry is numbered and started on a new line. In the case of the Slide Puzzle, at the end of the list is an entry telling what wrapping setting applies. In the case of the Hungarian Rings, there are two separate fields displaying the entries on the two rings, with the intersection points shown on both lists.

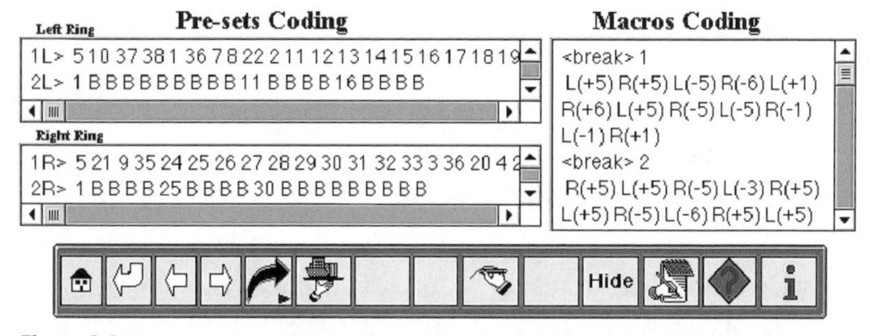

Figure A.4 The View Codes Display for the Hungarian Rings. This one has separate fields for the left and right rings. The displays for the other puzzles need only one field for presets.

The coding system for the macros differ according to which puzzle the macros are for. The coding system is the same as that within the puzzle's window when you are looking at a macro code there.

Most of the time, if you have written clear descriptions for all of the resources, you will have no need to read the details of their codes, but if a description is unclear, it is sometimes helpful to look at the code to determine what the configuration actually looks like, or to get some idea of what a macro program might do.

12. **The Note Pad Button** (note pad icon) works for Resources just like it does for the puzzles, bringing up a screen in which you can do limited word processing.
13. **The Documentation and Ideas Button** (question mark icon) takes you to the documentation window, opening it first if it is not already open or bringing it to the front if it is already open. The documentation for Puzzle Resources will be displayed initially when you come to the documentation window from Puzzle Resources, but you can navigate to the other pages.
14. **The Graphical Information Button** ("i" icon) functions like it does on the puzzles. It takes you to a graphically-based help screen, which looks like the Puzzle Resources screen, but which is designed to bring up explanations for objects when you click on them.

A.3 The Write/Edit Display

The Write/Edit display becomes visible when you press the writing hand icon on the button bar. Figure A.5 shows what this display looks like for the Resources of the Hungarian Rings. It includes a small data entry area that is laid out like the puzzle itself. For each of the other shapes of puzzle, this data entry has a corresponding shape.

Creating new configurations. If you want to create a totally new configuration for the puzzle, press the "Clear" button to empty out the fields for each of the locations. Then click into location 1 as you would in any field in any data entry screen. Type in the character you want for the first location.

You move between the entry fields for all of the locations of the puzzle using the tab key. Alternate presses on the tab key with presses on the keys for whatever characters you want in the locations of the puzzle.

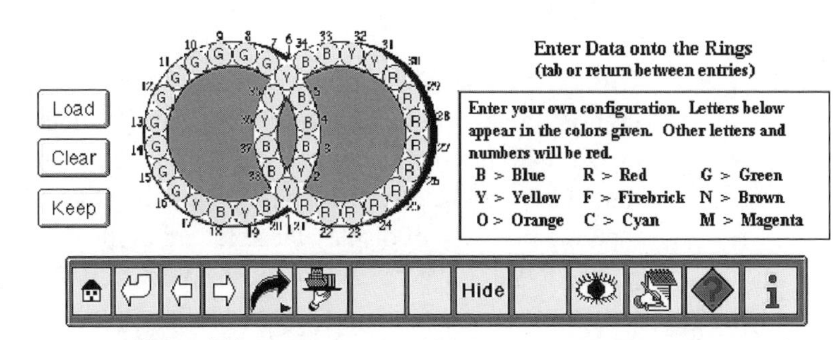

Figure A.5 The Write/Edit Display for the Hungarian Rings.

Once you have created a configuration, two buttons offer you options for doing something with it. If you press the "Load" key, the computer will load the configuration into the puzzle. However, it will first do some validity checks and will let you know if there is a problem. If there is none, it will ask you to write a name or description for the configuration that will be displayed on the puzzle when the configuration is loaded. The computer will supply a default name based upon the date and time if you wish.

The other option is to press the "Keep" button to have the configuration added to your collection, which is saved by Puzzle Resources. When you press this button, the computer again does a validity check and asks you for a name or description. If you choose not to name it, it will assign a default name based upon the date and time of creation.

Editing existing configurations. In addition to creating configurations by typing in all of the entries, you can also edit ones that you currently have in your collection. To do this, you click on the description of the configuration you want to edit. In the dialog box that opens, click on the "Edit" button. This loads the configuration into the data entry display. Once it is there, you can move your cursor into the various fields that you want to change and edit as you would in any data entry situation.

Once you have a modified configuration, you can load it or keep it. If you choose to keep it, be aware that what you will be doing is adding this modified configuration as a new entry at the end of your list. The entry that you loaded earlier for editing will remain in the list and unchanged.

Hungarian Rings Unique Features. Each of the different puzzles has a few unique features relating to its configuration process. For example, the Hungarian Rings configuration system allows you to create color versions of the puzzle with up to nine colors. The table at the right of the data entry area in Figure A.5 shows a mapping between nine letters of the alphabet and nine colors. If you prepare a configuration that uses only the letters from this set, then the puzzle will display these letters in the colors indicated. It will also initialize the color dots display on the puzzle to show where each of these colors starts out. Using these options, you can explore a very large number of variations on the basic four-color configuration. (See Exercises 16.6 through 16.9 for discussion of four additional configurations with colors.)

Slide Puzzle Unique Features. The Slide Puzzle needs at least one blank spot to function. Thus, if you are creating a configuration for Slide, you need to be sure to put in one or more of them. They are coded with the asterisk character, so be sure to enter at least one asterisk into the display. The computer will not let you load or keep the configuration without one.

If you enter at most two asterisks, then when you load the configuration into the puzzle, these asterisk-loaded locations will display as blanks. Slide Puzzle is programmed to display at most two asterisk-loaded spots with blanks. If there are more, then the extras will display as asterisks on disks.

The letter “X” has special significance for Slide Puzzle as the marker of an out of action spot. Thus, you cannot use X’s for symbols that you want to move around. Any other letter or number can be used, however, as a character on a disk.

As a time saver, you can leave the last locations on the data entry display blank if you want all of them to be out of action. The computer will interpret all locations that are blank from some location to the end as being out of action and will fill them with X’s.

Transpose Unique Features. The only thing to remember about creating Transpose configurations is that, like with Slide, the letter “X” should be used only for spots that are out of action. You may enter X’s wherever you want into the display, thus putting certain locations out of action.

The Transpose configure utility works like that of Slide in that it accepts blanks from some location to the end and interprets these as all being out of action spots and puts X’s in them.

Oval Track Unique Features. As with Slide and Hungarian Rings, on the Oval Track Puzzles, the character “X” means “out of action.” You may put a string of locations from some number up to 20 out of action, so long as you leave enough room for the entire “Do Arrows” action of the version of the puzzle you are configuring. However, the computer will not let you configure a puzzle so that an X-ed out location is followed by a higher numbered active location.

As with the previous two versions, on Oval Track a string of blanks in the highest numbered location is interpreted as meaning that these locations are out of action and should be loaded with X’s.

A.4 Working with Macro Programs

There is a difference in the way you create new configurations and new macros. The tools for creating new configurations for your puzzles are available via the Puzzle Resources window, whereas you create new macros using the tools available to you in the puzzle's window.

The creation of macros can be done in the puzzles using the "Program a Macro" button. You develop and test macros on the puzzle, and when you have something that you want to keep permanently, you move it to Puzzle Resources. To do this, first press the "Review a Macro" button and select the letter for the button that has the program you want to save. When the computer displays the program, it also makes visible a "Copy to Resources" button. Press this button and you start the process of saving your macro program. The computer will first open Puzzle Resources if it is not currently open, and then it will ask you to describe or name your macro, suggesting a default name based upon the time and date. Once you set the name, the macro will be added to your collection.

Another way to create macros is available when the "View Moves" display is visible. If you press the "View Moves" button (eye icon), the "Copy to Resources" button becomes visible as well. (The exception is for Transpose, which does not offer macros.) If you press this button, then the sequence of steps that currently shows will be converted into a macro program and saved.

There will be times when this feature will be handy for recording your results. For example, when you are exploring for useful procedures on a puzzle, a typical approach is to start with the puzzle in the solved configuration and try things like commutators, looking for procedures that will do controlled changes. If you find something useful this way, you can convert it directly into a macro and save it rather than having to first program it into a button.

Pressing the "Copy to Resources" button when the "View Moves" display is open causes the computer to convert your sequence of moves into the notation for macros, and readies it for saving. The computer will again ask you for a name, and once it has this, it will add this program to your Puzzle Resources macro collection. From there you can load the program into a button.

This feature of creating macros from an entire sequence of moves has another potential use that you might want to exploit from time to time. Suppose you have solved a scrambled puzzle, and the process raised some issues that you want to explore later. If you store your entire sequence of steps as a macro, you can come back at some later time and retrieve that scrambling of the puzzle. To do so, load the macro program into one of the macro buttons for a puzzle, be sure the puzzle is in a solved state, and then run the inverse of the macro. This will back the puzzle up through the steps of your solution procedure and return it to the scrambled state. At this point you can try a new approach to solving it, and perhaps improve your efficiency.

A.5 Saving Changes

It is important for you to be aware of how the Permutation Puzzle Package manages the saving of changes. This is particularly the case with respect to Puzzle Resources, since it is the component in which you will do most of your saving of changes.

The Permutation Puzzle Package manages the saving of changes more like a word processor than like a data base program. With a database, the program automatically updates the file on your hard drive to reflect the changes you make. In contrast, a word processor puts you in charge of when to update the documents you create.

This puzzle software follows the word processor model. This means that when you add things to your resources collection, they do not become a part of your permanent record until you do a save. The program will ask you if you want to save each component every time at shutdown, so you will have this opportunity to update your permanent Resources. However, depending on your level of concern about system crashes or power failures, in keeping with the old adage, "save early and save often," you may want to do a save after you have recorded some especially valuable find.

Exercises

This group of exercises is designed to make you aware of the different uses you can make of the Puzzle Resources.

Exercise A.1 The various buttons scrambles. Many of the preset configurations for the puzzles include one or more buttons scramble configurations. These are all scrambles that have been gotten by having the computer do buttons scrambles, with a small number of "spins," or rotations with an unknown number of applications of "Do Arrows" between successive rotations. Thus, they are all guaranteed to be solvable in a relatively small number of steps.

Explore the Puzzle Resources collections of preset configurations, and load some of these buttons scrambles. See if you can solve them using a number of rotations that is less than or equal to the number of spins that is indicated.

Exercise A.2 Conjugacy practice. Sometimes in solving the puzzle you need to focus on a few disks and try to ignore the neighboring ones. For example, there are times that you need to focus on three disks to get them into three specific locations, and not worry about what happens with anything else. (This is for setting up for performing conjugates of known tool permutations. See Chapter 9 for the background on this.)

Puzzle Resources gives a way to practice this skill. You can create configurations that distinguish only a few disks and label all the rest the same so that they cannot be distinguished. This makes it easy to focus on certain disks and ignore the others.

For each of the configurations described below, build each of them using the "create your own preset" feature accessed by pressing the hand and pencil icon button. Load it into the puzzle. Then have the computer scramble the disks, and get them back into place.

(a) On Oval Track 3, create a configuration that starts with disks labeled 1, 2, and 3 in their home spots, and has all of the rest labeled with the letter B.

(b) On Oval Track 4, create a configuration that starts with disks labeled 1, 2, 3, 4, and 5 in their home spots, and has all of the rest labeled with the letter B.

(c) On Oval Track 7, create a configuration that starts with disks labeled 1, 3, and 5 in their home spots, and has all of the rest labeled with the letter B.

Solutions and Hints for Selected Exercises

Introduction Exercises

0.1: (b) 11

(c) Hold down the alt key (Windows) or the option key (Mac).

0.5: 6.

0.7: (b) The background is not textured like the puzzle screens' backgrounds and is gray, there is a label which flashes "This is a help screen," and the "i" icon has changed to "Go back."

(c) A slow wipe to the left takes you to the help screen and a wipe to the right takes you back.

0.9: If you select a puzzle from the "Go" menu that is currently not open, Puzzle Driver will open the relevant window and go to the selected puzzle. If the prefs are set so that the musical introductory tour is to take place, it will run the tour before taking you to the desired puzzle. If you press the shift key down while the tour is in progress in order to stop it, this will also terminate the "Go" command and leave you where you stopped on the tour.

Chapter 1 Exercises

1.1: Ex. 1.2 describes a solution that is provided as a macro in Puzzle Resources.

1.3: Here is a 76 step opening game:

$$TR^3TR^2TR^8TR^{-1}TR^{-5}TR^2TR^2TR^{-2}TRTR^{-2}TRTR^{-4}TR^3TR^{-9}TR^2TR^3TR^3TR^{-1}$$
$$TR^{10}(TR^3)^3TR^{-7}TR^{-2}TR^3TR^2(TR)^2TR^{-2}TR^{-4}TR^3TR^{-1}TR^{-3}TR^2(TR^{-1})^2\,TRTR^{-1}$$

1.4: $RTR^4TR^5TR^{-9}$.

1.5: $R^{-8}TR^3TR^{-1}TR^9TR^6T$.

1.6: Ex. 1.7 describes a solution that is provided as a macro in Puzzle Resources.

1.8: Ex. 1.9 describes a solution that is provided as a macro in Puzzle Resources.

Chapter 2 Exercises

2.1: (a) A first solution: (1, 5) (1, 11) (2, 3) (2, 8) (2, 12) (2, 9) (2, 10) (2, 7).
A second solution: (11, 1) (7, 2) (7, 3) (11, 5) (10, 7) (10, 8) (12, 9) (12, 10).
(b) A first solution: (1, 5) (1, 8) (1, 10) (1, 9) (1, 6) (1, 4) (2, 11) (2, 3) (7, 12).
A second solution: (4, 1) (3, 2) (11, 3) (6, 4) (6, 5) (9, 6) (12, 7) (9, 8) (10, 9).

2.2: (a) (1, 2)(2, 3)(3, 4)(4, 5)
(b) (1, 2)(2, 3)(3, 4)(4, 5)(5, 6)
(c) (1, 2)(2, 3)(3, 4)(4, 5)(5, 6)(6, 7)
The highest numbered disk out of place is in spot 1. Keep dragging it up one place at a time until it is home.

2.4: Any series of steps for which each transposition restores at least one of the disks to its home spot will be as efficient as any other series that satisfies this same condition.

2.5: $T(\alpha) = m - c$, where m is the number of disks displaced by the scramble α, and c is the number of cycles in α. (See Chapter 5 for a definition of a cycle.)

2.6: Clearly $\mu_2 = 1/2$. Suppose you are considering a scramble with n disks active. If disk n is out of place, use your first transposition to put it into its home spot, and you now have a scramble involving disks 1 through $n-1$, and on average, it will take μ_{n-1} transpositions more. If disk n was already in place, the scramble can be regarded as an $n-1$ scramble. Show why these insights lead to the formula

$$\mu_n = \frac{n-1}{n}\left(\mu_{n-1}+1\right)+\frac{1}{n}\mu_{n-1} = \mu_{n-1}+\frac{n-1}{n}.$$

Get a formula for μ_n from this.

2.8: (b) Count how many A's are out of place. If you are efficient, you can restore the alternating pattern in this many transpositions.

(c) Make your own alternating configurations with 2, 4, 6, 8, etc. disks. Try many scrambles for each case, and see if it doesn't work out so that your average number of transpositions needed to restore the alternating pattern (with A first always) is near to $n/4$ if there are n disks active and n is even. (You need to be as efficient as possible to achieve this by always taking an A from a B spot to put into an A spot that has a B.) Then see if you can come up with a way to prove that $n/4$ is indeed the average. (One way might be to think about reducing the problem along the lines of the hint for Exercise 2.6.)

Chapter 3 Exercises

3.2: See Section 14.3 for further discussion on efficiency on the "15" puzzle. With practice, you should be able to get your average into the range of 120 to 130 moves.

Chapter 4 Exercises

4.3: See Chapter 16 for a discussion of what shuffles do. Basically, they run disks around the "outer loop" of disks outside of the intersection while running disks around the "inner loop" between the intersection points. What happens to the disks from the intersection points depends on whether you start the shuffle with the left or right ring, and on which way you do the turning.

4.4: (a) (LR), (RL), $(L^{-1}R^{-1})$, and $(R^{-1}L^{-1})$ all have order $9 \times 29 = 261$. After every 9 applications, the numbers on the inner loop return home, and after every 29 applications, the numbers on the outer loop get home. To get both things to happen at the same time, you need to have the number of applications be a multiple of these two numbers.

(b) $(L^2 R^2)$ has order 19.

(c) $(L^5 R^5)$ has order 20.

4.7: The procedure listed in (4.1) must be repeated ten times to return every disk home. That is, its order is 10.

Chapter 5 Exercises

5.1: (a) $\begin{pmatrix} 1 & 2 & 3 & 4 & 5 & 6 & 7 & 8 & 9 & 10 & 11 & 12 \\ 2 & 5 & 11 & 1 & 6 & 9 & 10 & 4 & 8 & 7 & 3 & 12 \end{pmatrix} = (1,2,5,6,9,8,4)(3,11)(7,10)$

(b) $\begin{pmatrix} 1 & 2 & 3 & 4 & 5 & 6 & 7 & 8 & 9 & 10 & 11 & 12 \\ 5 & 8 & 6 & 4 & 7 & 1 & 12 & 11 & 3 & 2 & 10 & 9 \end{pmatrix} = (1,5,7,12,9,3,6)(2,8,11,10)$

(c) $\begin{pmatrix} 1 & 2 & 3 & 4 & 5 & 6 & 7 & 8 & 9 & 10 & 11 & 12 \\ 10 & 4 & 2 & 1 & 9 & 6 & 7 & 12 & 8 & 5 & 3 & 11 \end{pmatrix} = (1,10,5,9,8,12,11,3,2,4)$

(d) $\begin{pmatrix} 1 & 2 & 3 & 4 & 5 & 6 & 7 & 8 & 9 & 10 & 11 & 12 \\ 12 & 1 & 10 & 2 & 5 & 7 & 3 & 11 & 8 & 4 & 6 & 9 \end{pmatrix} = (1,12,9,8,11,6,7,3,10,4,2)$

5.2: (a) (1, 12, 16, 11, 3, 2, 5, 14, 10, 13)(4, 8, 9, 15)
(b) (1, 13, 11, 3)(2, 12, 4, 15, 5, 14, 10, 16, 9, 8)

5.3: (a) (1, 17, 3, 6, 13, 18, 5, 11, 10, 15, 12, 4, 7, 9, 14, 8, 2)(16, 20)
(b) (1, 6, 3, 13, 5, 18)(4, 17, 15, 7)(8, 19, 12, 11, 9, 20)(10, 14)

5.4: (a) (1, 27, 5, 32, 37, 24, 30, 26, 22, 14, 19)(2, 10, 18, 34, 11, 9, 4)(3, 23, 29)(6, 13, 20, 12, 28, 8, 17, 21, 33, 25)(7, 38, 15, 31, 36)(16, 35)
This permutation is made up of cycles of length 11, 7, 3, 10, 5, and 2.

(b) (1,21,29,11,38,36,26,37,15,18,31,12,35,30,9,34,16,6,8,32,33,28,23,4,13,17,25,20,5,2,10,24,14,22)(3, 19, 27, 7)
This one consists of a 34-cycle and a 4-cycle.

5.5: (a) $\alpha = (1, 5, 2, 7)\,(3, 4)\,(8, 10, 9)$
(b) $\beta = (1, 10, 9, 7, 6, 5, 2, 4, 8)$
(c) $\gamma = (1, 2, 3, 4)\,(6, 10, 8, 7, 9)$
(d) $\delta = \begin{pmatrix} 1 & 2 & 3 & 4 & 5 & 6 & 7 & 8 & 9 & 10 \\ 5 & 9 & 6 & 1 & 8 & 3 & 2 & 4 & 10 & 7 \end{pmatrix}$
(e) $\mu = \begin{pmatrix} 1 & 2 & 3 & 4 & 5 & 6 & 7 & 8 & 9 & 10 \\ 1 & 6 & 10 & 2 & 5 & 8 & 4 & 3 & 7 & 9 \end{pmatrix}$
(f) $\nu = \begin{pmatrix} 1 & 2 & 3 & 4 & 5 & 6 & 7 & 8 & 9 & 10 \\ 2 & 3 & 4 & 1 & 6 & 5 & 8 & 10 & 9 & 7 \end{pmatrix}$

5.6: Here are the solutions in cycle notation. Convert them to array notation yourself.
(a) $\alpha \circ \beta = (1, 2, 6, 5, 4, 3, 8, 9)(7, 10)$
(b) $\beta \circ \alpha = (1, 9)(2, 3, 4, 10, 8, 5, 7, 6)$
(c) $\alpha \circ \alpha = (1, 2)(5, 7)(8, 9, 10)$
(d) $\beta \circ \beta = (1, 9, 6, 2, 8, 10, 7, 5, 4)$
(e) $\gamma \circ \delta = (1, 9, 3)(2, 6, 7, 10, 4, 5, 8)$
(f) $\delta \circ \gamma = (1, 5, 7, 3, 10, 9, 8)(2, 6, 4)$
(g) $(\mu \circ \nu) \circ \delta = (1, 9, 4, 6, 7, 5, 3, 2, 8)$
(h) $\mu \circ (\nu \circ \delta) = (1, 9, 4, 6, 7, 5, 3, 2, 8)$

5.7: (a) $\alpha = (1, 6)(2, 5, 4, 3, 7)$
(b) $\beta = (1, 2, 3, 4, 5, 6, 7, 8, 9)$
(c) $\gamma = (1, 5, 3, 6, 4, 2)$
(d) $\delta = (1, 2, 3, 4, 5, 6, 7, 8, 9)$
(e) $\mu = (1, 7)(2, 4, 5, 6, 9, 8)$
(f) $\nu = (1, 7, 9, 2, 4, 6, 8)$

5.8: Here are the solutions in cycle notation. Convert them to array notation yourself.

(a) $\alpha^{-1} = (9, 10, 8)\,(4, 3)\,(7, 2, 5, 1)$
(b) $\beta^{-1} = (8, 4, 2, 5, 6, 7, 9, 10, 1)$
(c) $\gamma^{-1} = (9, 7, 8, 10, 6)\,(4, 3, 2, 1)$
(d) $\delta^{-1} = (6, 3)\,(7, 10, 9, 2)\,(4, 8, 5, 1)$
(e) $\mu^{-1} = (4, 7, 9, 10, 3, 8, 6, 2)$
(f) $\nu^{-1} = (10, 8, 7)\,(6, 5)\,(4, 3, 2, 1)$

5.9: We saw in Ex. 5.6 that $\alpha \circ \beta \neq \beta \circ \alpha$ and that $\gamma \circ \delta \neq \delta \circ \gamma$. These cases show that not all pairs of permutations commute with each other.

5.10: If α is not the identity, then there is an i such that $j = i\alpha \neq i$. If $j\alpha = i$, then define β to be the permutation which maps i to j, j to some k other than i or j, (which must exist because $n \geq 3$), and which maps this k to i and all other elements to themselves. It is easy to show that $\alpha \circ \beta \neq \beta \circ \alpha$. If $j\alpha \neq i$, then we get a β which does not commute with α by having β map i and j to each other and everything else to itself.

5.11: The one case of a three-fold composition we looked at in Ex. 5.6, showed that $(\mu \circ \nu) \circ \delta = \mu \circ (\nu \circ \delta)$. This suggests that composition is associative, but of course does not prove it.

Chapter 6 Exercises

B

6.2: (a) An n-position counterclockwise rotation of the disks is the same as the inverse of an n-position clockwise rotation. That is, if one starts with the identity permutation, and then performs n-position rotations clockwise and counterclockwise in either order, one returns the disks to the identity permutation.

(b) Doing an n-position counterclockwise rotation gives the same result as doing n one-position counterclockwise rotations.

6.3: Since A and T are disjoint, they commute with each other. Thus, a positive integral exponent can be distributed over the product AT.

6.4: (a) $\{1, 3\}$

(b) $\{2, 4\}$

(c) $\{1, 2, 3, 4, 6, 7, 8, 9, 21, 22, 23, 35, 36, 37\}$

(d) $\{1, 2, 3, 4, 6, 7, 8\}$

(e) $\{11, 12, 13, 14, 15\}$

6.5 (c): On Oval Track 15 with $T = (1, 3, 4, 2, 5)$ and with 5 active disks, R and T do not commute, but $(RT)^5$, R^5, and T^5 all reduce to the identity, so $(RT)^5 = R^5T^5$.

6.6: (a) If $\alpha \circ \gamma = \beta \circ \gamma$, then $\alpha = \alpha \circ \varepsilon = \alpha \circ (\gamma \circ \gamma^{-1}) = (\alpha \circ \gamma) \circ \gamma^{-1} = (\beta \circ \gamma) \circ \gamma^{-1} = \beta \circ (\gamma \circ \gamma^{-1}) = \beta \circ \varepsilon = \beta$.

(b) Modify the proof of the right cancellation law to get one for the left cancellation law.

(c) Here is a puzzle interpretation for the left cancellation law. Suppose you do a series of moves which you program into macro button C. Suppose you also program different series of moves into buttons A and B, but discover that pressing A then C produces the same outcome as pressing B then C. You can conclude that you would get the same outcome from pressing button A as you would from pressing button B, assuming in each case starting from the identity permutation. The sequences of steps might be different, but the permutations they produce would be the same.

6.7: $\beta_1 = \beta_1 \circ \varepsilon = \beta_1 \circ (\alpha \circ \beta_2) = (\beta_1 \circ \alpha) \circ \beta_2 = \varepsilon \circ \beta_2 = \beta_2$

6.8: Since $(\alpha^{-1})^{-1}$ is an inverse for α^{-1}, this means that $(\alpha^{-1})^{-1} \circ \alpha^{-1} = \varepsilon$. But also $\alpha \circ \alpha^{-1} = \varepsilon$, so $(\alpha^{-1})^{-1} \circ \alpha^{-1} = \alpha \circ \alpha^{-1}$. By the right cancellation law, $(\alpha^{-1})^{-1} = \alpha$.

6.9: (a) If $(\alpha \circ \beta)^2 = \alpha^2 \circ \beta^2$, this means that $(\alpha \circ \beta) \circ (\alpha \circ \beta) = (\alpha \circ \alpha) \circ (\beta \circ \beta)$, so $\alpha \circ (\beta \circ \alpha) \circ \beta = \alpha \circ (\alpha \circ \beta) \circ \beta$. Applying both cancellation laws to this equation, we get that $\beta \circ \alpha = \alpha \circ \beta$.

(b) If $(\alpha \circ \beta)^{-1} = \alpha^{-1} \circ \beta^{-1}$, then

$$\alpha \circ \beta = ((\alpha \circ \beta)^{-1})^{-1} = (\alpha^{-1} \circ \beta^{-1})^{-1} = (\beta^{-1})^{-1} \circ (\alpha^{-1})^{-1} = \beta \circ \alpha.$$

(Note that we have used display (6.2) and Ex. 6.8 here.)

Chapter 7 Exercises

7.1: (a) (1, 3, 4, 6, 5, 2) = (1, 3) (1, 4) (1, 6) (1, 5) (1,2)
(b) (1, 6, 2) (5, 7) = (1, 6) (1, 2) (5,7)
(c) (1, 4, 2, 6, 3, 5) = (1, 4) (1, 2) (1, 6) (1, 3) (1,5)
(d) (1, 4) (3, 5)

7.2: (a) (1, 5) (1, 3) (1, 8) (2, 9) (2, 4) (2, 7) (2, 6)
(b) (1, 4) (1, 2) (3, 7) (3, 6) (5, 8) (5, 9)
(c) (1, 2) (1, 3) (1, 4) (1, 5) (1, 6) (7, 8)
(d) (1, 2) (3, 4) (5, 6) (5, 7) (8, 9)

7.3: Let T_i denote the "Do Arrows" permutation of Oval Track Puzzle i, and let $o(T_i)$ denote its order. Then $o(T_1) = 2$, $o(T_2) = 4$, $o(T_3) = 3$, $o(T_4) = 5$, $o(T_5) = 2$, $o(T_6) = 2$, $o(T_7) = 3$, $o(T_8) = 2$, $o(T_9) = 3$, $o(T_{10}) = 4$, $o(T_{11}) = 4$, $o(T_{12}) = 5$, $o(T_{13}) = 4$, $o(T_{14}) = 6$, $o(T_{15}) = 5$, $o(T_{16}) = 6$, $o(T_{17}) = 2$, $o(T_{18}) = 6$.

7.4: (a) $o(\alpha) = 6$
(b) $o(\beta) = 3$
(c) $o(\gamma) = 12$
(d) $o(\delta) = 15$
(e) $o(\mu) = 6$
(f) $o(\nu) = 12$
(g) $o(\rho) = 12$
(h) $o(\sigma) = 30$

7.5: Think of a k-cycle as k arrows cycling around a set of k dots. To follow the arrows from a given point until you get back to the same point will take you along each of the k arrows. No fewer applications will do.

7.6: Suppose that $\alpha = \sigma_1\sigma_2 \cdots \sigma_m$ where the σ_i's denote the pairwise disjoint cycles of α, and let k_i denote the length of σ_i. Since the cycles are disjoint, an integral exponent on α can be distributed across them. That is, $\alpha^k = \sigma_1^k \sigma_2^k \cdots$

B

σ_m^k. For each i, $\sigma_i^k = \varepsilon$ if and only if k is a multiple of k_i. To have $\sigma_i^k = \varepsilon$ for all i simultaneously, one needs to have k being a multiple of k_i for each i. This first happens when k is the least common multiple of the k_i's.

7.7: Suppose that $\alpha = \sigma_1\sigma_2\cdots\sigma_m$ where the σ_i's denote the pairwise disjoint cycles of α, and let k_i denote the length of σ_i. One can list these cycles in any of $m!$ possible orders, and list the cycle σ_i in exactly one way for each of the k_i elements as the first one. This leads to $m!\prod_{i=1}^{m} k_i$ ways to express α in terms of cycles.

Chapter 8 Exercises

8.2: (a) odd
(b) odd
(c) odd
(d) even

8.4: (a) The permutation is (1, 4, 16, 14, 6, 13, 12, 7, 5, 11, 9, 8, 15, 3, 2). This is a 15-cycle, which can be expressed using 14 transpositions, so it is even.

(b) This permutation is $(1,2)(3,7,16,14,13,9,15)(4,6,11) = (1,2)(3,7)(3,16)(3,14)(3,13)(3,9)(3,15)(4,6)(4,11)$. Since we have it expressed with 9 transpositions, it is odd.

8.5: Build a permutation from {1, 2,…, n} to itself by first defining what 1 maps to, then what 2 maps to, then 3, etc. You have n choices for mapping 1. Since the image of 1 cannot be reused, you have $n - 1$ choices for what 2 maps to. As you define the image of each number, that leads to one less choice. The result is $n \times (n-1) \times (n-2) \times \cdots \times 2 \times 1 = n!$ ways to define a permutation in S_n.

8.7: (a)
$$\begin{aligned}\alpha &= [(1,2)(1,3)]\,[(1,4)(1,6)]\,[(1,5)(3,4)]\,[(3,5)(2,5)]\,[(1,4)(5,2)] \\ &= (1,2,3)(1,4,6)(1,5,3)(3,1,4)(3,2,5)(1,4,5)(5,1,2)\end{aligned}$$

(b)
$$\begin{aligned}\beta &= [(1,2)(2,3)]\,[(4,5)(1,3)]\,[(6,7)(6,8)]\,[(9,10)(11,12)] \\ &= (1,3,2)(4,5,1)(1,4,3)(6,7,8)(9,10,11)(11,9,12)\end{aligned}$$

B

(c) $\gamma = [(1, 2)(1, 3)]\,[(1, 4)(2, 3)]\,[(2, 4)(2, 5)]\,[(4, 5)(4, 6)]\,[(4, 7)(8, 9)]$
$= (1, 2, 3)(1, 4, 2)(2, 1, 3)(2, 4, 5)(4, 5, 6)(4, 7, 8)(8, 4, 9)$

8.8: (a) $\alpha = (1, 2, 3)(1, 4, 5)(1,6)$

(b) $\beta = [(1, 2)(1, 3)]\,[(1,4)(5, 6)]\,[(5, 7)(8, 9)]\,(8, 10)$
$= (1, 2, 3)(1, 4, 5)(5, 1, 6)(5, 7, 8)(8, 5, 9)(8, 10)$

(c) $\gamma = (2, 5, 3)(2, 7, 6)(3, 5, 8)(3, 4)(6, 8, 2)(6, 1, 9)$

8.10: Here are the original sequences with their reductions directly below them. The First Box and Tagged Box numbers are identified with different typefaces.

(a) (**6**, ***12***) (11, 7) (12, ***10***) (9, **6**) (5, **6**) (7, **6**) (7, 5) (***10***, **6**) (9, 12) (10, 11) (10, 7) (10, 9)
(11, 7) (12, ***10***) (9, ***10***) (5, ***10***) (7, ***10***) (7, 5) (9, 12) (10, 11) (10, 7) (10, 9)

(b) (**4**, ***5***) (5, ***6***) (***10***, 6) (8, 6) (11, 6) (8, **4**) (5, 6) (8, 5) (11, 8) (***10***, **4**)
(5, ***6***) (***10***, 6) (8, 6) (11, 6) (8, ***10***) (5, 6) (8, 5) (11, 8)

a: This typeface denotes the First Box number. ***b***: This typeface denotes the Tagged Box number.

8.11: Generalizing from the formula $(1, 2, 3, 4)(1, 2, 3, 4)(1, 3, 2, 4) = (1, 2)$, one sees that if all 4-cycles are available, then one can get all transpositions, which means that one can get all permutations in S_n, with $n \geq 4$. Similar methods show that if $n \geq 5$, then one can express all 3-cycles in S_n using 5-cycles, and thus can get all even permutations expressed as a product of 5-cycles. In general, if $n \geq k$, then one can express all permutations in S_n in terms of k-cycles if k is even and one can express all even permutations in S_n in terms of k-cycles if k is odd.

8.14: (a) A solution of a scramble α on the Transpose Puzzle produces a sequence of transpositions that expresses α^{-1}. In the reverse order, we get a sequence that expresses α. If the expression for α^{-1} resulting from solving the puzzle is as short as possible, then so is the resulting expression for α.

B

8.15: If one applies the transposition (a_i, a_j) where $1 \le i < j \le r$, the formula

$$(a_1, a_2, \ldots, a_r)\,(a_i, a_j) = (a_1, \ldots, a_{i-1}, a_j, a_{j+1}, \ldots, a_r)\,(a_i, a_{i+1}, \ldots, a_{j-1})$$

applies. The r-cycle has been split by this transposition into a cycle of length $r - (j - i)$ and another of length $j - i$. The r-cycle needs $r - 1$ transpositions to express it, and the shortened cycles need $r - (j - i) - 1$ and $(j - i) - 1$ transpositions respectively. Adding these last two numbers, we get a sum of $r - 2$. By performing one transposition to achieve the split, we have reduced by one the total number of transpositions needed.

Chapter 9 Exercises

9.1: (a) $\alpha\beta\alpha^{-1} = (7, 4, 10)\,(1, 5)\,(2, 6, 3)$
$= (1\alpha^{-1}, 5\alpha^{-1}, 8\alpha^{-1})\,(2\alpha^{-1}, 6\alpha^{-1})\,(3\alpha^{-1}, 7\alpha^{-1}, 4\alpha^{-1})$

(b) $\alpha\beta\alpha^{-1} = (8, 6, 4, 7, 1, 2, 3)\,(5, 9, 10)$
$= (1\alpha^{-1}, 2\alpha^{-1}, 3\alpha^{-1}, 4\alpha^{-1}, 5\alpha^{-1}, 6\alpha^{-1}, 7\alpha^{-1})\,(8\alpha^{-1}, 9\alpha^{-1}, 10\alpha^{-1})$

(c) $\alpha\beta\alpha^{-1} = (1, 2)\,(8, 10, 5, 9, 6, 4, 12)$
$= (1\alpha^{-1}, 2\alpha^{-1})\,(5\alpha^{-1}, 11\alpha^{-1}, 7\alpha^{-1}, 9\alpha^{-1}, 12\alpha^{-1}, 4\alpha^{-1}, 10\alpha^{-1})$

9.3: (a) A 12-step setup procedure that brings the disks 5, 11, and 14 into spots 6, 7, and 10 respectively with the blank in spot 11, is

$$A = (16,12)(12,8)(8,7)(7,11)(11,10)(10,6)(6,5)(5,9)(9,10)(10,14)(14,15)(15,11).$$

If $B = (11, 7)(7, 6)(6, 10)(10, 11)$, then $ABA^{-1} = (5, 11, 14)$.

(b) A 17-step setup that uses the 6, 7, 10, 11 square is:

$$A = (16,15)(15,11)(11,7)(7,3)(3,2)(2,1)(1,5)(5,9)(9,10)(10,6)(6,2)(2,3)(3,7)(7,11)(11,12)(12,8)(8,7).$$

(c) This 19-step setup also uses the 6, 7, 10, 11 square:

$$A=(16,15)(15,14)(14,10)(10,9)(9,13)(13,14)(14,15)(15,11)(11,10)(10,9)(9,5)(5,1)(1,2)(2,3)(3,7)(7,6)(6,2)(2,3)(3,7).$$

B

9.5: Form a train putting the three target disks in consecutive positions. Park a target disk in spot 10 as you circle the train to where the target disk can be coupled in where it needs to go.

9.7: (a) The disks alternate in parity in the initial identity configuration. Rotations do not change this, nor does the permutation (5, 3, 1). Since you can only use these two methods to move disks, you cannot break out of the alternating parity pattern. (b) We'll come back to the second question in Chapter 13. (c) You need to get the three target disks so that the distances between them are two. The "Do Arrows" can reduce distances by two or four. Consider target disks a and b. If the distance between them clockwise is an odd number, then the distance between them counterclockwise will be even, and vice-versa. Whichever way the distance is an even number, reduce it that way to get the right spacing.

Chapter 10 Exercises

10.1: (a) (1, 2, 5) (b) (1, 2, 3, 6, 5) (c) (1, 5, 3)(2, 6, 4) (d) (1, 3, 5) (e) (1, 2, 6) (f) (1, 3, 6) (g) (1, 2, 4, 5, 3) (h) (1, 2)(5, 6) (i) (1, 2)(3, 4, 5)(6, 7) (j) (1, 6, 11)(2, 5)(3, 4)(7, 10)(8, 9)

10.2: (a) (1, 21, 2)(6, 35, 7) (b) (1, 38, 2)(6, 34, 7) (c) (1, 2, 21)(6, 7, 35) (d) (1, 20, 21)(5, 35, 6) (e) (1, 21, 35, 11, 6) (f) (1, 6, 11, 35, 21) (g) (1, 38, 34, 11, 6) (h) (1, 6, 30, 7, 2) (i) (1, 25)(6, 11) (j) (1, 25)(6, 11) (k) (1, 16)(6, 30) (l) (1, 6)(11, 30)

B

10.3: On OT 8 with $n \geq 5$, $[R^{-1}, T]^2 = (5, 3, 1)$. On OT 10 with $n \geq 9$, $[R^{-2}, T] = (1, 7, 9)$. On OT 11 with $n \geq 8$, $[R^{-1}, T^2]^2 = (7, 5, 2)$.

10.4 (b): The "commutator" $[L^2, U^2]$ on the Slide Puzzle in the basic "15" configuration is the 7-cycle (15, 12, 8, 7, 6, 10, 14).

10.5: Multiply $\alpha\beta\alpha^{-1}\beta^{-1} = [\alpha, \beta]$ by $\beta\alpha\beta^{-1}\alpha^{-1} = [\beta, \alpha]$ and see how the product simplifies.

10.6: Use the fact that α and β commute with each other if and only if α^{-1} and β commute with each other.

10.7: (a) On OT 1 with $n \geq 7$, $[[R^{-1}, T], T] = (1, 2, 3, 4, 5)$ while $[R^{-1}, [T, T]]$ is the identity.

(b) If β commutes with both α and γ, then, according to Ex. 10.6, $[\alpha, \beta]$ and $[\beta, \gamma]$ both reduce to ε.

10.8: (a) On Oval Track 1, $[(RT), T] \neq \varepsilon$ while $[R, (TT)] = \varepsilon$.

(b) $[(\alpha\beta), \gamma] = [\alpha, (\beta\gamma)]$ iff $(\alpha\beta)\gamma(\alpha\beta)^{-1}\gamma^{-1} = \alpha\,(\beta\gamma)\alpha^{-1}\,(\beta\gamma)^{-1}$ iff $\alpha\beta\gamma\beta^{-1}\alpha^{-1}\gamma^{-1} = \alpha\beta\gamma\alpha^{-1}\gamma^{-1}\beta^{-1}$ iff (left cancellation) $\beta^{-1}\alpha^{-1}\gamma^{-1} = \alpha^{-1}\gamma^{-1}\beta^{-1}$ iff $(\beta^{-1}\alpha^{-1}\gamma^{-1})^{-1} = (\alpha^{-1}\gamma^{-1}\beta^{-1})^{-1}$ iff $\gamma\alpha\beta = \beta\gamma\alpha$.

10.9: $[\gamma\alpha\gamma^{-1}, \gamma\beta\gamma^{-1}] = (\gamma\alpha\gamma^{-1})\,(\gamma\beta\gamma^{-1})(\gamma\alpha\gamma^{-1})^{-1}\,(\gamma\beta\gamma^{-1})^{-1} = (\gamma\alpha\gamma^{-1})\,(\gamma\beta\gamma^{-1})\,(\gamma\alpha^{-1}\gamma^{-1})\,(\gamma\beta^{-1}\gamma^{-1}) = (\gamma\alpha)\,(\gamma^{-1}\gamma)\beta\,(\gamma^{-1}\gamma)\alpha^{-1}(\gamma^{-1}\gamma)(\beta^{-1}\gamma^{-1}) = \gamma(\alpha\beta\alpha^{-1}\beta^{-1})\gamma^{-1} = \gamma[\alpha, \beta]\gamma^{-1}$.

10.10: $[\alpha, \beta]\,(\beta[\alpha, \gamma]\beta^{-1}) = (\alpha\beta\alpha^{-1}\beta^{-1})\,(\beta(\alpha\gamma\alpha^{-1}\gamma^{-1})\beta^{-1}) = \alpha\beta\gamma\alpha^{-1}\gamma^{-1}\beta^{-1} = \alpha(\beta\gamma)\alpha^{-1}(\beta\gamma)^{-1} = [\alpha, (\beta\gamma)]$. The rest are proven similarly.

10.11: We have that α commutes with $[\alpha, \beta]$ iff $\alpha(\alpha\beta\alpha^{-1}\beta^{-1}) = (\alpha\beta\alpha^{-1}\beta^{-1})\,\alpha$ iff $\alpha\beta\alpha^{-1}\beta^{-1} = \beta\alpha^{-1}\beta^{-1}\alpha$ iff $[\alpha, \beta] = [\beta, \alpha^{-1}]$. The rest are proven similarly.

10.12: Prove first that under these assumptions, $\beta^m\alpha = \alpha\beta^m\,[\beta, \alpha]^m$ for all positive integers m. Then the proof goes by induction, and makes use of the equation $(\alpha\beta)^{m+1} = (\alpha\beta)^m\,(\alpha\beta)$.

Chapter 11 Exercises

11.1: (a) If β is a k-cycle, then since by Ex. 7.5, $\beta^k = \varepsilon$, then $\beta\beta^{k-1} = \varepsilon = \beta^{k-1}\beta$. Thus, we have $\beta^{-1} = \beta^{k-1}$. By display (5.4), β^{-1} is also a k-cycle.

(b) First observe that if $\beta = (b_1, b_2, \ldots, b_k)$, then $b_j\,(\beta^i)^r = b_{j+ri}$, where the addition is modulo k. Thus, $b_j(\beta^i)^r = b_j$ means that $j + ri \equiv j \pmod{k}$, which means that $ri \equiv 0 \pmod{k}$. If k is a prime number and $1 \leq i \leq k - 1$, the smallest value of r for which this can happen is $r = k$.

B

(c) As in (b), $b_j(\beta^r)^i = b_j\beta^{ri} = b_j$ for all j means that $ri \equiv 0 \pmod{k}$. The smallest value of i that makes this possible is $i = s$.

11.2: (a) (1, 3, 4), (1, 4), (1, 3, 2, 4), and (1, 3, 2).

(b) (i) Perform this 3-cycle as a conjugate of the tool 3-cycle (7, 4, 1) by first getting disks 1, 3, and 4 into spots 1, 4, and 7. (ii) Perform this transposition as a conjugate of the tool transposition (1, 3) by first getting disk 4 into spot 3. (iii) An application of the "Do Arrows" reduces the permutation to (1, 2). From here, follow a strategy like that used for (ii). (iv) Use a strategy like that used for (i).

11.3: The 3-cycles in (i) and (iv) can be solved using the same strategy for any number n of active disks as long as $n \geq 7$. The odd permutations in (ii) and (iii) cannot be solved if n is odd, but can be solved if n is even and $n \geq 6$.

11.4: (i) First perform the tool transposition (2, 3) to exchange disks 1 and 4. Then apply the "Do Arrows" twice. (ii) Swap disks 1 and 2 using the tool transposition (2, 3). Then apply T^2 to get disks 3 and 4 into spots 2 and 3 and perform the tool transposition again to swap disks 3 and 4. Then apply T^{-1}. (iii) First apply the tool transposition to swap disks 1 and 3. Then apply T^{-1} to get disks 2 and 1 into spots 2 and 3. Then apply the tool transposition to swap disks 2 and 1. Then apply T. (iv) Swap the disks 2 and 1 with the tool transposition, then apply T.

11.5: (b) One can show that any three disks a, b, and c, can be put into spots 5, 4, and 3. As a result, any 3-cycle (a, b, c) can be obtained as a conjugate of (5, 4, 3). Since we can get all of the 3-cycles, we can produce—and thus also solve—any even permutation on the puzzle. If n, the number of active disks, is odd, then the rotation R is an even permutation, as is the "Do Arrows," T. This means that even permutations are the only ones obtainable. On the other hand, if n is even, then R is an odd permutation. Since we can get at least one odd permutation along with all the even ones, we can also get all of the odd permutations.

11.6: By Ex. 8.8, we can express any odd scramble α in the form $\tau\sigma_1\sigma_2\cdots\sigma_k$, where the σ_i's are 3-cycles and τ is a transposition. Since we can perform all 3-cycles, we can, in succession, perform the 3-cycles $\sigma_k^{-1}, \sigma_{k-1}^{-1}, \ldots, \sigma_1^{-1}$, thereby reducing the scramble α to the transposition τ.

11.7: (a) On the first pass through the list, as one compares adjacent numbers, one will eventually come across the largest number, and the subsequent steps of that pass will bubble this largest number to the top of the list. The second pass bubbles the second largest number up to the second from the top position. Etc.

(b) Locate the largest disk, and establish it as the top end of the list. Apply the bubble sort algorithm by rotating the pair of adjacent disks to be compared into the two adjacent locations where you have a tool transposition for performing a swap if it is needed.

(c) The bubble sort is well known as one of the least efficient sorting algorithms. Its virtue is not efficiency but conceptual simplicity. A little puzzle experience ought to convince you that the same is true when this algorithm is used as a puzzle solving strategy.

Chapter 12 Exercises

12.1: $[R^{-3}, T]^2 = (5, 4, 3)$. This is an easy conjugate, $R^{-2}TR^2$, of the "Do Arrows," $T = (3, 2, 1)$, of Oval Track 3. Thus, everything that is possible to do on Oval Track 3 is also possible to do on Oval Track 5. Since we know we can completely solve OT 3 when n is even and sufficiently large, we now know we can completely solve OT 5 for such n's.

12.2: (b) See the explanation for Ex. 12.3 below.

12.3: When you can produce the "Do Arrows" of a first Oval Track Puzzle on a second one and also produce the "Do Arrows" of the second on the first, then anything that is possible (or impossible) on either is also possible (or impossible) on the other.

12.4: (a) On OT 5, $(1, 4)\,(2, 3) = TR^{-1}TR^{-1}TRT^{-1}RT^{-1}RT^{-1}R^{-5}TR^5TR^{-1}TR^{-1}TR^{-3}TR^4TR^{-1}TR$.

(b) On OT 1, $(1, 2)\,(3, 4) = R^{-1}TR^{-2}TRTRTRTR^{-2}TRTR^{-4}TRTR^2TR^{-1}TRTR^{-1}TR^{-2}TRTR^4$.

(c) On OT 11, $(4, 3, 2, 1) = ATA^{-1}$, where $A = R^{-2}T^{-1}R^{-3}T^{-1}R^5$.

B

12.5: (a) On OT 14, $(6, 5, 4, 3, 2, 1) = ATA^{-1}$, where $A = R^2T^{-1}RT^2RT^2RT^2R^{-5}$.

(b) On OT 17, $(3, 2, 1) = R^{-1}TR^{-5}TR^{-1}TRTR^{-1}TRTR^5T^{-1}R^{-1}TRTRTR^{-6}TR^{-1}TRTR^{-1}TRTR^5TR$.

(c) On OT 19 with $T = (1, 4, 2, 5, 3)$, $(5, 4, 3, 2, 1) = ATA^{-1}$, where $A = (RT^{-1})^4R^{-4}$.

12.6: On OT 7, we get (3, 2, 1) with this formula for the given values of n active disks:

$n = 19$: $(R^{-2}T)(R^{-4}T)^3(R^{-2}T^{-1})(R^4T^{-1})^4$ $\quad$ $n = 17$: $(R^{-2}T)(R^{-4}T)^3(R^2T^{-1})(R^4T^{-1})^3$

$n = 15$: $(R^{-2}T)(R^{-4}T)^2(R^{-2}T^{-1})(R^4T^{-1})^3$ $\quad$ $n = 13$: $(R^{-2}T)(R^{-4}T)^2(R^2T^{-1})(R^4T^{-1})^2$

$n = 11$: $(R^{-2}T)(R^{-4}T)^1(R^{-2}T^{-1})(R^4T^{-1})^2$ $\quad$ $n = 9$: $(R^{-2}T)(R^{-4}T)^1(R^2T^{-1})(R^4T^{-1})^1$

$n = 7$: $(R^{-2}T)(R^{-4}T)^0(R^{-2}T^{-1})(R^4T^{-1})^1$

We can write a single formula for this if we use the greatest integer function, $\lfloor x \rfloor$ = the greatest integer k such that $k \le x$. We get that with n active disks, n odd, and $n \ge 7$,

$$(3, 2, 1) = (R^{-2}T)(R^{-4}T)^{m_1}(R^{(-1)^k \cdot 2}T^{-1})(R^4T^{-1})^{m_2}, \text{ with } m_1 = \left\lfloor \frac{n-4}{4} \right\rfloor,\ m_2 = \left\lfloor \frac{n-3}{4} \right\rfloor,\ k = \left\lfloor \frac{n-1}{2} \right\rfloor.$$

12.7: On OT 8, we get (3, 2, 1) with this formula for the given values of n active disks:

$n = 19$: $R^{-1}(TR^{-2})^8$ $\quad$ $n = 17$: $R^{-1}(TR^{-2})^7$ $\quad$ $n = 15$: $R^{-1}(TR^{-2})^6$

We see that in general, if n is odd, then $(3, 2, 1) = R^{-1}(TR^{-2})^k$, where $k = (n - 3)/2$.

B

12.8: Clearly $R \in OT_i(n)$, so if also $T_j \in OT_i(n)$, then every combination of R and T_j is in $OT_i(n)$. But the set of all combinations of R and T_j is $OT_j(n)$. Thus, $OT_j(n) \subseteq OT_i(n)$. Since the permutations obtainable on Oval Track Puzzle k are exactly the permutations in $OT_k(n)$, this containment relationship says that every permutation obtainable on Oval Track j is also obtainable on Oval Track i.

Chapter 13 Exercises

For these exercises, let $OT_i(n)$ denote the group associated with Oval Track Puzzle i with n active disks, and let $|OT_i(n)|$ denote the size of this group.

13.1: (a) $|OT_4(6)| = 720 = 6!$, $|OT_4(7)| = 2{,}520 = 7!/2$, $|OT_4(8)| = 40{,}320 = 8!$, $|OT_4(9)| = 181{,}440 = 9!/2$, $|OT_4(10)| = 3{,}628{,}800 = 10!$.

(b) $|OT_9(4)| = 24 = 4!$, $|OT_9(5)| = 60 = 5!/2$, $|OT_9(6)| = 720 = 6!$, $|OT_9(7)| = 2{,}520 = 7!/2$, $|OT_9(8)| = 40{,}320 = 8!$.

13.2: Here is a table that compares the sizes of $OT_7(n)$ and $OT_8(n)$.

n	$\|OT_7(n)\|$	$\|OT_8(n)\|$
10	7,200 = ?	14,400 = 2 ×?
11	19,958,400 = 11!/2	19,958,400 = 11!/2
12	518,400 = ?	518,400 = ?
13	3,113,510,400 = 13!/2	3,113,510,400 = 13!/2
14	12,700,800 = ?	25,401,600 = 2 ×?
15	653,837,184,000 = 15!/2	653,837,184,000 = 15!/2
16	1,625,702,400 = ?	1,625,702,400 = ?
17	177,843,714,048,000 = 17!/2	177,843,714,048,000 = 17!/2
18	65,840,947,200 = ?	131,681,894,400 = 2 ×?
19	60,822,550,204,416,000 = 19!/2	60,822,550,204,416,000 = 19!/2
20	13,168,189,440,000 = ?	13,168,189,440,000 = ?

13.3: (a) We have (1, 3) in $OT_1(6)$, and clearly we can get all transpositions as conjugates of this one. Thus, we have all transpositions in $OT_1(6)$, so $OT_1(6)$ is all of S_6.

B

(b) With $n = 5$, the disks start out with the neighbors of disk 1 being disks 5 and 2, the neighbors of disk 2 being 1 and 3, etc. Rotations clearly do not change any disk's neighbors. Also, the "Do Arrows" makes no such changes either, since if the disks in spots 1 and 4 were the original neighbors of the disk in spot 5 before an application of T, then applying T preserves the neighbor relationships. We can clearly put any of the five disks into spot 1. After picking one, we have only two choices of what can go into spot 2, namely one of the two original neighbors of whatever disk we put into spot 1. After picking one of them, we have only one choice for each of the remaining spots, namely the neighbor not yet used. This leads to ten possible arrangements of this puzzle.

(c) Similar to the analysis in (b).

13.4: (a) In $OT_8(n)$ with $n \geq 5$, $(5, 3, 1) = [T, R^{-1}]$.

(b) The "Do Arrows" of Oval Track 7, namely (5, 3, 1), performs an even permutation on the odd numbered disks if there are odd numbered disks in the odd numbered spots. Similarly it performs an even permutation on the even numbered disks if they are the ones in the odd numbered spots. If there are an odd number of odd numbered disks, that is, if n, the total number of disks, is not a multiple of 4, then a rotation that returns odd numbered disks to odd numbered spots is also an even permutation. Since (1, 3) is an odd permutation, we cannot get it under these circumstances, and the same applies to (2, 4). On the other hand, if $n = 4m$, and $n \geq 8$, then $(TR^{-4})^{m-1}(TR^{-3})(TR^{-4})^{m-1}(TR^{-2})(TR^{-1}) = (1, 3)(2, 4)$.

(c) These observations show that $OT_7(4m) = OT_8(4m)$ and $|OT_7(4m + 2)| < |OT_8(4m + 2)|$. One needs to explore further to show why $|OT_8(4m + 2)| = 2 \times |OT_7(4m + 2)|$.

B

13.5: (a) The disks must remain in an odd-even alternating pattern, and since a "place" has two adjacent spots, it must contain one disk of each parity.

(b) (i) $\alpha = (1, 4, 2, 8, 6, 7, 5, 3)$, $\beta = (1, 8, 7, 5, 3)(4, 6, 9)$, $\gamma = (1, 2)$.

(ii) $\alpha = (1, 6, 9, 2, 4, 3)(5, 7)$, $\beta = (2, 9, 5, 4, 6, 3, 7)$, $\gamma = (1, 2)$.

(iii) $\alpha = (1, 3, 5, 8, 2, 4)$, $\beta = (1, 3, 5, 8, 2, 7, 6)$, $\gamma = (1, 2)$.

(iv) $\alpha = (1, 2, 6, 3)\,(4, 7)$, $\beta = (1, 6, 3, 7, 4, 2, 5)$, $\gamma = \varepsilon$.

(c) The triples (α, β, γ) translate into these elements of S_n:

(i) (1, 7, 9, 3, 13) (5, 15) (2, 6, 16, 12, 20) (4, 8, 14) (10, 18)

(ii) (1, 10) (2, 9, 4, 13, 14, 3, 6, 7, 12, 5) (8, 11)

(iii) (1, 7, 9) (2, 6) (3, 11) (4, 12, 8, 10)

13.6: (a), (b) The triple that describes the initial state of the puzzle is $(\varepsilon, \varepsilon, \varepsilon)$. From that state, a one-position clockwise rotation moves the odd-numbered disks to the other spots within their places, thus leaving the first component of the triple at ε, while the even disks are moved to the next place, thus adjusting the second component by the $n/2$-cycle $(1, 2,\ldots, n/2)$. A second one-position clockwise rotation shifts the even disks within their places, and moves the odd disks to the next place, thus adjusting the first component by the $n/2$-cycle. Each one position rotation adjusts the third component by the transposition (1, 2) since it moves all disks from spots with one parity to spots of the other parity. The "Do Arrows" permutation clearly adjusts one or the other of the first two components of the triple by the 3-cycle (3, 2, 1) and leaves the other alone.

13.7: (a) By Exercise 13.4, you can produce (1,3)(2,4) on this puzzle when $n/2$ is even. Use this tool together with the methods developed for Oval Track 3 with "Do Arrows" (3,2,1) to handle the odd and even disks separately.

(b) Use the methods of Oval Track 3 to handle the odd and even disks separately.

13.9: Here is a table listing the sizes of $OT_{19}(n)$ where $T = (7, 4, 1)$. Can you explain these numbers?

n	$\lvert OT_{19}(n)\rvert$	n	$\lvert OT_{19}(n)\rvert$
7	$2{,}520 = 7!/2$	8	$40{,}320 = 8!$
9	$81 = (3 \times (3!)^3)/8$	10	$3{,}628{,}800 = 10!$
11	$19{,}958{,}400 = 11!/2$	12	$10{,}368 = (3 \times (4!)^3)/4$
13	$3{,}113{,}510{,}400 = 13!/2$	14	$87{,}178{,}291{,}200 = 14!$
15	$648{,}000 = (3 \times (5!)^3)/8$	16	$20{,}922{,}789{,}888{,}000 = 16!$

B

Chapter 14 Exercises

14.1: The permutations $(14,15,16) = (16,15)\ (15,14)$ and $(16,12)$ are both in *SP*. Their composition is $(14,15,16)$ $(16,12) = (14,15,12,16)$. This permutation puts the blank into spot 14, which is an even spot. However the permutation is a 4-cycle which is odd. Only even permutations in *SP* can put the blank into even spots. Thus, this composition of two members of *SP* cannot be in *SP*, and thus, *SP* is not a subgroup.

14.5: Because the dimensions of the puzzle grid are both even, having wrapping turned on makes no difference in the parity of the locations. If it takes an even (or odd) number of moves to get the blank from a given location, moving horizontally only, to the rightmost column by going directly, then it will take an even (or odd) number of moves if one goes there over the left edge. A similar argument shows the same parity for vertical movements to the bottom row.

14.6: These particular obstacles do not make the puzzle any less solvable. There is still sufficient maneuverability to get any three disks together with the blank in a two by two square. The obstacles greatly increase the number of moves needed to solve the puzzle, however.

14.7: Note that the transposition $(4, 8) = (12, 8)\ (8, 4)\ (4, 12)$ is possible on this puzzle if one takes advantage of the vertical wrap. Clearly one can get any two disks into spots 4 and 8 and thus can get any transposition as a conjugate of this one. Since one can produce all the transpositions on this puzzle, one can produce (and therefore also solve) all permutations.

B

14.10: Since we can get a 4-cycle, namely $\alpha = (1, 5, 9, 13)$, on this puzzle, as stated in the hint, and since a 4-cycle is odd, we have at least one odd permutation that is obtainable. Clearly we have enough flexibility to get all of the 3-cycles, and thus all of the even permutations β. As β runs through all the even permutations, $\alpha\beta$ runs through all of the odd ones. All of these combinations of solvable permutations are solvable.

Chapter 15 Exercises

15.6: The end game permutation is (3, 8, 34, 19, 38). To solve the puzzle, we use its inverse (3, 38, 19, 34, 8). Another way to think of this is that we need to have disk 3 go where disk 38 is now, disk 38 go where disk 19 is now, etc. An 11-step setup procedure which gets these five disks into spots 1, 25, 30, 11, and 16 is $A = R^{-6}L^2R^{-5}L^{-9}RL^{-3}R^{-1}L^{-8}RL^3R^{-1}$. Following this by the tool 5-cycle $(L^5R^5)^4$ = (1, 25, 30, 11, 16), and following that by A^{-1}, we have the puzzle solved.

15.7: The end game permutation is (4, 37, 35, 20). To set up for inverting this, we get disks 4, 20, 35, and 37 into spots 1, 35, 11, and 21, respectively. An 8-step procedure that will do this is $A = R^{-9}L^3RL^6R^6LR^{-3}L^{-1}$. Follow this by the tool 4-cycle (1, 35, 11, 21) of display (15.5), and then by A^{-1}, and the puzzle is solved.

15.8: The end game permutation is (3, 7, 36). To set up for inverting this, we get disks 3, 36, and 7 into spots 6, 11, and 12, respectively. A 6-step procedure that does this is $A = LRL^{-3}RL^{-6}R^{-2}$. Follow this by the tool 3-cycle (6, 11, 12) of display (15.1) and this by A^{-1}, and the puzzle is solved.

15.9: The end game permutation is (2, 38, 5) (20, 34, 36). A 16-step setup procedure is $A = R^4L^{-1}R^5L^6R^{-4}L^{-10}R^{-1}L^4R^{-1}L^{-8}RL^{-1}R^{-2}L^{-1}R^{-7}L^6$. Follow this by the tool double 3-cycle $[L, R^{-1}]$ = (34, 7, 6) (38, 2, 1) and that by A^{-1}, and the puzzle is solved.

Chapter 16 Exercises

B

16.6: One strategy is: (1) Get about 10 green disks together on the left ring and about 10 red disks together on the right ring; (2) Get additional green and red disks together with their main groups on their respective rings as opportunities present themselves; (3) Use jump rotations to get any remaining green disks onto the left ring and red disks onto the right ring, not necessarily contiguous with the main groups; (4) Use jump rotations to get all the green disks and red disks together. This strategy makes use of methods that are used regularly and thus that you might have already memorized. On a test of ten scrambles using it, the author had an average of 75.8 moves, with a range from 51 to 107.

16.7: Here is a strategy for the five color version: (1) Get the firebrick (F) and red (R) disks together and on the correct rings; (2) Get as many as possible of the green (G) disks near the top as possible and as many as possible of the blue (B) disks near the bottom; (3) Use jump rotations to get the yellow disks separated out from among the green and blue disks. You will likely need more moves for this five color version than for the four color version.

16.8: A similar strategy to the one described for the five color version (Ex. 16.7 hint above) works for the six-color version. (1) Get the blue (B) and most of the magenta (M) disks in order and together on the left ring and the firebrick (F) and most of the yellow (Y) disks together on the right ring; (2) Get the green (G) disks up near the top and the rest of the magenta, yellow, and red (R) disks near the bottom; (3) Use jump rotations to do the final separation of the greens, reds, yellows, and magentas. This seems to not take too many more steps than the five-color version does.

16.9: The two-color version of the Hungarian Rings is not as easy as the inexperienced intuition would guess it to be. The same basic strategy that works for the four-color version continues to work. With only two colors, however, there are more occasions when one can get "two for the price of one" and place two disks on the correct ring at the same time, one at each intersection point. Also, one should be on the lookout for making use of the easy pairs of transpositions such as $[L^5, R^5] = (1, 25)\,(6, 11)$ and $[L^5, R^{-5}] = (1, 6)\,(11, 30)$, as there are more frequent opportunities to use them. On a test of ten scrambles with this version, the author averaged 63.3 moves to solve the scrambles, with a range from 29 to 79.

B

Chapter 17 Exercises

17.1, 17.2: If you get any improved results for transpositions or 3-cycles on any of the puzzles, please contact me by e-mail at `kiltinen@nmu.edu`.

17.3: The permutation that produces a complete reversal of the disks corresponds to the triple (α, β, γ) as described in Ex. 13.5, where α and β are both $(1,10)(2,9)(3,8)(4,7)(5,6)$ and $\gamma = (1,2)$. According to Ex. 13.6(c), for a permutation to be possible on Oval Track 7 with n and $n/2$ both even, α and β must have opposite parities if γ is odd. Since γ is indeed odd here, and since α and β, being equal, clearly have the same parity, this permutation is not possible.

17.4: For any possible permutation, the units digits of the disks must stay in sequence around the track. Consider how many ways one can build a permutation satisfying this condition. Pick a units digit for the disk to go into spot 1. (Say we picked 5.) There are ten ways to make this choice. Once we have made it, we have two choices of a number with this units digit. (For our example, they are 5 or 15.) Make one of these two choices. Now the units digit of the number in spot 2 is determined, (For our example, it is 6) but we again have two numbers with this units digit. For spots 2 through 10, we can choose either of the two disks with the prescribed units digit, and beyond that for spots 11 through 20, we must place the one remaining disk with the right units digit. This gives us 10×2^{10} ways to complete this process. However on the puzzle, we cannot obtain all of these permutations because the "Do Arrows" (1, 11) (4, 14) always switches the parity of the permutations on both the even and odd disks, thus constraining them to always be of the same parity. Only half of the 10×2^{10} permutations that satisfy the units-digits-in-sequence constraint will also satisfy this additional condition. For example, the single transposition (1, 11) is impossible on this puzzle.

17.6: (a) Pick the modulo 5 reduction for the number on the disk to go into spot 1. There are five ways to do this. Once the choice is made, the sequence of mod 5 reductions is completely determined. However, you start out with three disks with each mod 5 reduction, for example {1, 6, 11}, {2, 7, 12}, etc., and you can pick any of the three for spots 1 through 5, and any of the remaining two for spots 6 through 10, and then must put in the remaining one in each of spots 11 through 15. This gives $5 \times 3^5 \times 2^5 \times 1^5 = 5 \times 6^5 = 38{,}880$ ways to do the task.

(b) *Maple* or *GAP* return the number 19,440 as the number of elements in the subgroup associated with this puzzle. That is half of our number. The reason that we can get only half is that since R, a 15-cycle, and $T = (1,11)(4,14)$ are both even, we can only get even permutations. It will still take some work to show why we can indeed get all of the even permutations that satisfy the cycling condition we have described. Explore the squared commutator $[R^{-2}, T]^2$.

17.7: *Maple* or *GAP* returns 1,625,702,400 = $(8!)^2$ as the order of the group $OT_6(16)$ and returns 131,681,894,400 = $(9!)^2$ as the order of the group $OT_6(18)$.

17.13: (a) If the sum of the exponents is 1 modulo n, then each of the $n-2$ disks that ultimately returns home must get a net total of one counterclockwise nudge from the "Do Arrows," T. If each of these disks is acted upon by T at least twice, then there must be at least two times that T gives each of them a counterclockwise nudge. This means there are at least $2(n-2)$ counterclockwise nudges administered by T. Since T can only do this to two disks at a time, and since $m/2$ steps of the m-step procedure P are T's, this means that $2(m/2) \geq 2(n-2)$.

(b) One can show that P must administer a total of $n-2$ clockwise nudges to the two disks a and b that are moved by the transposition. If it does not perform two of these nudges at the same time, then there must be at least $n-2$ applications of T in the procedure P.

B

Annotated List of Web and Print References

Web Sites

Here is a collection of web sites that have materials related to permutation puzzles. In each case, a brief description of the nature of the site is given.

Our Permutation Puzzles web site. `www.nmu.edu/mathpuzzles`
Check out this web site from time to time to learn of developments regarding the puzzles, including contributions from readers of this book.

Innovations in Teaching Abstract Algebra web site. `www.central.edu/MAANotes/`
This web site supplements a book with the same title. (See below for a citation of the book.) The web site gives abstracts of the articles in the book and links to many other web sites from which resources related to learning about abstract algebra are available.

The *GAP* web site. `www-gap.dcs.st-andrews.ac.uk/~gap/`
You can download the group theory software package, *GAP*, from this web site.

JOMA Parity Theorem article. `www.joma.org/vol1-3/articles/kiltinen/ParityJOKMain.htm`
This is the URL for the author's Parity Theorem article in JOMA, the *Journal of Online Mathematics and its Applications*. It includes a bibliography and a review of pedagogical approaches to the proof of this theorem. One can also download the Parity Theorem Demo software from this site.

Binary Arts puzzles web site. `www.puzzles.com`
Binary Arts is the company that made the *Top Spin* puzzle upon which our Oval Track Puzzles are modeled. At their web site, they describe their current products and have on-line puzzles and links.

Interactive Mathematics Miscellany and Puzzles. `www.cut-the-knot.com/content.html`
This web site has a wide variety of mathematics related stuff. The description given above is the title on the introductory page. Click on the Games and Puzzles link to get to a variety of puzzles, many of which can be played from within the web browser.

Buying *Rubik's Cube* and related puzzles on-line. `www.rubiks.com`
This is the web site of the company licensed to market *Rubik's Cube* and other Erno Rubik puzzles in the United States. You can order these puzzles from there. There are links to related web sites.

Another source for buying puzzles on-line: `www.puzzle-shop.de/`
Hendrik Haak of Kuhs, Germany maintains this "Puzzle-Shop" site from which one can purchase puzzles. He also has a "Puzzle-Museum" with photos of items in his extensive puzzle collection.

A computerized version of *Rubik's Cube* sold on a CD ROM. `www.rubiksgames.com`
This web site promotes a CD ROM product called Rubik's Games. The CD ROM contains several puzzles including the classic Rubik's Cube. You can download a demo from the site.

Jaap's Puzzle Page. `www.geocities.com/jaapsch/puzzles/`
This site, maintained by Jaap Scherphuis, has analyses of dozens of types of permutation puzzles and links to lots of other resources. See the next entry for one of the topics he covers.

C

Entrée to the *Lights Out* puzzle world. `www.geocities.com/jaapsch/puzzles/lights.htm`
Lights Out is a mathematical puzzle that has lighted squares on the faces of a cube. By pressing on one of the squares, you switch that light one way and the neighboring squares the other way. The object is to get all the

lights turned out starting from a random arrangement.

The group theory of permutation puzzles. web.usna.navy.mil/~wdj/rubik_nts.htm
This web site is maintained by Professor W. D. Joyner at the United States Naval Academy, and has his lecture notes for a course on Rubik's Cube. See entry 3 in the list of books for one by Joyner.

A variety of mathematical puzzles. www.mathpuzzle.com
Ed Pegg, Jr., of Champaign, Illinois, maintains this site. It has a variety of mathematical puzzle information, but is not particularly focused on permutation puzzles. There are links and a search engine. You can also sign up to be on a recreational mathematics e-mail list that he maintains.

An online *Rubik's Cube* bibliography. webplaza.pt.lu/public/geohelm/myweb/cubbib.htm
Georges Helm in Germany maintains this site. It has an extensive bibliography of material written about Rubik's Cube and related puzzles.

A source for the theory of permutation groups. www.maths.qmw.ac.uk/~pjc/permgps/
Professor Peter J. Cameron of Queen Mary, University of London maintains this site, which has links to information about the theory of permutation groups. There are lots of links.

Journal Articles

Here are a few journal articles that are of interest.

1. C. Bennett, "Top Spin on the symmetric group," *Math Horizons* v. 8, no. 2 (2000) 11–15.
2. D. Hofstadter, "The Magic Cube's Cubies Are Twiddled by Cubists and Solved by Cubemeisters," *Scientific American* 244 (1981) 20–39.
3. J. O. Kiltinen, "How Few Transpositions Suffice? … You Already Know!," *Mathematics Magazine* 67 (1994) 45–47.

4. J. O. Kiltinen, "A Tactile, Real-time Proof of the Parity Theorem for Permutations," *Journal of Online Mathematics and its Applications* 1(2001). (See Web Sites listing above for the URL.)

Books

This is a short list of books that present information about puzzles or related recreational mathematics and the formal mathematics that allows for detailed understanding.

1. J. A. Gallian, *Contemporary Abstract Algebra*, fifth edition, Houghton Mifflin, Boston, 2002.
   (*This is a good reference for the group theory we have developed, as are many other introductory abstract algebra texts, but Gallian also has extensive annotated references to other literature.*)
2. A. Hibbard and E. Maycock, *Innovations in Teaching Abstract Algebra*, Notes #60, MAA, Washington, DC, 2002.
   (*A collection of articles about innovative ways of teaching abstract algebra.*)
3. David Joyner, *Adventures in Group Theory: Rubik's Cube, Merlin's Machine, and Other Mathematical Toys*, Johns Hopkins Univ. Press, Baltimore, 2002
   (*This is another new book that presents group theory via puzzles.*)
4. S. B. Morris, *Magic Tricks, Card Shuffling and Dynamic Computer Memories*, MAA Spectrum Series, MAA, Washington, DC, 1997.
   (*The theory of permutation groups also provides a foundation for understanding card shuffling problems. This book deals with that topic.*)

C

The next group consists of a variety of books that deal with recreational mathematics.

5. W. W. R. Ball, *Mathematical Recreations and Essays*, Macmillan, New York, 1962.
6. C. Bandelow, *Inside Rubik's Cube and Beyond*, Birkhäuser, Boston, 1982.

7. E. R. Berlekamp, J. H. Conway, R. K. Guy, *Winning Ways for Your Mathematical Plays*, Academic Press, London, 1982.
8. A. P. Domoryad, *Mathematical Games and Pastimes*, Macmillan, New York, 1964.
9. M. Gardner, *New Mathematical Diversions from Scientific American*, Simon and Schuster, New York, 1966.
10. M. Gardner, *Martin Gardner's Sixth Book of Mathematical Games from Scientific American*, W. H. Freeman, San Francisco, 1971.
11. M. Gardner, *Entertaining Mathematical Puzzles*, Dover, New York, 1986, c1961.
12. M. Gardner, *The Colossal Book of Mathematics: Classic Puzzles, Paradoxes, and Problems: Number Theory, Algebra, Geometry, Probability, Topology, Game Theory, Infinity, and Other Topics of Recreational Mathematics*, Norton, New York, 2001.
13. M. Kraitchik, *Mathematical Recreations*, Dover, New York, 1953.
14. D. B. Lewis, *Eureka*, Putnam, New York, 1983.
15. G. Mott-Smith, *Mathematical Puzzles*, Dover, New York, 1954.
16. T. Rice, *Mathematical Games and Puzzles*, St. Martin's Press, New York, 1973.
17. E. Rubik, T. Varga, K. Gerzson, G. Marx, T. Vekerdy, *Rubik's Cubic Compendium*, Oxford University Press, Oxford, 1987.
18. W. L. Schaaf, *A Bibliography of Recreational Mathematics*, National Council of Teachers of Mathematics, Reston, 1978.
19. F. Schuh, *The Master Book of Mathematical Recreations*, Dover, New York, 1968.

Index

I

I

I

R

S

T

U

V

W

I

Mathematical Notation

$A\gamma$	For a set A and a permutation γ, $A\gamma = \{x\gamma \mid x \in A\}$.
$\lvert A\rvert$	If A is a set, $\lvert A\rvert$ is the number of elements in A.
A_n	The subgroup of S_n consisting of the even permutations (*alternating* group).
$\alpha, \beta, \gamma \ldots$	Arbitrary permutations in S_n.
$[\alpha, \beta]$	The *commutator* of permutations α and β, or $\alpha\beta\alpha^{-1}\beta^{-1}$.
D_n	The *dihedral* subgroup of S_n consisting of those permutations which rigidly map a regular n-sided polygon onto itself when the permutation acts on sequentially numbered vertices.
ε	The *identity* permutation in S_n.
F_α	The elements of $\mathbb{Z}_n$ that are left *fixed* by the permutation α in S_n.
$k!$	For a positive integer k, $k! = k \cdot (k-1) \cdot (k-2) \cdot \cdots \cdot 2 \cdot 1$. This is called k factorial.
L	A one-position clockwise rotation of the disks on the left ring of the Hungarian Rings Puzzle.
M_α	The elements of $\mathbb{Z}_n$ that are *moved* by the permutation α in S_n.
min	The *minimum* function. If $m, n, \ldots$ are integers, $\min\{m, n, \ldots\}$ is the smallest among them.
mod n	$i \equiv j \bmod n$ means i and j are *equivalent modulo n*, or $i - j$ is a multiple of n.
O_n	The subgroup of S_n consisting of all the odd permutations.
$OT_i(n)$	The subgroup S_n associated with Oval Track Puzzle i with n active disks.

R	A one-position clockwise rotation of the disks on an Oval Track Puzzle, or of the right ring on the Hungarian Rings Puzzle.
S_n	The set of all permutations on the set $\mathbb{Z}_n$.
SP	For a version of the *Slide Puzzle*, the set of all permutations that are possible to achieve on that puzzle.
SP^*	For a version of the *Slide Puzzle*, the set of all permutations that are possible to achieve on that puzzle and that return the blank to its initial box.
$\sum_{j=1}^{k} i_j$	Sigma (summation) notation. $\sum_{j=1}^{k} i_j = i_1 + i_2 + \cdots + i_k$.
T	An application of the "Do Arrows" (*Turntable*) permutation on an Oval Track Puzzle.
$\mathbb{Z}_n$	The set of integers $\{1, 2,\ldots, n\}$.
$\cap$	Intersection symbol. $X \cap Y$ is the set of elements in both X and Y.
$\cup$	Union symbol. $X \cup Y$ is the set of elements in either X or Y.
$\subseteq$	Subset symbol. $X \subseteq Y$ means that X is a subset of Y.
$\varnothing$	Empty set symbol. $\varnothing$ is the set that contains no elements.
$\in$	The element symbol. $x \in A$ means that x is an element of the set A.
$\circ$	Composition symbol. For functions α and β, $\alpha \circ \beta$ is their composition defined by the equation $x\alpha \circ \beta = (x\alpha)\beta$.
$\times$	The product symbol. If m and n are numbers, $m \times n$ is their arithmetic product. If A and B are sets, $A \times B$ is the set $\{(x, y) \mid x \in A \text{ and } y \in B\}$.

About the Author

John O. Kiltinen is a native of Marquette, Michigan. He earned his Bachelor of Arts degree at Northern Michigan University in 1963 and his Doctor of Philosophy in mathematics at Duke University in 1967. After teaching at the University of Minnesota for four years, he returned in 1971 to his undergraduate *alma mater*, where he has been a faculty member ever since.

Dr. Kiltinen's dissertation dealt with topological rings and fields, and, after publishing several research papers in that specialty, his interests shifted toward pure algebra. He has authored over twenty research and expository articles. He particularly enjoys pursuing questions that are accessible to his undergraduate students of abstract algebra. Many of his recent papers have grown out of questions that arose in undergraduate classes and he is especially proud of several that he has co-authored with his undergraduate students.

He spent a 1978–79 sabbatical leave as a visiting professor in Finland, supported in part by a Fulbright Lectureship.

Dr. Kiltinen has been active in the Michigan Section of the MAA, having served on the examination committee for the Section's high school competition, as newsletter editor, and as Governor. He chaired the committee that oversaw the establishment of the Glenn T. Seaborg Center for Teaching and Learning Science and Mathematics at Northern Michigan University, and served as the Center's first acting director.

Outside of his profession, he has been deeply involved in the promotion of Finnish-American culture and is also involved with music as a choral singer and supporter of the performing arts. He and his wife have commissioned several compositions for chorus and orchestra. He has served on the Board of Trustees of Carthage College.